Optimal Traffic Control

URBAN INTERSECTIONS

Optimal Traffic Control

URBAN INTERSECTIONS

Slobodan Guberinić
Gordana Šenborn
Bratislav Lazić

CRC Press
Taylor & Francis Group
Boca Raton London New York

CRC Press is an imprint of the
Taylor & Francis Group, an **informa** business

CRC Press
Taylor & Francis Group
6000 Broken Sound Parkway NW, Suite 300
Boca Raton, FL 33487-2742

International Standard Book Number-13: 978-1-4200-6268-7 (Hardback)

Library of Congress Cataloging-in-Publication Data

Library of Congress Cataloging-in-Publication Data
Guberinić, Slobodan.
Optimal traffic control : urban intersections / Slobodan Guberinić, Gordana Šenborn, and Bratislav Lazić.
p. cm.
Includes bibliographical references and index.
ISBN 978-1-4200-6268-7 (alk. paper)
1. Signalized intersections. 2. Traffic signs and signals--Control systems. 3. Traffic engineering. I. Šenborn, Gordana. II. Lazić, Bratislav. III. Title.

TE228.G82 2008
388.4'1312--dc22 2007029044

Visit the Taylor & Francis Web site at
http://www.taylorandfrancis.com

and the CRC Press Web site at
http://www.crcpress.com

CONTENTS

PART IV

PART V

APPENDICES

Preface

One of the main features of modern cities is the permanent growth of population in a relatively small area. The consequence of this fact is the increase in the number of cars and also the necessity of movement and transport of people and goods in urban city networks.

The increase of the capacity of the street network can't, as a rule, follow the increase of the necessities for transport. It has unwanted consequences, such as the increase in time losses of traffic participants, the increase of environmental pollution and noise, and also the increase in the number of traffic accidents.

Different measures are applied for elimination of these unwanted consequences of intensive development of modern cities. Some of them are, for instance, the land use planning, the improvement of traffic control, etc. Significant attention is paid just to the improvement of the automatic traffic control systems. The reason for that is the fact that this measure doesn't imply significant changes of infrastructure.

Traffic control in modern cities, however, is one of the most complex control problems in the sense of the theoretical problem statement as well as in the sense of practical realization of traffic control systems. The transportation system in the modern city has all features of so-called complex systems: the great number of state and control variables, the presence of uncertainty and indeterminism, the complex interactions between subsystems, the necessity to optimize several optimization criteria, active behavior of the controlled process, etc.

The control systems of such degree of complexity couldn't be realized without significant progress achieved in the development of information technologies, especially computer and telecommunication techniques, during recent decades.

The necessary condition, however, for realization of such complex control systems is also the development of solution methods of control problems, especially the optimization methods.

A signalized intersection can be isolated, with negligible influence of other signalized intersections to its own performance. On the other hand, it may be close to other signalized intersections so that the interactions between them can be very significant. Because of that, the optimal control problems have to be formulated separately: for isolated intersections, for a sequence of signalized intersections along an artery (green wave), and for a street network with greater number of signalized intersections with strong mutual interactions.

This monograph is concerned with the traffic control problem on a single, isolated signalized intersection. This problem is still of current interest because the existing algorithms don't enable to take advantage of all capabilities of modern, microprocessor-based traffic signal controllers. The fact that a great number of intersections (in many countries over 50% of all signalized intersections) are isolated points out the importance of the traffic control problem on isolated intersections.

The approach to the optimal traffic control presented in this book and also the optimization methods based on the graph theory and combinatorial optimization are results of a long-term work of authors in the field of traffic control, in "Automation & Control Systems" department, within Mihajlo Pupin Institute in Belgrade.

The authors are very grateful to all colleagues who contributed to the realization of this book. We are especially grateful to Mgr. Miomir Šegović for his participation regarding the influence of the choice of the complete set of signal groups to the intersection performance and also regarding the solution of the optimization problems presented as the problems of nonlinear programming.

The authors owe gratitude also to Dr. Snežana Mitrović-Minić for her participation in the statement and solution of some optimization problems regarding the choice of the complete set of signal groups.

The authors express distinct gratitude to Professor Radivoj Petrović for his continuous support and encouragement during the work that resulted in appearance of this book.

Introduction

The most common means of traffic control in modern cities is traffic control by traffic signals. Traffic signals made it possible to "solve" conflicts between traffic flows at intersections. This goal, however, can be achieved in different ways, and application of particular methods has different consequences regarding the intersection capacity, delay, or environmental pollution. The main problem of traffic engineering is choosing the way of conflict prevention that has the lowest unwanted consequences, such as minimal delay or minimal pollution, etc.

A significant development of traffic control systems using traffic lights has been achieved since the first traffic controller was installed in London in 1868. Starting from an isolated signalized intersection, the area covered by a traffic control system extended to a series of signalized intersections along an artery ("green wave") out to street networks with several hundred signalized intersections ("area traffic control systems"). The first green wave was realized in Salt Lake City in 1918 [85], and the first area traffic control was introduced in Toronto in 1960.

Traffic control equipment has followed technology development. At the very beginning, traffic control had been performed by electromechanical devices. Then, semiconductor-based controllers were introduced, and nowadays microprocessor-based controllers are used in traffic control systems.

The development of area traffic control systems, especially since 1960, has led to introduction of other equipment in traffic control systems, such as computers, telecommunication devices, vehicle detectors, etc.

Traffic control strategies have also improved since the installation of the first traffic controller. The strategies can be classified in respect to different features. The most important features are as follows:

- *The influence of real-time traffic data to traffic control*

  Regarding this feature, there exist two main types of strategies:

  - *Fixed-time* (FT) strategies. The control (signal plan) is calculated in advance, using statistical data.
  - *Real-time* (RT) strategies. The real-time data about traffic processes are used to determine control or its modification.

- *The performance indices*

  The most frequently used performance indices are:

  - For traffic control on an isolated intersection:
    The total rate of delay (the sum of the rate of delays on all intersection approaches during a determined time interval, usually one cycle time); the number of stops; the weighted sum of the rate of delay and number of stops; the sum of all green times during a cycle; the total flow through a congested intersection during a cycle; the number of accidents; the cycle time, etc.
  - For arterial traffic control:
    The bandwidth, i.e., the interval in which it is possible to enter the "green wave" and pass through the sequence of intersections without stopping.
  - For traffic control in a network of signalized intersections:
    The total rate of delay on all intersections in the network; the total number of stops in the network; the weighted sum of the rate of delay and number of stops on all links in the network; the total fuel consumption of all vehicles in the network; the air pollution level, the noise level, the number of accidents, etc.

- *The state of the traffic process*

  According to this feature, traffic control strategies can be classified as:

  - Strategies for weak traffic
  - Strategies for normal traffic
  - Strategies for congestion in the network
  - Strategies for special purposes (e.g., giving priority to mass transit vehicles, setting of fire brigade routes, etc.)

- *Distribution of functions between subsystems of one traffic control system*
  In respect to this feature, there exist two types of traffic control strategies:
  - Strategies used in centralized traffic control systems where all control functions are performed by the control center computer
  - Strategies based on control problem decomposition—one part of a control problem is solved in microprocessor-based controllers on intersections, and the other part by the control center computer [76], [80]
- *The influence of traffic control to traffic assignment in the network*
  On the basis of this feature, two types of strategies can be noted:
  - Strategies which assume that traffic assignment is independent of traffic control by traffic lights
  - Strategies assuming that both signal settings parameters and link flows are not fixed, i.e., the strategies that optimize the chosen performance index and influence to the traffic assignment

The development of traffic control strategies and information technologies enabled the realization of complex traffic control systems in modern cities, including hundreds of signalized intersections. These systems enable coordination between intersection control subsystems, which is necessary in dense street networks.

However, there are many intersections that are isolated, i.e., not included in complex traffic systems, like "green waves" or area traffic control systems. For example, more than 60% of the total number of signalized intersections in Sweden are isolated [53]. Because of that it is very important to develop good algorithms for optimal traffic control on isolated intersections. This is significant especially when bearing in mind that modern microprocessor-based traffic light controllers are capable of applying very complex control strategies. Moreover, the control on a signalized intersection has to be determined even in the case the intersection belongs to a complex control system.

Optimal traffic control on an isolated intersection is a very complex problem, especially because of the combinatorial nature of the problem. In this book, the problem is treated from the very beginning as a complex combinatorial problem and is formulated as the problem of finding the best closed path on a certain graph.

The traffic control on a signalized intersection is performed by means of traffic lights of different colors (green, amber, and red) that are repeating periodically. Conflicts between traffic participants are prevented by dividing the cycle time in intervals allocated to traffic flows so that the conflicting flows don't get the right-of-way in the same interval. The control in one interval is defined by one control vector (so-called phase) whose components are control variables that control traffic by means of traffic lights. Several traffic flows that are not mutually conflicting can get the right-of-way during the same interval.

In classical controllers the composition of phases and their sequence are fixed. In modern, microprocessor-based controllers, these restrictions are eliminated. However, due to the elimination of mentioned restrictions control problems became much more complex. The traffic control problem not only includes the problem of of splitting cycle time into particular phases, but the composition and sequence of the phases. Because of that, the traffic control problem is transformed to a complex problem of combinatorial optimization.

The necessity of phase sequence change is illustrated by the following example. The traffic flow through the intersection presented in Fig. 1 is much greater in the West-East than in the East-West direction during the morning peak hour. During the evening peak hour the greatest traffic flow is in the East-West direction. The number of left-turning vehicles is also significantly different in these two intervals. (This example is similar to the Hank Van Zylen example in [92].)

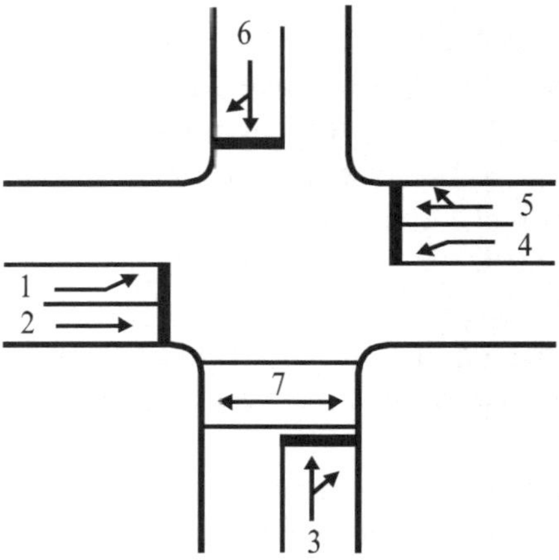

Figure 1

The phase sequences suitable for traffic control in these 2 peak hours are presented in Fig. 2. The number of phases giving the right-of-way to left turners fits the demand. This right, in case (a), is given to vehicles coming from the West, by control vectors 3, 4, and 5. In case (b) the right of left turning is given to vehicles coming from the East, by control vectors 4, 5, and 6. Arrows in Fig. 2 mark the traffic flows getting the right-of-way, and dashed lines and transversal dashes instead of arrows mark traffic flows not having the right-of-way.

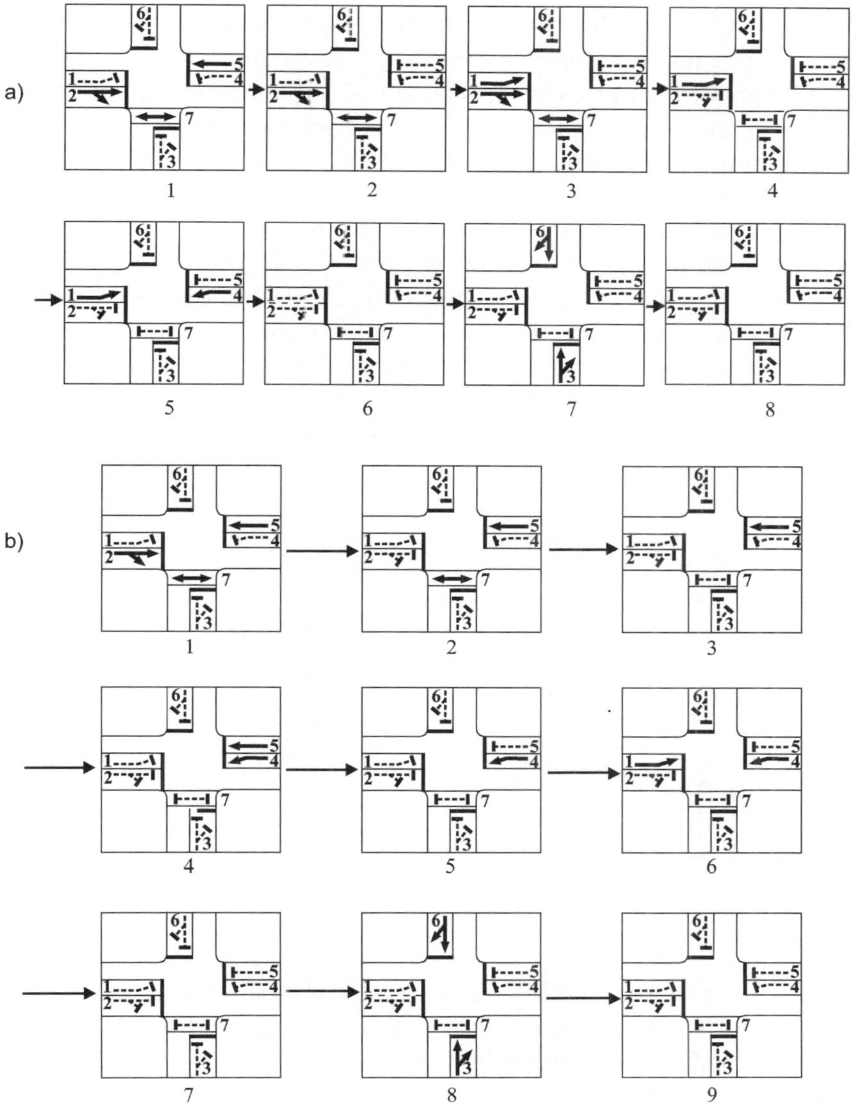

Figure 2

This example illustrates the combinatorial nature of the problem. It is obvious that the number of possible sequences of control vectors is very large. The determination of the relation between these sequences and the values of a chosen performance index and finding the sequence corresponding to the optimal value of the performance index is a complex problem of combinatorial optimization.

The development of the methods of combinatorial optimization, besides the progress in equipment development, was also an unavoidable condition for the solution of such problems. A method of combinatorial optimization, of the branch-and-bound type, is used in this book for solving the optimal traffic control problem. This method was developed at the beginning of the 1960's [54], [56].

The combinatorial approach to the optimal traffic control problem on isolated intersection was founded by Stoffers, K. [77] by introduction of the compatibility graph of traffic streams. Stoffers also noted that the traffic stream sets with maximal number of nonconflict traffic streams, which can get the right-of-way simultaneously, can be determined by extracting cliques from the compatibility graph. The compatibility graph for the intersection given in Fig. 1, together with the set of complete graphs whose sets of nodes are cliques of the compatibility graph, are presented in Fig. 3.

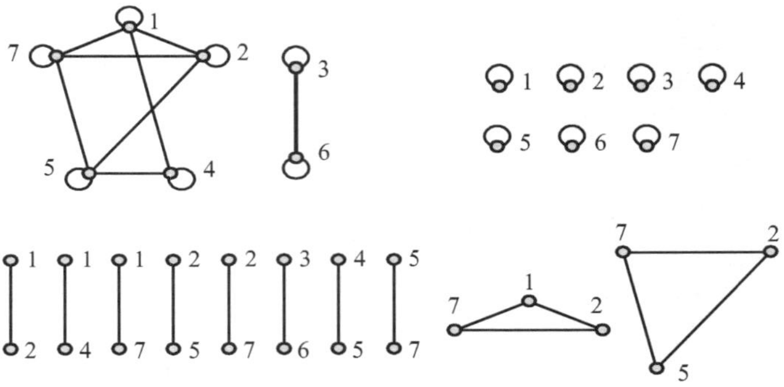

Figure 3

Cliques of the compatibility graph are used for determining the set of feasible control vectors. However, the information about feasible control vectors is not sufficient for determination of their sequence. Control vector sequences have to satisfy certain conditions. The most important conditions are that each traffic stream has to get the right-of-way during one cycle, and the right-of-way can be given only once to any traffic stream. Because of that, the control vectors giving the right-of-way to some traffic stream have to follow one after the other. To stream 1 (in case [a], in Fig. 2) the right-of-way is given by control vectors 3, 4, and 5, and to streams 2 and 7 by control vectors 1, 2, and 3.

The relations of green indication successions and control vector transitions are introduced for the sake of the exact control problem statement. There are also certain constraints related only to the control vector composition and to their sequence. Because of that the concept of "structure" is introduced. The structure is the sequence of control vectors. A feasible structure of control vectors can be presented as a closed path on the control vectors transition graph.

The optimal control problem is transformed to the problem of optimal closed path on the graph of control vectors transition. The method of the branch-and-bound type is developed for solving this problem. The computer program is developed also, enabling determination of the optimal structure and optimal cycle time split to control vectors.

Introduction of the control vectors transition graph and development of the combinatorial optimization method enabled a solution to several optimal traffic control problems. These problems have different performance indices or different constraints defining the set of feasible controls.

The book is organized in the following way:

Part I gives the description of the dynamical process on isolated signalized intersection. The elements of the mathematical model are: the space of inputs, the space of outputs, the state space, the state transition function, and the output function. Mathematical models of uncontrolled inputs—arrival flows, which are stochastic processes, are presented in this part.

The definition of the state, as the vector whose components are vehicle queues on particular approaches, is presented and also the function of state transition. The relations in the input set, such as the conflictness and compatibility relations, are defined. These relations are very significant for the statement and solution of the optimal traffic control problem on isolated, signalized intersections. The problem of assignment of control variables to subsets of traffic flows—signal groups is formulated in this part. The concept of signal plan structure is also defined—the signal plan structure is a sequence of control vectors.

In Part II the problem of control (signal plan) determination is formulated as an optimization problem. The set of constraints and different optimization criteria are defined. Mathematical expressions are given for the constraints:

- Control vectors sequence constraints
- The constraints of minimal effective green times for signal groups
- The constraints of maximal effective red times for signal groups
- The flow balance constraints
- The constraints of minimal effective intergreen times
- The constraints of single interval of green indication for a signal group during one cycle
- The constraints of control vectors composition
- The constraints on the sum of control vectors duration

In Part III the mathematical expressions for different optimization criteria are presented. The criteria of capacity type are: the capacity of a traffic stream, signal group, or the whole intersection, the capacity factor, the sum of the squares of differences of saturation degrees of particular traffic streams. The mathematical expression of the total vehicles delay on all intersection approaches is given also. The limiting values of the signal plan elements can be determined by the solution of certain optimization problems. In that case, the optimization criteria are: green time of a signal group, the cycle time, the total number of control vectors.

Part III is devoted to the solution methods of the problems formulated in Part II. In this part it is shown that the optimization problems formulated in Part II can be presented as the problems of finding the best path in a graph of control vectors transitions. The method, based on the branch-and-bound algorithm, is developed for solving these problems. The branching function, dividing the solution set, is defined in the same way, regardless of the chosen optimization criterion. The bounding function definition depends on the type of the chosen optimization criterion.

In Part IV the developed algorithm is applied to find solutions to several problems formulated in Part II. These problems differ in regard to the optimization criteria or the set of constraints that define the feasible set of solutions.

In Part V the influence of the choice of the complete set of signal groups on intersection performances is analyzed. It is concluded that this influence is very significant, and one heuristic is proposed for the choice of the complete set of signal groups.

Part I

MATHEMATICAL MODEL OF TRAFFIC PROCESS ON A SIGNALIZED INTERSECTION

The process that takes place on a signalized intersection consists of transformation of input flows into output ones, with queuing and queue discharging; thus, it represents a dynamic process.

According to the general systems theory [58], [23], the following objects can be associated with any dynamic process:

- An input space $\mathscr{X}$ (i.e., the set of all inputs)
- An output space $\mathscr{Y}$ (i.e., the set of all outputs)
- A state space $\mathscr{W}$ (i.e., the set of all states)

The input space, $\mathscr{X}$, and the output space, $\mathscr{Y}$, are two sets of time functions. Their elements $x(\cdot)$ and $y(\cdot)$ are vector functions of time, defined in the time domain $\mathscr{T} \subseteq \mathrm{R}$, where $\mathscr{T} = (-\infty, \infty)$ or $[0, \infty)$ in the case of continuous time functions, and $\mathscr{T} = \{n\,\Delta t \mid n \in \mathrm{N}\}$ or $\mathscr{T} = \{n\,\Delta t \mid n \in \mathrm{N}_{+}\}$ in the discrete case.

A state of the process is a vector that depends on time t and whose components are real numbers.

The fundamental property of a dynamic system is that given any "initial" time $t_0 \in \mathscr{T}$, any "initial" state $w_0 \in \mathscr{W}$, and any input $x(\cdot) \in \mathscr{X}$, both $w(t)$ (the resulting state at some later time t) and $y(t)$ (the resulting output at some later time t) are uniquely specified. Also, $w(t)$ and $y(t)$ depend only on $w_0 = w(t_0)$ and the values of the input $x(\cdot)$ in interval $[t_0, t]$.

Therefore, in order to fully describe a dynamic process, the functions have to be defined by which $w(t)$ and $y(t)$ are determined.

1. GENERAL MATHEMATICAL DESCRIPTION OF THE DYNAMIC PROCESS ON A SIGNALIZED INTERSECTION

The dynamic process on a signalized intersection is fully defined by the quintuple $(\mathscr{X}, \mathscr{W}, \mathscr{Y}, \varphi^1, \varphi^2)$.

An element of set $\mathscr{X}$ comprises uncontrolled inputs in the system—traffic flows and inputs that influence the process—control inputs. Traffic flows are influenced by traffic signals, controlled by *control variables*. Thus the input can be described by the set of ordered pairs, i.e.:

$$\mathscr{X} = \mathscr{U}' \times \Sigma . \tag{1.1}$$

Set $\mathscr{U}'$ is the set of controls, i.e., vectors

$$u'(\cdot) = (u_1'(\cdot), u_2'(\cdot), \ldots, u_p'(\cdot), \ldots, u_P'(\cdot)) .$$

A control $u'(\cdot) = \{u'(t) \mid t \in \mathscr{T}\}$ is a periodic time function by which the process of traffic signal changes is described (green, amber, red, red and amber is the standard control sequence in many countries). The smallest part of this process, the repeating of which makes up the whole process, is called the *control cycle*, or, most often the *cycle*, and its duration, c, is called the *cycle duration* or *cycle time*.

Since control components $u_p'(\cdot)$ are periodic functions of time, there holds:

$$u_p'(t) = u_p'(t + kc), \quad (k = 0, \pm 1, \pm 2, \ldots \quad p = 1, 2, \ldots P) . \tag{1.2}$$

A control variable can assume either 0 or 1 value. Value 0 corresponds to red (effective red) and 1 to green (effective green) signal indication, i.e.,

$$u_p'(\cdot) = \{u_p'(t) \mid t \in \mathscr{T}\} . \quad (p = 1, 2, \ldots P) , \tag{1.3}$$

where

$$u_p'(t) \in \mathscr{B} , \quad (t \in \mathscr{T}), \text{and } \mathscr{B} \in \{0,1\} .$$

The green, amber, red, red-amber sequence is here transformed to the effective green, effective red sequence (see Section 4.1 and Appendix V).

Σ – the set of vectors $\sigma = (\sigma_1, \sigma_2, \ldots, \sigma_i, \ldots, \sigma_{\mathrm{I}})$ is made up of elements that are arrival flows or traffic streams (see Section 2.1). Traffic streams are defined, as well as traffic flows in general, by their volume, speed, density, headway or spacing interval, composition, the percentage of straight-through or left-turning and right-turning volume, and the paths they use to traverse the intersection. The quantitative traffic stream characteristic most frequently used in algorithms for optimum signal plan determination is the traffic flow volume, i.e., the number of vehicles that pass a given point in a unit of time; in this case it is the number of vehicles that arrive on an intersection approach during a time unit.

There exist arrival flow models [7], [31], in which it is assumed that flows are deterministic, but these models are not realistic enough, except in very specific cases. However, in expressions used for delay calculation, the delay resulting from deterministic, average flow volume values represents a part of more complex expressions that take into account the stochastic nature of traffic flows.

Trajectories traversed by different traffic streams through the intersection have to be known in order to determine whether a pair of traffic streams can simultaneously gain the right-of-way, i.e., whether the streams are compatible.

To each set Σ there corresponds a set $\mathscr{Q}$—the set of flow volume vectors $q(\cdot) = (q_1(\cdot), q_2(\cdot), \ldots, q_i(\cdot), \ldots, q_{\mathrm{I}'}(\cdot))$, whose elements are volumes of traffic streams $\sigma_1, \sigma_2, \ldots$. The set of indices

$$\mathscr{I} = \{1, 2, \ldots, i, \ldots, \mathrm{I}', \ldots, \mathrm{I}\} \tag{1.4}$$

is such that vehicle flows are assigned indices from 1 to I', and pedestrian and other flows from $\mathrm{I}'+1$ to I. Thus, the set of indices corresponding to vehicle flows is:

$$\mathscr{I}' = \{1, 2, \ldots, \mathrm{I}'\}.$$

$\mathscr{W}$ is the set of system states. The state at time t is described by the vector

$$w(t) = (w_1(t), w_2(t), \ldots, w_i(t), \ldots, w_{\mathrm{I}'}(t)),$$

whose components represent queue lengths formed by arrival traffic flows, i.e., traffic streams.

The set of outputs, $\mathscr{Y}$, is the set of vectors of output flows, i.e., outputs

$$y(\cdot) = (y_1(\cdot), y_2(\cdot), \ldots, y_h(\cdot), \ldots, y_{\mathrm{H}}(\cdot)).$$

Output flows are characterized, in the same way as input flows, by volume, speed, etc., and by the surface used for leaving the intersection. To each set of outputs, $\mathscr{Y}$, there corresponds set $\mathscr{Q}^e$—the set of vector functions $q^e(\cdot) = (q_1^e(\cdot), q_2^e(\cdot), \ldots, q_h^e(\cdot), \ldots, q_{\mathrm{H}}^e(\cdot))$, whose elements are volumes of output elements $y_1(\cdot), y_2(\cdot), \ldots$. Vector $q^e(\cdot)$ represents the function defined by the following expression:

$$q^e(\cdot) = \{q^e(t) \mid t \in \mathscr{T}\}, \tag{1.5}$$

where

$$q_h^e(\cdot) = \{q_h^e(t) \mid t \in \mathscr{T}\}, \quad (h = 1,2,\ldots,\mathrm{H}).$$

These volumes represent the volumes of traffic streams that are transformed by traffic signals or streams comprising several transformed traffic streams.

The state transformation (transition) function φ^1 can now be described by the following expression:

$$\varphi^1 : \; \mathscr{W}_{t_0} \times (\mathscr{Q}_{[t_0,t]} \times \mathscr{U}_{[t_0,t]}) \to \mathscr{W}_t \,. \tag{1.6}$$

In this expression,

$$u_{[t_0,t]} \in \mathscr{U}_{[t_0,t]} \tag{1.7}$$

represents the restriction of function $u'(\cdot)$ to interval $[t_0, t] \cap \mathscr{T}$, and $q_{[t_0,t]} \in \mathscr{Q}_{[t_0,t]}$ represents the restriction of function $q(\cdot)$ to the same interval.

Expression (1.6) shows the fact that for determining the state in time t it is necessary to know the state at a previous time t_0 and the input, which is an element of set $(\mathscr{Q}_{[t_0,t]} \times \mathscr{U}_{[t_0,t]})$, i.e., the vector of traffic stream volumes and control in every instant of interval $[t_o, t]$.

The reaction (readout or output) function, φ^2, can be represented as:

$$\varphi^2 : \; \mathscr{W}_t \times (\mathscr{Q}_t \times \mathscr{U}'_t) \to \mathscr{Q}_t^e \,, \tag{1.8}$$

where:

$\mathscr{W}_t$ – the set of states at time t

$\mathscr{Q}_t$ – the set of vectors of traffic stream volumes at time t

$\mathscr{Q}_t^e$ – the set of output volume vectors at time t

$\mathscr{U}'_t$ – the set of controls at time t

Figure 1.1 represents an intersection and a part of the process on the intersection.

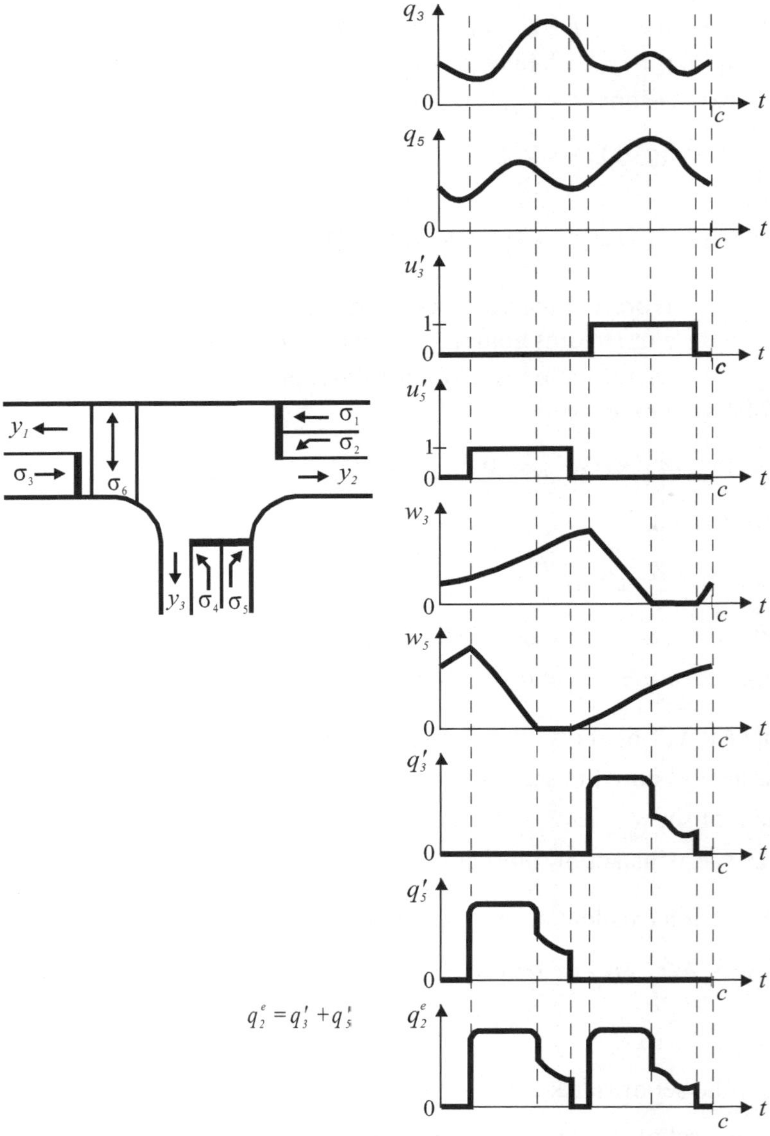

Figure 1.1

The process is described using the following variables:

(a) Volumes $q_3(\cdot)$ and $q_5(\cdot)$, corresponding to flows σ_3 and σ_5. These volumes represent "uncontrolled" input variables.

(b) Control variables $u_3'(\cdot)$ and $u_5'(\cdot)$, i.e., "controlled" input variables. These variables can assume either 0 or 1 value. Value 0 denotes red signal indication (effective red), and 1 denotes green indication (effective green).

(c) Queue lengths $w_3(t)$ and $w_5(t)$, representing process state components, with $t \in [0,c]$.

(d) Volumes $q_3'(\cdot)$ и $q_5'(\cdot)$ that resulted from transformation of variables $q_3(\cdot)$ and $q_5(\cdot)$ under influence of control variables $u_3'(\cdot)$ and $u_5'(\cdot)$.

(e) The volume $q_2^e(\cdot)$, which corresponds to output flow $y_2(\cdot)$ and represents the sum of $q_3'(\cdot)$ and $q_5'(\cdot)$ volumes.

All these variables are shown in the $[0,c]$ interval, where c is the basic period of functions $u_3(\cdot)$ and $u_5(\cdot)$, which are periodic time functions.

2. UNCONTROLLED SYSTEM INPUTS

The system input, as already described, are vectors whose components are traffic flows on intersection approaches, and sequences of different signal indications by which the right-of-way is given or taken away, i.e., by which traffic control is performed. This chapter describes the way to define and represent arrival flows.

Several models have been used for describing arrival flows on an intersection [7], [31]. The simplest is the "uniform arrivals" model. This model is a rather rough approximation of the real process, which is stochastic by its nature, and all other models account for this fact. Arrival flow parameters, such as volume, speed, density, spacing, etc., represent stochastic processes. Namely, these quantities are random numbers at a given time, i.e., defined by the set of values they can assume and the probabilities of taking these values.

Characteristics of these parameters, such as mathematical expectation, dispersion, *et al.* are not constant—they change during a day. However, it can be assumed that these processes are stationary in limited time intervals, e.g., morning or evening peaks or between peaks. Therefore, in further discussion, when stating control problems and developing methods for their solution, it will be assumed that we consider only intervals in which these processes are stationary.

2.1. Input components—traffic streams

Vehicles approaching an intersection prepare themselves to perform a certain "maneuver," i.e., to drive straight through, turn left, or turn right at the intersection. The vehicles that perform the same maneuver and form the same queue on an approach, in one or several lanes, represent a flow component that can be considered separately from other flow components, which perform other maneuvers [4], [5]. Such an arrival flow component is termed a *traffic stream*. In fact, this is the smallest flow component that can be controlled by a separate traffic signal, i.e., by a sequence of signal indications different than the sequences on other signals.

Traffic streams on an intersection can represent passenger vehicle flows, pedestrian flows, flows of public transport vehicles, etc. Traffic streams σ_i are components of vector $\sigma = (\sigma_1, \sigma_2, \ldots, \sigma_i, \ldots, \sigma_I)$.

The queue formed by a vehicle traffic stream, during the red signal indication, occupies one or more lanes. Horizontal signalization is the same in all lanes used by one traffic stream. Vehicles joining the queue can choose any lane on the approach, expecting to leave the queue in the same order as when joining it [5].

An *approach* is a part of the street, comprising one or more lanes, along which vehicles arrive to the intersection and on which the queue of a particular traffic stream forms. Traffic stream σ_i uses approach T_i, $(i \in \mathscr{I}')$.

Volume unit (pcu)—Traffic stream volume is equal to the number of vehicles that pass a given point in a unit time. Since traffic streams are composed of various vehicles, the volume is expressed as the number of average passenger cars per time unit (pcu/s). For each vehicle type there can be determined an equivalent number of passenger cars [88], [45]. This number is determined on the basis of the fact that various vehicles need different time to pass through the intersection. Heavier vehicles need more time than light ones. Vehicles of the same type need more time when turning left or right than when going straight. The equivalent number is equal to the number of passenger cars that would pass through an intersection in the same time as the given vehicle type. Table 2.1 presents an example of equivalent passenger car units for some vehicles [15] when going straight or turning.

Table 2.1

Vehicle Type	**Straight (pcu/s)**	**Turning (pcu/s)**
Passenger car	1	1
Bus	1.7	2
Heavy truck	1.7	2

2.2. Mathematical models of arrival flows —traffic streams

Several models have been used to describe traffic streams, as already mentioned. The simplest model is the uniform flow model.

A traffic stream is uniform [7] (with uniform arrivals) if its approach volume is constant, q [pcu/s]. The spacing interval is then also constant, equal to $1/q$ [s/pcu].

This model is a rough approximation of the real flow. However, for certain purposes, such as determination of vehicle delay in realistic, stochastic models, the uniform flow model can be used for determination of the delay component that results from average arrival volume values.

All other models used for arrival flow modeling, i.e., traffic stream modeling, are based on the theory of stochastic systems. In this case, $\mathcal{Q}_t$ is the set of traffic streams volume vectors at time *t*, and its element represents the vector whose components are random variables.

Most frequently it is assumed that vehicles belonging to traffic streams arrive according to the Poisson process [1], [89]. The real process is more complex so that other models have been used in which some limitations inherent to the Poisson process were eliminated [7], [31].

2.2.1. Modeling arrival flow with the Poisson process

Several authors pointed out that the stochastic Poisson process could be used for modeling arrivals of vehicles at an intersection. Webster used this model in his expression for calculation of the delay of vehicles at an intersection [89]. The expression was verified in practice and is still used.

With Poisson flow, the probability of arrival of m_i vehicles of traffic stream σ_i to an intersection, during interval τ, is expressed by the following formula:

$$p_{m_i}(\tau) = p[m_i \text{ vehicles arrive in interval } \tau] = \frac{(q_i\tau)^{m_i}}{m_i!}e^{-q_i\tau}, \qquad (2.1)$$

where q_i is constant.

The mathematical expectation of the number of vehicle arrivals, $\overline{m}_i$, in interval τ is, according to [86]:

$$\begin{aligned} \mathrm{M}m_i = \overline{m}_i &= \sum_{m_i=0}^{\infty} m_i\, p_{m_i}(\tau) = \sum_{m_i=0}^{\infty} m_i \frac{(q_i\tau)^{m_i}}{m_i!}e^{-q_i\tau} \\ &= (q_i\tau)e^{-q_i\tau}\sum_{m_i=0}^{\infty} m_i \frac{(q_i\tau)^{m_i-1}}{m_i!} = q_i\tau . \end{aligned} \qquad (2.2)$$

The average volume (flow intensity, or flow rate) of traffic stream σ_i is thus equal to

$$\frac{\mathrm{M}m_i}{\tau} = \frac{\overline{m}_i}{\tau} = \frac{q_i\tau}{\tau} = q_i \quad \text{(pcu/s)} . \qquad (2.3)$$

The spacing interval, Θ_i, between two consecutive vehicles belonging to stream σ_i is a random variable [86]. If a segment τ' is considered, the beginning of which coincides with the beginning of interval Θ_i, the probability that in τ' at least one vehicle will arrive is:

$$F_i(\tau') = p_i[\Theta_i < \tau'] = 1 - p_0, \tag{2.4}$$

where p_0 is the probability that $m_i = 0$, i.e., that during τ' no vehicle of stream σ_i will arrive. This probability is obtained by setting $m_i = 0$ in formula (2.1). Then we have:

$$p_i^0(\tau') = \frac{(q_i\,\tau')^0}{0!} e^{-q_i\,\tau'} = e^{-q_i\,\tau'}$$

and

$$p_1[\Theta_i < \tau'] = 1 - e^{-q_i\,\tau'}. \tag{2.5}$$

This is, in fact, the probability distribution function of the random quantity Θ_i. The probability distribution density $h_i(\tau')$ of Θ_i is obtained as the derivative of the distribution function over τ', i.e.,

$$h_i(\tau') = q_i e^{-q_i\,\tau'}, \quad (\tau' > 0). \tag{2.6}$$

The mathematical expectation of the spacing interval is:

$$\mathrm{M}\Theta_i = \overline{\Theta}_i = \int_0^\infty \tau' q_i e^{-q_i \tau'} d\tau' = q_i \int_0^\infty \tau' e^{-q_i \tau'} d\tau' = \frac{1}{q_i}. \tag{2.7}$$

Thus, if the probability distribution of the number of vehicles that arrive during time segment τ' to an approach is Poisson distribution, then the spacing interval distribution is negative exponential.

The probability that during a short interval δt one vehicle arrives, $p_i^1(\delta t)$, is approximately given by:

$$p_i^1(\delta t) \approx 1 - p_i^0(\delta t),$$

where $p_i^0(\delta t)$ is the probability that in interval δt no vehicle arrives. This probability is determined as

$$p_i^0(\delta t) = \frac{(q_i\,\delta t)^0}{0!} e^{-q_i\,\delta t} = e^{-q_i\,\delta t}. \tag{2.8}$$

Therefore,

$$p_i^1(\delta t) \approx 1 - e^{-q_i\,\delta t}.$$

Evolving $e^{-q_i \delta t}$ into a series, and neglecting higher order terms, results in

$$p_i^1(\delta t) \approx 1-(1-q_i\delta t) = q_i\delta t\,, \qquad (2.9)$$

i.e., the probability of one vehicle arrival to the intersection in a short interval δt is approximately equal to $q_i\delta t$.

The Poisson model is good for describing the real arrival process on an intersection approach in the case when the ratio between volume and approach capacity is not high.

2.2.2. Modeling arrival flows with more complex stochastic processes

For traffic streams with larger volumes, when interactions between vehicles cannot be neglected, other models are used, which give better results than the simple Poisson process. Some of them actually represent certain generalizations of the simple Poisson process.

The simplest generalization of the simple Poisson process is the *Compound Poisson process* [40], [57]. According to this process, vehicles of one traffic stream arrive on an approach in groups. Group arrivals are Poisson arrivals with intensity λ, and the number of vehicles in a group, *z*, is an integer random variable with seed function $\Phi(z)$. If $N(T)$ is the number of arrivals in interval of duration *T*, then

$$\mathrm{M}[z^{N(T)}] = e^{-\lambda T(1-\Phi(z))}\,. \qquad (2.10)$$

The average volume is

$$q = \lambda\,\Phi'(1)\,. \qquad (2.11)$$

The mathematical expectation of the number of vehicle arrivals is

$$\mathrm{M}[N(t)] = \lambda T\,, \qquad (2.12)$$

and the dispersion is

$$D[N(t)] = I\lambda T\,, \qquad (2.13)$$

where *I* is the dispersion index, defined by expression

$$I = 1 + \frac{\Phi''(1)}{\Phi'(1)}\,, \qquad (2.14)$$

where

$$\Phi'(1) = \left.\frac{d\Phi(z)}{dz}\right|_{z=1}, \ldots, \Phi''(1) = \left.\frac{d^2\Phi(z)}{dz^2}\right|_{z=1}.$$

The Poisson model does not impose an upper bound on the number of vehicles that can arrive during a given time interval. In reality this limit always exists. If the upper bound on the number of vehicles that can arrive during interval τ equals p, then the probability of n vehicle arrivals in interval τ is given by the following expression:

$$p(n) = \binom{N}{n} p^n v^{N-n}, \tag{2.15}$$

where $v = 1 - p$.

This expression, in fact, defines *binomial distribution* of vehicle arrivals on an approach.

Many other models ([67], [21], [60], [46]) have been used for describing the arrival process of vehicles on an approach. All of them were developed in order to provide a better model of real process than the Poisson model. However, most of them are much more complex, and the Poisson model is still used for most practical purposes.

2.2.3. Traffic stream parameters

The path used by a traffic stream to traverse an intersection is called the *trajectory*. A trajectory connects an approach on which vehicles enter the intersection to the intersection leg on which these vehicles leave the intersection. Fig. 2.1 shows trajectories of all traffic streams that pass through a four-leg intersection.

Vehicles belonging to some streams may use more than one trajectory when traversing the intersection (e.g., streams σ_3 and σ_7 in Fig. 2.1), and accordingly, several exit approaches. Therefore, these streams can be regarded as multicomponent streams, where the distinction between components lies in the fact that they use different trajectories when traversing the intersection. Traffic streams representing these components are called *partial traffic streams* [77].

Vehicles belonging to different partial traffic streams use the same lane when arriving to the intersection and different paths for leaving it. For example, in Fig. 2.1, stream σ_3 comprises two partial traffic streams: σ_3^1 and σ_3^2.

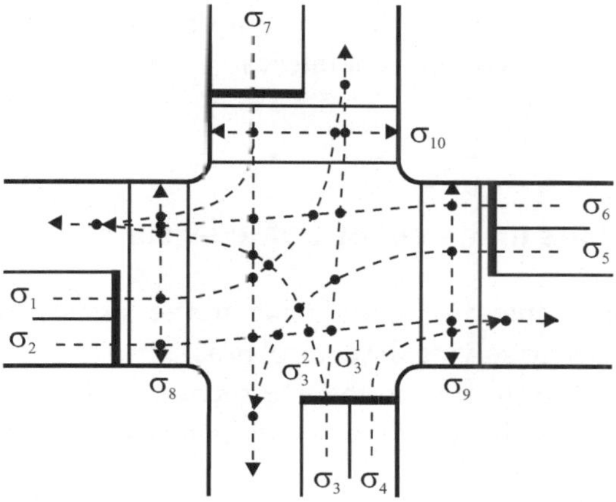

Figure 2.1

For every traffic stream σ_i on an intersection, it is necessary to have the following information:

a) The *approach* T_i used by vehicles belonging to stream σ_i, with defined lanes assigned to the stream, and traffic signs that regulate direction of movement through the intersection.

b) The *mathematical model* of the random process, which describes vehicle arrivals. The most important characteristic of the model is the volume, expressed in pcu per time unit. The Poisson process is most often used.

c) The *average arrival volume* of the traffic stream, and if it is a multicomponent stream, then also the average volumes of partial streams, expressed as percentages of the arrival volume.

d) The *type* of the traffic stream. Traffic streams represent different flows: vehicles, pedestrians, trams, etc.

e) The *trajectory* of the traffic stream through the intersection. If the stream comprises partial traffic streams, their trajectories have to be known also.

Traffic streams on an intersection, as mentioned in Section 2.1, can be represented as components of a vector (the vector of arrival flows or uncontrolled input):

$$\sigma = (\sigma_1, \sigma_2, \ldots, \sigma_i, \ldots, \sigma_I).$$

The uncontrolled input, σ, is transformed, by control, into the output, i.e., the output flows in interval $[t_0, t]$. Vector σ is an element of set $\Sigma_{[t_0,t]}$, whose elements are all uncontrolled inputs in interval $[t_0, t]$, i.e.,

$$\sigma \in \Sigma_{[t_0,t]} .$$

2.3. Basic relations in the set of traffic streams

For solving the problem of introducing traffic signals on an isolated intersection, as well as for solving the control problem, it is necessary to know relations between traffic streams on the intersection. A thorough analysis of these relations is very significant if their combinatorial effects are to be taken into account.

Since the main objective of traffic control by traffic lights is to give the right-of-way to some traffic streams, and to stop others, it is necessary to find, in the set of traffic streams of an intersection, the traffic streams that can simultaneously get the right-of-way. Therefore, a *traffic stream compatibility relation* is introduced, defined by a set of traffic stream pairs, such that elements of the pair can simultaneously get the right-of-way.

The traffic stream compatibility relation plays an important role in solving traffic control problems related to:

- Deciding whether traffic control by traffic lights should be introduced at an intersection
- Assigning control variables to traffic streams, or to subsets of traffic streams
- The traffic control process on an intersection

The factors that have to be considered when defining the compatibility relation are:

- The intersection geometry
- Factors related to traffic process safety, for which expert estimations of traffic engineers are needed

The analysis of intersection geometry considers mutual relations of trajectories of traffic streams. Obviously, when trajectories of two traffic streams do not cross, these streams can simultaneously get the right-of-way, i.e., they are *compatible*. On the other hand, when trajectories of two traffic streams do cross, the streams are in a *conflict*, and their simultaneous movement through the intersection should not be permitted.

If volumes are not high, a "filtering" of one stream through another can be permitted in some cases. However, when determining the compatibility relation, some special requirements should be taken into account, e.g., some streams are required to pass through the intersection without any disturbance, although filtering could be permitted if only their volumes were considered. These requirements are usually achieved by so-called *directional signals*.

When only geometrical factors are considered, the *relation of conflictness* and the *relation of nonconflictness* can be defined.

Traffic streams on an intersection are elements of the set of traffic streams, $\mathscr{S}$, i.e.,

$$\mathscr{S} = \{\sigma_1, \sigma_2, \ldots, \sigma_i, \ldots, \sigma_\mathrm{I}\}, \tag{2.16}$$

where $i \in \mathscr{I}$, and $\mathscr{I}$ is the set of traffic stream indices (1.4):

$$\mathscr{I} = \{1,2,\ldots,i,\ldots,\mathrm{I}\} = \{1,2,\ldots,i,\ldots,\mathrm{I}',\ldots,\mathrm{I}\}.$$

Indices $i = 1,2,\ldots,\mathrm{I}'$ are assigned to vehicle traffic streams, and indices $i = \mathrm{I}'+1,\ldots,\mathrm{I}$ to pedestrian and other traffic streams.

Elements of set $\mathscr{S}$ are also components of vector $\sigma = (\sigma_1, \sigma_2, \ldots, \sigma_\mathrm{I})$, which describes the uncontrollable system input.

For exact statement and solution of traffic control problems, it is necessary to study the relations of conflictness, nonconflictness, and compatibility.

2.3.1. Conflictness relation of traffic streams

The conflictness relation is illustrated by the intersection presented in Fig. 2.1, with the set of traffic streams

$$\mathscr{S} = \{\sigma_1, \sigma_2, \ldots, \sigma_{10}\}.$$

Some pairs of traffic stream use, along a part of their trajectories, the same space on the intersection, i.e., the conflict area. These are the streams whose trajectories cross or merge. A *conflict* exists between such streams.

The set of all pairs of traffic streams that creates a conflict between elements of the pair represents the conflictness relation. Thus, the conflictness relation, C_1, can be defined in the following way:

$$C_1 \subset \mathscr{S} \times \mathscr{S}, \tag{2.17}$$

i.e.,

$$C_1 = \{(\sigma_i, \sigma_j) \mid \text{the trajectories of } \sigma_i \text{ and } \sigma_j \text{ cross or merge}, \; \sigma_i, \sigma_j \in \mathscr{S}\}. \tag{2.18}$$

The graph of conflictness, G_k, is defined by set $\mathscr{S}$ and relation C_1:

$$G_k = (\mathscr{S}, C_1)\,. \tag{2.19}$$

The incidence matrix, B', of graph G_k for the intersection given in Fig. 2.1 is:

$$B' = [b'_{ij}]_{10\times 10} = \begin{bmatrix} 0 & 0 & 1 & 0 & 0 & 1 & 1 & 1 & 0 & 1 \\ 0 & 0 & 1 & 1 & 1 & 0 & 1 & 1 & 1 & 0 \\ 1 & 1 & 0 & 0 & 1 & 1 & 1 & 1 & 0 & 1 \\ 0 & 1 & 0 & 0 & 0 & 0 & 0 & 0 & 1 & 0 \\ 0 & 1 & 1 & 0 & 0 & 0 & 1 & 0 & 1 & 0 \\ 1 & 0 & 1 & 0 & 0 & 0 & 1 & 1 & 1 & 0 \\ 1 & 1 & 1 & 0 & 1 & 1 & 0 & 1 & 0 & 1 \\ 1 & 1 & 1 & 0 & 0 & 1 & 1 & 0 & 0 & 0 \\ 0 & 1 & 0 & 1 & 1 & 1 & 0 & 0 & 0 & 0 \\ 1 & 1 & 1 & 0 & 0 & 0 & 1 & 0 & 0 & 0 \end{bmatrix}. \tag{2.20}$$

Elements of this matrix, b'_{ij}, are defined as follows:

$$b_{ij} = \begin{cases} 1, (\sigma_i, \sigma_j) \in C_1 \\ 0, (\sigma_i, \sigma_j) \notin C_1 \end{cases}. \tag{2.21}$$

All conflict points of traffic streams are indicated in Fig. 2.1. It can be noted that the conflictness between two traffic streams means that they can have one or more than one conflict point. For instance, streams σ_2 and σ_3 have two conflict points.

Since the conflict exists between any two streams whose trajectories cross or merge, it is obvious that the conflictness relation is symmetrical:

$$\sigma_i C_1 \sigma_j \Rightarrow \sigma_j C_1 \sigma_i\,, \quad ((\sigma_i, \sigma_j) \in C_1)\,. \tag{2.22}$$

This fact can also be noted observing expression (2.20).

Relation C_1 is not reflexive (a stream cannot be in conflict by itself). Therefore, $(\sigma_i, \sigma_i) \notin C_1$, $(i \in \mathscr{J})$.

2.3.2. Nonconflictness relation of traffic streams

The nonconflictness relation of traffic streams represents a set of ordered traffic stream pairs, such that the trajectories of the pair elements do not cross nor merge. Thus, this relation is the set of all pairs of traffic streams that are not mutually in conflict:

$$C_2' = \overline{C}_1 = (\mathscr{S} \times \mathscr{S}) \setminus C_1 . \tag{2.23}$$

The graph of nonconflictness is defined by set $\mathscr{S}$ and relation C_2', as

$$G_k' = (\mathscr{S}, C_2') .$$

2.3.3. Compatibility relation of traffic streams

As already mentioned, when determining the compatibility relation of traffic streams, besides data on geometrical features of traffic stream trajectories, it is necessary to consider some other factors, i.e., it is necessary to determine:

- The pairs of conflicting traffic streams that can simultaneously get the right-of-way
- The traffic streams required to pass through the intersection without any disturbance (the streams to which the right-of-way is given by directional signals)

Some pairs of conflicting traffic streams can at the same time be pairs of compatible streams, although the streams are conflicting. Therefore, it is necessary to divide the conflicts into allowed and forbidden [26]. Forbidden conflicts can be regulated only by traffic lights, while allowed conflicts are solved by traffic participants themselves, respecting priority rules prescribed by traffic regulations. Without traffic lights, conflicts are solved by "filtering" one stream through another. Obviously, the possibility of filtering depends on vehicle spacing interval, which depends on volume of traffic streams. The volumes change during a day. There are intervals with very high volume like morning peak, afternoon peak, and intervals with significantly lower volume like off-peak and night periods. Hence, situations may arise that two conflicting traffic streams may simultaneously have the right-of-way in one period and not in another. Thus, it might be necessary to change the compatibility relation during a day.

The set of traffic stream pairs, which comprise *conditionally compatible* streams, i.e., conflicting streams allowed to simultaneously pass through an intersection, can be, thus, defined as follows:

$$C_2'' = \{(\sigma_i, \sigma_j) \mid (\sigma_i, \sigma_j) \in C_1 \text{, streams } \sigma_i \text{ and } \sigma_j \text{ can simultaneously get the right-of-way} \}. \tag{2.24}$$

The problem of introducing traffic signals for traffic control on an intersection is actually a problem of the same kind. Namely, it is necessary to determine when traffic lights have to be introduced in order to remove conflicts, i.e., which are the values of traffic stream volumes when filtering is not possible anymore. Before traffic signals are introduced, traffic participants themselves, using filtering and respecting priority rules, solve all conflicts.

When volumes of conflicting traffic streams reach some level that filtering becomes impossible, the introduction of traffic lights becomes unavoidable because traffic participants themselves cannot solve the conflicts. The values of traffic stream volumes that justify signalization of an intersection are given in tables in various traffic-engineering handbooks. Avoiding introducing traffic lights when these levels are reached can lead to many negative effects, such as enormous delay and number of stops, increase in the number of traffic accidents, etc. Therefore, conflicts at all conflict points on an unsignalized intersection are prevented by traffic participants respecting priority rules, while at a signalized intersection traffic lights are used in order to avoid conflicts at most of the conflict points, with a possibility of some conflict points still left for "self-regulation" by traffic participants.

The compatibility relation of traffic stream pairs whose elements can simultaneously get the right-of-way is:

$$C_2 = C_2' \cup C_2'' . \tag{2.25}$$

In some cases it may be necessary to control the traffic in such a way that certain streams can pass through an intersection without conditional conflicts. Then they cannot gain the right-of-way simultaneously with any conflicting streams, although that would be justified if only volumes were considered. For controlling these streams, directional signals are used.

If the set of streams that passes through the intersection without any conflict is denoted by $\mathscr{S}'$, where $\mathscr{S}' \subset \mathscr{S}$, then the set of pairs of traffic streams that can simultaneously get the right-of-way is defined by the following expression:

$$C_3 = C_2 \setminus \{(\sigma_i,\sigma_j) \mid (\sigma_i,\sigma_j) \in C_2'', \quad (\sigma_i \text{ or } \sigma_j \in \mathscr{S}')\} . \tag{2.26}$$

Assuming that each traffic stream is compatible with itself, in order to define the set of pairs that define the compatibility relation, the set of pairs C_3 should be extended by the diagonal Δ_S in set $\mathscr{S}$.

Therefore, the compatibility relation can be defined as:

$$C = C_3 \cup \Delta_S , \tag{2.27}$$

where

$$\Delta_S = \{(\sigma_i,\sigma_i) \mid \sigma_i \in \mathscr{S}\}, \quad (i \in \mathscr{I}) . \tag{2.28}$$

Relation C is symmetric and reflexive.

The compatibility graph of traffic streams is defined by the set of traffic streams, $\mathscr{S}$, and compatibility relation C:

$$G_c = (\mathscr{S}, C)\,. \tag{2.29}$$

Since set $\mathscr{S}$ is finite, and relation C symmetric and reflexive, graph G_c is a finite, nonoriented graph, with a loop at each node. The incidence matrix of this graph is $B = [b_{ij}]_{\mathrm{I}\times\mathrm{I}}$, where $\mathrm{I} = card\,\mathscr{S}$. Elements of the incidence matrix are defined as

$$b_{ij} = \begin{cases} 1, & (\sigma_i, \sigma_j) \in C \\ 0, & (\sigma_i, \sigma_j) \notin C \end{cases} \quad (i, j \in \mathscr{I})\,. \tag{2.30}$$

A compatibility graph does not have to be a connected graph. In Example 2.1.γ, the compatibility graph has two connected components. In some cases a connected component can consist of only one node.

Example 2.1

α) For the intersection presented in Fig. 2.1 determine the compatibility relation and compatibility graph if the set of permitted conflicts is defined by the following relation:

$$C_2'' = \{(\sigma_1, \sigma_6), (\sigma_1, \sigma_{10}), (\sigma_2, \sigma_4), (\sigma_5, \sigma_7), (\sigma_7, \sigma_8)\}\,,$$

and the set of traffic streams that shall have no disturbance in passing through the intersection (controlled by directional signals) is:

$$\mathscr{S}' = \{\sigma_1\}\,.$$

All conflict points (the point at which trajectories of traffic streams cross or merge) are shown in Fig. 2.1.

The conflict points for stream σ_1 are points A, B, C, D, E, and F (Fig. 2.2). The allowed conflicts are marked by B and C (circled points), i.e., $(\sigma_1, \sigma_{10}) \in C_2$, $(\sigma_1, \sigma_6) \in C_2$.

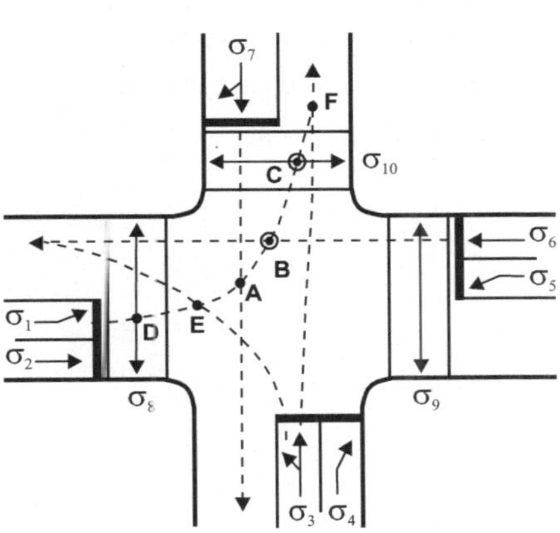

Figure 2.2

Traffic stream σ_1 could be compatible with σ_6 and σ_{10}. However, stream σ_1 is controlled by a directional signal, and the vehicles belonging to this stream shall not meet any obstacle. Therefore, stream σ_1 cannot be compatible with streams σ_6 and σ_{10}.

The compatibility graph incidence matrix is:

$$B = [b_{ij}]_{10\times 10} = \begin{bmatrix} 1 & 1 & 0 & 1 & 1 & \cancel{\textcircled{1}} & 0 & 0 & 1 & \cancel{\textcircled{1}} \\ 1 & 1 & 0 & \textcircled{1} & 0 & 1 & 0 & 0 & 0 & 1 \\ 0 & 0 & 1 & 1 & 0 & 0 & 0 & 0 & 1 & 0 \\ 1 & \textcircled{1} & 1 & 1 & 1 & 1 & 1 & 1 & 0 & 1 \\ 1 & 0 & 0 & 1 & 1 & 1 & \textcircled{1} & 1 & 0 & 1 \\ \cancel{\textcircled{1}} & 1 & 0 & 1 & 1 & 1 & 0 & 0 & 0 & 0 \\ 0 & 0 & 0 & 1 & \textcircled{1} & 0 & 1 & \textcircled{1} & 1 & 0 \\ 0 & 0 & 0 & 1 & 1 & 0 & \textcircled{1} & 1 & 1 & 1 \\ 1 & 0 & 1 & 0 & 0 & 0 & 1 & 1 & 1 & 1 \\ \cancel{\textcircled{1}} & 1 & 0 & 1 & 1 & 0 & 0 & 1 & 1 & 1 \end{bmatrix}.$$

The circled 1s in matrix B correspond to pairs of signal groups that can simultaneously have the right-of-way in spite of their conflictness. The circled and crossed 1s represent the conflicts of stream σ_1 that would be allowed if stream σ_1 were not controlled by the directional signal. Since $\sigma_1 \in \mathscr{S}'$, the stream pairs (σ_1, σ_6) and (σ_1, σ_{10}) cannot belong to compatibility relation C, and therefore $b_{1,6} = 0$ and $b_{1,10} = 0$ (also, $b_{6,1} = b_{10,1} = 0$).

The compatibility graph (G_c') of traffic streams for the intersection in Fig. 2.1 is presented in Fig. 2.3 for the case when stream σ_1 is controlled by an ordinary, nondirectional signal, and Fig. 2.4 presents compatibility graph G_c for the same intersection if stream σ_1 is controlled by the directional signal.

The edges of graph G_c' between nodes σ_1 and σ_6, as well as between nodes σ_1 and σ_{10}, do not exist in graph G_c because an edge between two nodes representing traffic streams indicates that these streams are in the compatibility relation.

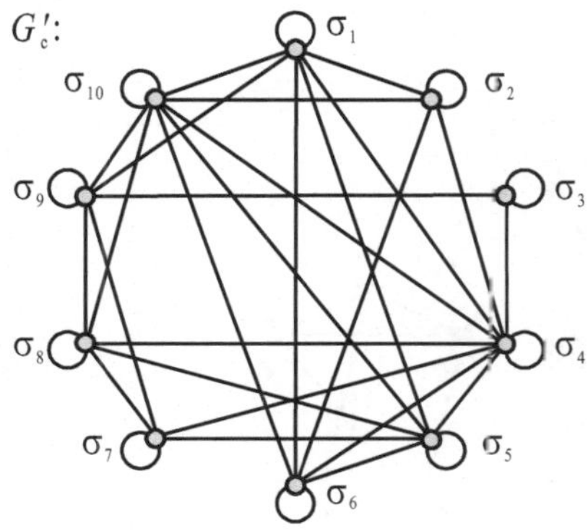

Figure 2.3

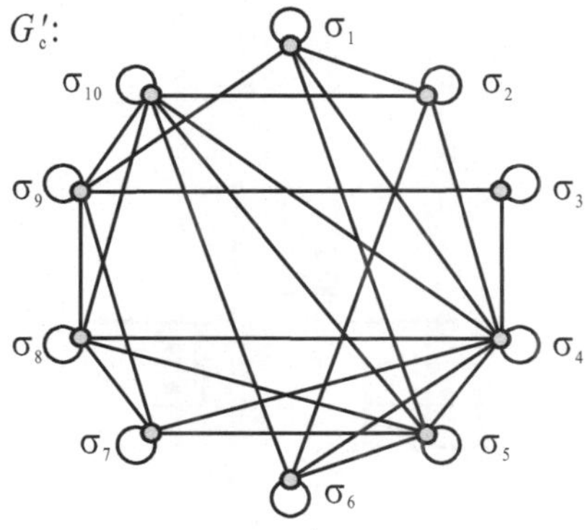

Figure 2.4

β) An intersection with six traffic streams is presented in Fig. 2.5. Determine the relation and graph of compatibility if no conflicts are allowed.

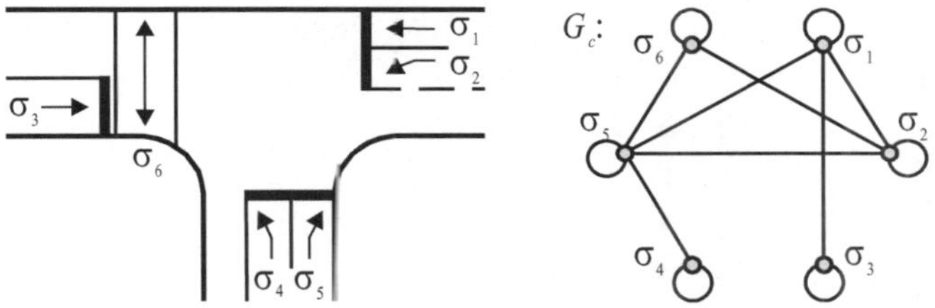

Figure 2.5

The compatibility graph is given in Fig. 2.5, and the compatibility relation is represented by matrix B, i.e., the incidence matrix of graph G_c.

$$B = \begin{bmatrix} 1 & 1 & 1 & 0 & 1 & 0 \\ 1 & 1 & 0 & 0 & 1 & 1 \\ 1 & 0 & 1 & 0 & 0 & 0 \\ 0 & 0 & 0 & 1 & 1 & 0 \\ 1 & 1 & 0 & 1 & 1 & 1 \\ 0 & 1 & 0 & 0 & 1 & 1 \end{bmatrix}.$$

γ) For the intersection presented in Fig. 2.6 determine the relation and graph of compatibility if no conflicts are allowed.

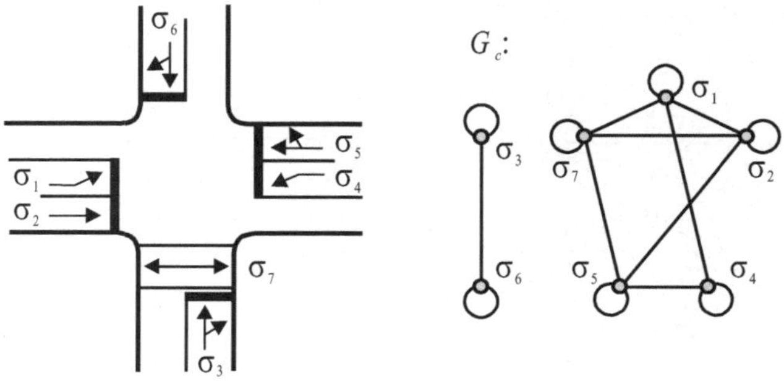

Figure 2.6

The compatibility graph is given in Fig. 2.6, and the compatibility relation is represented by matrix B, i.e., the incidence matrix of graph G_c.

$$B = \begin{bmatrix} 1 & 1 & 0 & 1 & 0 & 0 & 1 \\ 1 & 1 & 0 & 0 & 1 & 0 & 1 \\ 0 & 0 & 1 & 0 & 0 & 1 & 0 \\ 1 & 0 & 0 & 1 & 1 & 0 & 0 \\ 0 & 1 & 0 & 1 & 1 & 0 & 1 \\ 0 & 0 & 1 & 0 & 0 & 1 & 0 \\ 1 & 1 & 0 & 0 & 1 & 0 & 1 \end{bmatrix}.$$

3. SIGNAL GROUP

Introduction of a traffic control system on an intersection consists of installation of signals that will control traffic streams by different light indications. The basic intention of traffic signals introduction is, of course, prevention of simultaneous movement of incompatible traffic streams.

Traffic control at an intersection consists of giving and canceling the right-of-way to particular traffic streams. Giving/canceling the right-of-way is performed by different signal indications. Meaning of indications is assigned by a convention. A green indication for vehicles means *allowed* passage, while red means *forbidden* passage. An amber indication, appearing after a green indication, as well as a red–amber after a red indication, informs drivers that the right-of-way will be changed. The duration of amber and red–amber intervals in some countries is determined by traffic regulations, and it is most frequently specified as 3 s for amber and 2 s for red–amber indication.

Signals that control pedestrian streams usually have only two indications: red (“stop”) and green (“walk”).

The most frequently used sequence of signal indications for vehicles and for pedestrians is presented in Fig. 3.1. However, in some countries other sequences exist, such as flashing amber before a steady amber indication, or direct switching from red to green, etc.

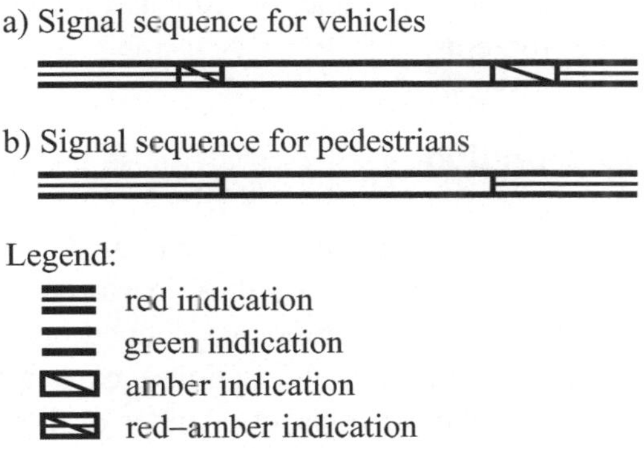

Figure 3.1

The control of traffic lights, i.e., forming and implementing specified signal sequences, is performed by an electronic device—a traffic controller. A controller changes signal indications using sequence of pulses.

Changes of signal indications are described by a mathematical variable, so-called *control variable*. Control variables can be assigned to *every* traffic stream. However, the fact that compatible traffic streams can simultaneously gain/lose the right-of-way makes it possible that a subset of traffic streams, comprising several compatible streams, can be controlled by a single control variable [34].

Therefore, one of the first problems to be solved when introducing traffic lights control at an intersection is to establish the correspondence between traffic streams and traffic signal sequences, i.e., the control variables that control these traffic streams.

The simplest assignment of control variables to traffic streams is to assign one control variable to each traffic stream. However, there exist some practical reasons that make this assignment not generally applicable.

Technical and economic considerations in earlier years of traffic control systems development had caused a tendency to minimize the number of control variables. Namely, in this case the traffic controller could be simpler, with a smaller number of modules that form control variables, and thus it would give a cheaper solution. This reasoning is nowadays not as important as it was before. Modern, microprocessor-based traffic controllers can control almost any number of signal groups, and the price of their components (modules) is not so high any more.

Modern traffic controllers can implement more complex control algorithms than was possible before their introduction. However, increasing the number of control variables significantly emphasizes the combinatorial nature of traffic control problems.

3.1. Signal group definition

Various intersection performance indices depend on the choice of traffic control systems for an intersection. Among these performance indices are: total delay or total number of vehicle stops in a defined interval, total flow through the intersection (for saturated intersections), capacity factor, linear combination of delay and number of stops, etc. Values of these performance indices also depend on the assignment of control variables to traffic streams. The best results are, obviously, obtained if each traffic stream is controlled by one control variable.

When the number of control variables is less than the number of traffic streams, certain constraints have to be introduced that will express the requirement that several traffic streams simultaneously get/lose the right-of-way. The consequence of these constraints is "corruption" of optimum values of performance indices compared to the case when each traffic stream is at the same time a signal group.

The reduction in the number of control variables results in simplification of traffic control problems and also in possibility to use cheaper and simpler traffic controllers.

In real-time traffic control systems, in which data on current values of traffic stream parameters are used for determining values of control variables, a particular attention has to be devoted to choosing appropriate set of control variables and their assignment to traffic streams.

Determination of the set of control variables is very complex due to all the mentioned reasons. This problem, in fact, is the problem of partitioning the set of traffic streams, $\mathscr{S}$, in subsets of traffic streams, such that a single control variable can be used to control a subset.

A subset of traffic streams that simultaneously gains/loses the right-of-way, i.e., that is controlled by a single control variable, is called a *signal group*.

Another way to define a signal group is as follows: A signal group is the set of traffic streams that are controlled by identical traffic signal indications. Some authors define a signal group as the set of signals on various traffic lights that always show a same indication [69]. For traffic equipment manufacturers, a signal group is a controller module, which always produces one sequence of traffic signal indications.

It is obvious that traffic streams belonging to one signal group have to be mutually compatible. However, it is not enough. Namely, signals used for control of traffic streams of various types—vehicles, pedestrians, trams, etc., cannot always have the same indications, which is necessary if they were to belong to a same signal group. Vehicle streams are, for example, controlled by signal sequences with four indications, while for pedestrian streams only two indications are used. Therefore, signal groups are formed to contain only same types of traffic streams and the set of traffic streams $\mathscr{S}$ has to be partitioned in several subsets: the subset of *vehicle traffic streams*, the subset of *pedestrian traffic streams*, etc.

According to the signal group definition, for the intersection presented in Fig. 2.5, together with its compatibility graph, the signal groups are the following subsets: $D_1 = \{\sigma_1, \sigma_2, \sigma_5\}$, $D_2 = \{\sigma_1, \sigma_3\}$, $D_3 = \{\sigma_6\}$, etc.

A signal group D_p represents a subset of the set of traffic streams $\mathscr{S}$ and can be represented as follows:

$$D_p = \{\sigma_{p1}, \sigma_{p2}, \ldots, \sigma_{pe}, \ldots, \sigma_{p\mathrm{E}(p)}\}, \tag{3.1}$$

where $\sigma_{pe} \in \mathscr{S}$, $e \in \mathscr{E}_p$ and $\mathscr{E}_p$ is the set of traffic stream indices in signal group D_p, i.e.,

$$\mathscr{E}_p = \{1, 2, \ldots, e, \ldots, \mathrm{E}(p)\}.$$

3.2. The relation of identical signal indications (Identity relation)

In order to form signal groups, it is necessary to determine for each pair of compatible traffic streams whether they can be controlled by traffic lights that always have identical indications. The set of such signal group pairs represents a relation in the set of traffic streams $\mathscr{S}$. Since this relation determines whether identical traffic light indications can be used for controlling signal group pairs, it is called the relation of identical signal indications, or the identity relation.

The identity relation C_α is defined as:

$$\begin{aligned} C_\alpha = \{(\sigma_i, \sigma_j) \mid & \text{ traffic streams } \sigma_i \text{ and } \sigma_j \text{ can be controlled} \\ & \text{by a single control variable,} \\ & \sigma_i, \sigma_j \in \mathscr{S}\}. \end{aligned} \tag{3.2}$$

Relation C_α can be represented as:

$$C_\alpha = C \setminus C_4,$$

where

$$\begin{aligned} C_4 = \{(\sigma_i, \sigma_j) \mid & ((\sigma_i, \sigma_j) \in C) \wedge \\ & \wedge (\sigma_i \in \mathscr{S}^f, \sigma_j \in \mathscr{S}^l, f, l \in \mathscr{F}, f \neq l), \\ & \sigma_i, \sigma_j \in \mathscr{S})\}. \end{aligned} \tag{3.3}$$

The subsets $\mathscr{S}^1, \mathscr{S}^2, \ldots, \mathscr{S}^f, \ldots, \mathscr{S}^\mathrm{F}$ represent subsets of signal groups that are of the same type (vehicle, pedestrian, tram, etc.). Traffic streams of one type are controlled by signals that have same sequences of indications. For vehicle traffic streams, for example, this sequence is: green, amber, red, red–amber.

The set $\mathscr{F}$ is the index set of signal group types, i.e., signal types:

$$\mathscr{F} = \{1,2,\ldots,f,\ldots \mathrm{F}\}. \tag{3.4}$$

The collection

$$\overline{\mathscr{S}} = \{\mathscr{S}^1,\mathscr{S}^2,\ldots,\mathscr{S}^f,\ldots,\mathscr{S}^{\mathrm{F}}\} \tag{3.5}$$

represents a partition of set $\mathscr{S}$.

Hence, we have:

$$\bigcup_{f=1}^{\mathrm{F}} \mathscr{S}^f = \mathscr{S} \tag{3.6}$$

$$\mathscr{S}^f \cap \mathscr{S}^l = \varnothing, \quad (f \in \mathscr{F}, \quad l \in \mathscr{F}, \quad f \neq l). \tag{3.7}$$

The relation of identical traffic signal indications C_α is:

a) Reflexive, i.e.,

$$(\sigma_i,\sigma_i) \in C_\alpha, \quad (\sigma_i \in \mathscr{S}) \tag{3.8}$$

b) Symmetric, i.e.,

$$(\sigma_i,\sigma_j) \in C_\alpha \Rightarrow (\sigma_j,\sigma_i) \in C_\alpha, \quad ((\sigma_i,\sigma_j) \in C_\alpha) \tag{3.9}$$

To an identity relation there corresponds the identity graph:

$$G_\alpha = (\mathscr{S},C_\alpha) = (\mathscr{S},\Gamma_\alpha), \tag{3.10}$$

where Γ_α is

$$\Gamma_\alpha : \mathscr{S} \to \mathcal{P}(\mathscr{S}).$$

Besides the definition $G_\alpha = (\mathscr{S},C_\alpha)$ in further discussion the definition $G_\alpha = (\mathscr{S},\Gamma_\alpha)$ is also used (Appendix I).

The identity graph given in Fig. 3.2 refers to the intersection with ten traffic streams, presented in Fig. 2.1, with its compatibility graph of traffic streams given in Fig. 2.3. This identity graph corresponds to the identity relation in the case when no directional signals are used.

If traffic stream σ_1 is controlled by a directional signal, the identity graph will have the form presented in Fig. 3.3 and the corresponding compatibility graph is given in Fig. 2.4.

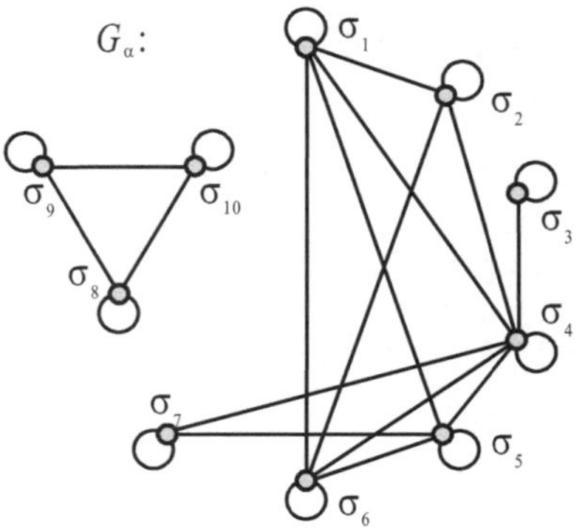

Figure 3.2

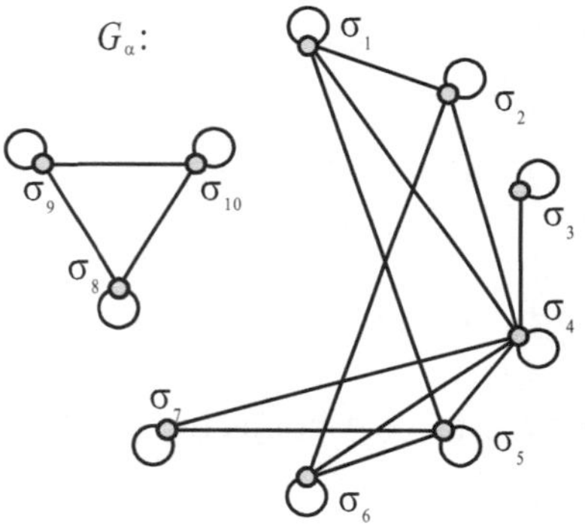

Figure 3.3

The identity graph given in Fig. 3.4 refers to the intersection with six traffic streams presented in Fig. 2.5, together with its compatibility graph.

If traffic streams of various types pass through an intersection ($F > 1$), the identity graph G_α is a nonconnected graph. The number of connected components is equal to or greater than the number of stream types F. Graph G_α is a nonoriented graph with a loop in each node.

Since graphs $G_c = (\mathscr{S}, C)$ and $G_\alpha = (\mathscr{S}, C_\alpha)$ have the same set of nodes, and $C_\alpha \subseteq C$, then the identity graph G_α is a subgraph of the compatibility graph G_c.

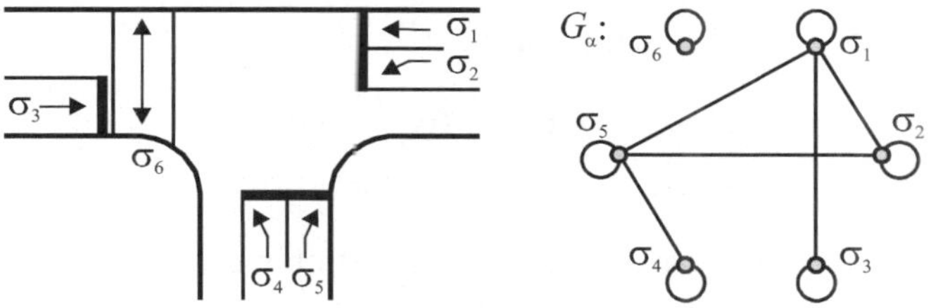

Figure 3.4

3.3. The complete set of signal groups

The identity relation C_α defines the set of signal group pairs that can be controlled by identical signal indications, and the identity graph G_α enables determination of all set $\mathscr{S}$ subsets that represent signal groups.

A set of nodes of any subgraph of identity graph G_α, such that the subgraph is a complete graph, represents, in fact, a signal group. Since the set of nodes of a complete subgraph of a graph is a clique (Appendix I), a signal group can be also defined in the following way:

A signal group is a clique (in Berge's sense [9]) of the graph of identical signal indications G_α.

For traffic control at an intersection, therefore, it is necessary to determine a set of signal groups such that each element of set $\mathscr{S}$ belongs to one and only one signal group, i.e., clique of graph G_α. Such set of signal groups is called a *complete set of signal groups* and it represents a partitioning of set $\mathscr{S}$.

For one graph of identical signal indications there can exist several complete sets of signal groups. This means that one intersection can be controlled in several ways, based on the choice of the complete set of signal groups. Introducing an appropriate measure for comparison of complete sets of signal groups, the choice of the complete set can be formulated as an optimization problem, which could be stated as follows: Find the complete set of signal groups such that the value of the chosen performance index is optimized. The set of feasible solutions for this problem is the collection of all complete sets of signal groups.

3.3.1. Collection of all complete sets of signal groups

A complete set of signal groups represents one partitioning of set $\mathscr{S}$. Therefore, a complete set of signal groups represents an equivalence relation

C_ε^m in set $\mathscr{S}$ [47], which is defined as:

$$C_\varepsilon^m = \{(\sigma_i,\sigma_j) \mid (((\sigma_i,\sigma_j)\in C_\alpha \wedge ((\sigma_j,\sigma_k)\in C_\alpha)) \\ \Rightarrow ((\sigma_i,\sigma_k)\in C_\alpha)\ \ i,j,k \in \mathscr{J})\}. \tag{3.11}$$

The signal groups that are elements of the complete set of signal groups are equivalence classes of this relation, and the complete set of signal groups is the quotient set $\mathscr{S}/C_\varepsilon^m$ of set $\mathscr{S}$ by equivalence relation C_ε^m (Appendix II). In this definition, $m \in \mathscr{M} = \{1,2,\ldots,\mathrm{M}\}$, and M is the total number of equivalence relations defined in set $\mathscr{S}$, such that their equivalence classes are cliques of the graph of identical signal indications G_α.

Therefore, the problem of determining all complete sets of signal groups for a given graph of identical signal indications $G_\alpha = (\mathscr{S},\Gamma_\alpha)$ can be formulated as follows: Find quotient sets by all equivalence relations defined in set $\mathscr{S}$ such that the equivalence classes of these relations are cliques of graph G_α, i.e., in another words, find all partitionings of set $\mathscr{S}$ such that the subsets into which $\mathscr{S}$ is partitioned are cliques of graph G_α.

The procedure for determining all complete sets of signal groups for a given graph G_α consists of the following steps:

- Determining the set of all cliques of graph G_α
- Forming the collection of quotient sets of all equivalence relations defined in set $\mathscr{S}$, with the equivalence classes of these relations being cliques of graph G_α, i.e., elements of the set determined in step 1

These steps are determined by functions d' and d'', which are defined below, in the scope of the steps explanation.

a) Determination of the set of all cliques of graph $G_\alpha = (\mathscr{S},\Gamma_\alpha)$

Graph G_α cliques, which are in fact signal groups, represent subsets of set $\mathscr{S}$, such that for any clique D_{l_c} of graph G_α there holds $D_{l_c} \in \mathcal{P}(\mathscr{S})$. The set of all cliques $\mathfrak{D}$ of graph G_α is a collection of subsets of set $\mathscr{S}$, i.e., $\mathfrak{D} \in \mathcal{P}(\mathcal{P}(\mathscr{S}))$, where $\mathcal{P}(\mathscr{S})$ is a partitive set of set $\mathscr{S}$.

Function $d'(G_\alpha)$, by which set $\mathfrak{D}$ is determined, is defined as follows:

$$d'(G_\alpha) = \mathfrak{D} = \{ D_{l_c} \mid (\sigma_i \in D_{l_c}) \wedge (\sigma_j \in D_{l_c}) \Rightarrow (\sigma_j \in \Gamma_\alpha \sigma_i), l_c \in \mathscr{L}_c \} \\ = \{ D_1, D_2, \ldots, D_{l_c}, \ldots, D_{\mathrm{L}_c} \}, \tag{3.12}$$

where

$$\mathscr{L}_c = \{1,2,\ldots,l_c,\ldots,\mathrm{L}_c\},$$

because each clique is the set of nodes of a complete graph.

Relation C_α is reflexive (a loop exists in each node of graph Γ_α), so that the sets containing just a single node are also cliques of this graph, i.e., they are signal groups also. This can be expressed as $\sigma_i \in \Gamma_\alpha \sigma_i$, $(\sigma_i \in \mathscr{S})$.

The set of all cliques, $\mathfrak{T}$, of a graph is obtained using a CLIQ program, the pseudocode of which is given in Appendix III. Other algorithms used for finding graph cliques are listed in Appendix VI.

b) Forming the collection of quotient sets by all equivalence relations defined in set $\mathscr{S}$, with the equivalence classes of these relations being cliques of graph $G_\alpha = (\mathscr{S}, \Gamma_\alpha)$

Collection $\mathfrak{D}$ contains all cliques of graph G_α. Thus, it is obvious that one quotient set, $\mathfrak{D}_a^m$, i.e., a complete set of signal groups, is a subset of collection $\mathfrak{D}$, namely:

$$\mathfrak{D}_a^m = \mathscr{S} / C_\varepsilon^m \in \mathcal{P}(\mathfrak{D}), \text{ or } \mathfrak{D}_a^m \subset \mathfrak{D}, (m \in \mathscr{M}) \tag{3.13}$$

$$\mathfrak{D}_a^m = \mathscr{S} / C_\varepsilon^m = \{D_1^m, D_2^m, \ldots, D_p^m, \ldots, D_{\mathrm{P}_m}^m\}, \quad (m \in \mathscr{M}) \tag{3.14}$$

and the collection of all complete sets of signal groups $\mathfrak{D}_b$ is a subset of set $\mathcal{P}(\mathfrak{D})$, i.e.,

$$\mathfrak{D}_b \in \mathcal{P}(\mathcal{P}(\mathfrak{D})). \tag{3.15}$$

Collection $\mathfrak{D}_b$ is the collection of quotient sets by all equivalence relations C_ε^m, i.e.,

$$\mathfrak{D}_b = \{\mathfrak{D}_a^1, \mathfrak{D}_a^2, \ldots, \mathfrak{D}_c^m, \ldots, \mathfrak{D}_a^{\mathrm{M}}\}, \tag{3.16}$$

where:

$\mathscr{M}$ – the index set of all equivalence relations C_ε^m, i.e.,

$$\mathscr{M} = \{1, 2, \ldots, m, \ldots, \mathrm{M}\}, \tag{3.17}$$

$\mathscr{P}_m$ – the index set of all classes of equivalence C_ε^m, i.e.,

$$\mathscr{P}_m = \{1, 2, \ldots, p, \ldots, \mathrm{P}_m\}. \tag{3.18}$$

Since each equivalence relation and its quotient set, $\mathfrak{D}_a^m$, determine one partitioning of set $\mathscr{S}$, function d'' can be defined as follows:

$$d''(\mathfrak{D}) = \mathfrak{D}_b = \{ \mathfrak{D}_a^m \mid ((D_p^m \in \mathfrak{D}_a^m)(D_r^m \in \mathfrak{D}_a^m)((D_p^m \cap D_r^m = \varnothing), \; p \in \mathscr{P}_m, r \in \mathscr{P}_m, p \neq r, \; \bigcup_{p=1}^{\mathrm{P}_m} D_p^m = \mathscr{S}), m \in \mathscr{M}\}. \tag{3.19}$$

Collection $\mathfrak{D}_b$ of all complete sets of signal groups for a given graph G_α is determined by composition of functions d' and d'':

$$\mathfrak{D}_b = d(G_\alpha) = d' \circ d'' . \tag{3.20}$$

The following example illustrates determination of all complete sets of signal groups for an intersection with six signal groups.

Example 3.1

Determine the collection of all complete sets of signal groups for the graph of identical signal indications given in Fig. 3.4.

The graph of identical signal indications, $G_\alpha = (\mathcal{S}, \Gamma_\alpha)$ is defined as follows:

$$\mathcal{S} = \{\sigma_1, \sigma_2, \sigma_3, \sigma_4, \sigma_5, \sigma_6\}$$

$$\Gamma_\alpha \sigma_1 = \{\sigma_1, \sigma_2, \sigma_3, \sigma_5\} \quad \Gamma_\alpha \sigma_2 = \{\sigma_2, \sigma_1, \sigma_5\} \quad \Gamma_\alpha \sigma_3 = \{\sigma_3, \sigma_1\}$$

$$\Gamma_\alpha \sigma_4 = \{\sigma_4, \sigma_5\} \quad \Gamma_\alpha \sigma_5 = \{\sigma_5, \sigma_1, \sigma_2, \sigma_4\} \quad \Gamma_\alpha \sigma_6 = \{\sigma_6\}$$

1) Determining the set of all cliques

The set of all cliques, $\mathfrak{D}$ of graph G_α, i.e., the set of all signal groups, determined by CLIQ program, is:

$$\begin{aligned}\mathfrak{D} = d'(G_\alpha) &= \{\{\sigma_1\}, \{\sigma_2\}, \{\sigma_3\}, \{\sigma_4\}, \{\sigma_5\}, \{\sigma_6\}, \{\sigma_1, \sigma_2\},\\ &\quad \{\sigma_1, \sigma_3\}, \{\sigma_1, \sigma_5\}, \{\sigma_2, \sigma_5\}, \{\sigma_4, \sigma_5\}, \{\sigma_1, \sigma_2, \sigma_5\}\}\\ &= \{D_1, D_2, \ldots, D_{12}\}.\end{aligned}$$

The signal groups are:

$$D_1 = \{\sigma_1\},\ D_2 = \{\sigma_2\},\ D_3 = \{\sigma_3\},\ D_4 = \{\sigma_4\},\ D_5 = \{\sigma_5\},\ D_6 = \{\sigma_6\},$$

$$D_7 = \{\sigma_1, \sigma_2\},\ D_8 = \{\sigma_1, \sigma_3\},\ D_9 = \{\sigma_1, \sigma_5\},\ D_{10} = \{\sigma_2, \sigma_5\},$$

$$D_{11} = \{\sigma_4, \sigma_5\},$$

$$D_{12} = \{\sigma_1, \sigma_2, \sigma_5\}.$$

2) Forming the collection $\mathfrak{D}_b$ of the complete sets of signal groups

The elements $\mathfrak{D}_a^m$ of collection $\mathfrak{D}_b$ are the following partitionings of set $\mathcal{S}$:

$$\begin{aligned}\mathfrak{D}_a^1 &= \mathcal{S} / C_\varepsilon^1 = \{\ \{\sigma_1\}, \{\sigma_2\}, \{\sigma_3\}, \{\sigma_4\}, \{\sigma_5\}, \{\sigma_6\}\ \}\\ &= \{D_1^1, D_2^1, \ldots, D_6^1\} = \{D_1, D_2, D_3, D_4, D_5, D_6\}\end{aligned}$$

$$\mathfrak{D}_a^2 = \mathscr{S} / C_\varepsilon^2 = \{\{\sigma_3\},\{\sigma_4\},\{\sigma_5\},\{\sigma_6\},\{\sigma_1,\sigma_2\}\}$$
$$= \{D_1^2, D_2^2, \ldots, D_5^2\} = \{D_3, D_4, D_5, D_6, D_7\}$$

$$\mathfrak{D}_a^3 = \mathscr{S} / C_\varepsilon^3 = \{\{\sigma_1\},\{\sigma_4\},\{\sigma_5\},\{\sigma_6\},\{\sigma_1,\sigma_3\}\}$$
$$= \{D_1^3, D_2^3, \ldots, D_5^3\} = \{D_2, D_4, D_5, D_6, D_8\}$$

$$\mathfrak{D}_a^4 = \mathscr{S} / C_\alpha^4 = \{\{\sigma_2\},\{\sigma_3\},\{\sigma_4\},\{\sigma_6\},\{\sigma_1,\sigma_5\}\}$$
$$= \{D_1^4, D_2^4, \ldots, D_5^4\} = \{D_2, D_3, D_4, D_6, D_9\}$$

$$\mathfrak{D}_a^5 = \mathscr{S} / C_\varepsilon^5 = \{\{\sigma_1\},\{\sigma_3\},\{\sigma_4\},\{\sigma_6\},\{\sigma_2,\sigma_5\}\}$$
$$= \{D_1^5, D_2^5, \ldots, D_5^5\} = \{D_1, D_3, D_4, D_6, D_{10}\}$$

$$\mathfrak{D}_a^6 = \mathscr{S} / C_\varepsilon^6 = \{\{\sigma_1\},\{\sigma_2\},\{\sigma_3\},\{\sigma_6\},\{\sigma_4,\sigma_5\}\}$$
$$= \{D_1^6, D_2^6, \ldots, D_5^6\} = \{D_1, D_2, D_3, D_6, D_{11}\}$$

$$\mathfrak{D}_a^7 = \mathscr{S} / C_\varepsilon^7 = \{\{\sigma_3\},\{\sigma_4\},\{\sigma_6\},\{\sigma_1,\sigma_2,\sigma_5\}\}$$
$$= \{D_1^7, D_2^7, \ldots, D_4^7\} = \{D_3, D_4, D_6, D_{12}\}$$

$$\mathfrak{D}_a^8 = \mathscr{S} / C_\varepsilon^8 = \{\{\sigma_3\},\{\sigma_6\},\{\sigma_1,\sigma_2\},\{\sigma_4,\sigma_5\}\}$$
$$= \{D_1^8, D_2^8, \ldots, D_4^8\} = \{D_3, D_6, D_7, D_{11}\}$$

$$\mathfrak{D}_a^9 = \mathscr{S} / C_\varepsilon^9 = \{\{\sigma_4\},\{\sigma_6\},\{\sigma_1,\sigma_3\},\{\sigma_2,\sigma_5\}\}$$
$$= \{D_1^9, D_2^9, \ldots, D_4^9\} = \{D_4, D_6, D_8, D_{10}\}$$

$$\mathfrak{D}_a^{10} = \mathscr{S} / C_\varepsilon^{10} = \{\{\sigma_2\},\{\sigma_6\},\{\sigma_1,\sigma_3\},\{\sigma_4,\sigma_5\}\}$$
$$= \{D_1^{10}, D_2^{10}, \ldots, D_4^{10}\} = \{D_2, D_6, D_8, D_{11}\}$$

The collection of complete sets of signal groups, $\mathfrak{D}_b$, is thus

$$\mathfrak{D}_b = d''(\mathfrak{D}) = \{\mathfrak{D}_a^1, \mathfrak{D}_a^2, \ldots, \mathfrak{D}_a^{10}\}.$$

3.3.2. Number of signal groups in a complete set of signal groups

The number of signal groups in a complete set of signal groups is in fact the number of control variables used to control traffic on the intersection. The complexity and price of the device that controls traffic, i.e., the controller, as well as complexity of the control algorithm, significantly depend on the number of signal groups. Because of these reasons, practitioners often tend to choose the minimal number of signal groups, although it is obvious that such a choice deteriorates performance indices of the traffic control system compared to the case when each traffic stream makes one signal group.

Complete sets of signal groups, as it can be seen in Example 3.1 in the preceding subsection, contain various numbers of elements, i.e., signal groups. The number of signal groups in complete sets of signal groups in this example is as follows:

$$card\,\mathfrak{D}_a^1 = 6$$

$$card\,\mathfrak{D}_a^2 = card\,\mathfrak{D}_a^3 = \ldots = card\,\mathfrak{D}_a^6 = 5$$

$$card\,\mathfrak{D}_a^7 = card\,\mathfrak{D}_a^8 = \ldots = card\,\mathfrak{D}_a^{10} = 4\,.$$

Thus, among 10 complete sets of signal groups, 1 set contains 6 signal groups, 5 of them contain 5 signal groups, and 4 contain 4 signal groups.

The number of complete sets of signal groups grows very fast with the number of signal groups. Table 3.1 [34] presents data on the number of signal groups and the number of complete sets of signal groups for intersections having five to eleven traffic streams. These data are obtained for randomly chosen intersections in Belgrade.

Table 3.1

Number of traffic streams - $I = card\,\mathcal{S}$	5	6	7	8	9	11
Number of signal groups - $card\,\mathfrak{D}$	10	12	14	13	30	48
Number of complete sets of signal groups - $card\,\mathfrak{D}_b$	10	10	20	16	235	2490

In choosing the complete set of signal groups, i.e., the control variables that will be used for controlling an intersection, it is interesting to determine the complete sets with the minimal and the maximal number of signal groups.

a) Complete sets of signal groups with maximal number of signal groups

Obviously, the maximal number of signal groups is equal to the number of traffic streams in set $\mathscr{S}$. In this case, each signal group controls exactly one traffic stream. This means that a control variable is assigned to each traffic stream. The maximal number of signal groups in a complete set of signal groups is, thus:

$$\mathrm{P}'_g = \max_m \mathrm{P}_m = \max\{card\, \mathfrak{D}_a^m \mid \mathfrak{D}_a^m \in \mathfrak{D}_b\} = card\, \mathscr{S} = \mathrm{I}, \tag{3.21}$$

where I is the number of traffic streams in set $\mathscr{S}$.

b) Complete sets of signal groups with minimal number of signal groups

The complete set of signal groups containing the minimal number of elements can be determined by solving the problem of optimal partitioning of the set of traffic streams, $\mathscr{S}$. The problem can be stated as follows:

If

$$\mathrm{P}''_g = \min_m \mathrm{P}_m = \min\{card\, \mathfrak{D}_a^m \mid \mathfrak{D}_a^m \in \mathfrak{D}_b\}, \tag{3.22}$$

find the set $\mathfrak{D}_a^*$ defined as:

$$\mathfrak{D}_a^* = \{\mathfrak{D}_a^m \mid card\, \mathfrak{D}_a^m = \mathrm{P}''_g\ , \quad \mathfrak{D}_a^m \in \mathfrak{D}_b\}. \tag{3.23}$$

The problem can be formulated as the problem of optimal partitioning of set $\mathscr{S}$ [37]. To each complete set of signal groups $\mathfrak{D}_a^m$, i.e., to each partitioning of set $\mathscr{S}$, a "selection vector" x can be assigned, defined as

$$x = [x_1, x_2, \ldots, x_{l_c}, \ldots, x_{\mathrm{L}_c}]^T,$$

where $x_{l_c} \in \mathscr{B}$, $(l_c \in \mathscr{L}_c)$, and

$$x_{l_c} = \begin{cases} 1, \text{ if } D_{l_c} \in \mathfrak{D}_a^m \\ 0, \text{ otherwise} \end{cases}, \quad (l_c \in \mathscr{L}_c).$$

The following notations can be introduced:

$$E = [e_{il_c}]_{\mathrm{I}\times\mathrm{L}_c}$$

$$e_{il_c} = \begin{cases} 1, \text{ if } \sigma_i \in D_{l_c},\ D_{l_c} \in \mathfrak{D} \\ 0, \text{ otherwise} \end{cases}$$

$$\mathrm{L}_c = card\, \mathfrak{D}, \quad \mathrm{I} = card\, \mathscr{S}$$

$$a_c = [a_{c1}, a_{c2}, \ldots, a_{cl_c}, \ldots, a_{c\mathrm{L}_c}]^T = [1,1,\ldots,1,\ldots,1]^T$$

$$b_c = [b_{c1}, b_{c2}, \ldots, b_{ci}, \ldots, b_{c\mathrm{I}}]^T = [1,1,\ldots,1,\ldots,1]^T$$

Then, the problem of determining the collection of sets of complete signal groups $\mathcal{D}_a^*$ with the minimal number of elements P_g'' can be stated as follows:

Find all vectors x so as to minimize the function

$$\mathrm{P}_m = a_c^T x = \sum_{l_c=1}^{\mathrm{L}_c} x_{l_c} \tag{3.24}$$

subject to the following constraints:

$$E\,x = b_c \tag{3.25}$$

$$x_{l_c} \in \{0,1\}, \quad (l_c \in \mathcal{L}_c)\,.$$

If there exists only one solution to the stated problem, x^*, it is obvious that

$$\mathrm{P}_g'' = \min \mathrm{P}_m = a_c\, x^* = \sum_{l_c=1}^{\mathrm{L}_c} x_{l_c}^*\,. \tag{3.26}$$

Several algorithms can be used for solving problems of such type [74].

Appendix II contains the pseudocode of MINA program, which is based on Garfinkel and Nemhauser algorithm [28].

Example 3.2

Determine complete sets of signal groups with the minimal number of elements for the intersection whose graph of identical signal indications is given in Fig. 3.4.

The set of all traffic streams is:

$$\mathcal{S} = \{\sigma_1, \sigma_2, \sigma_3, \sigma_4, \sigma_5, \sigma_6\}\,.$$

The set of all cliques $\mathcal{D}$ of graph G_α , i.e., the set of signal groups is:

$$\begin{aligned}\mathcal{D} &= \{\{\sigma_1\},\{\sigma_2\},\{\sigma_3\},\{\sigma_4\},\{\sigma_5\},\{\sigma_6\},\{\sigma_1,\sigma_2\},\{\sigma_1,\sigma_3\}\\ &\qquad \{\sigma_1,\sigma_5\},\{\sigma_2,\sigma_5\},\{\sigma_4,\sigma_5\},\{\sigma_1,\sigma_2,\sigma_5\}\}\\ &= \{D_1, D_2, \ldots, D_{12}\}\end{aligned}$$

$$E = \begin{bmatrix} 1&0&0&0&0&0&1&1&1&0&0&1\\ 0&1&0&0&0&0&1&0&0&1&0&1\\ 0&0&1&0&0&0&0&1&0&0&0&0\\ 0&0&0&1&0&0&0&0&0&0&1&0\\ 0&0&0&0&1&0&0&0&1&1&1&1\\ 0&0&0&0&0&1&0&0&0&0&0&0 \end{bmatrix}$$

$$a_c = [a_{c1}, a_{c2}, \ldots, a_{c12}]^T = [1,1,\ldots,1]^T$$

$$b_c = [b_{c1}, b_{c2}, \ldots, b_{c6}]^T = [1,1,\ldots,1]^T .$$

The problem can be, thus, stated as follows:

Find all vectors x^{*n} for which the function

$$\mathrm{P}_g = a_c\, x = x_1 + x_2 + \ldots + x_{12}$$

obtains minimal value, subject to constraints

$$E\, x = b_c ,$$

which can be written as:

$$\begin{array}{llllllllllllll}
x_1 & & & & & & +x_7 & +x_8 & +x_9 & & & +x_{12} & =1 \\
& x_2 & & & & & +x_7 & & & +x_{10} & & +x_{12} & =1 \\
& & x_3 & & & & & +x_8 & & & & & =1 \\
& & & x_4 & & & & & & & +x_{11} & & =1 \\
& & & & x_5 & & & & +x_9 & +x_{10} & +x_{11} & +x_{12} & =1 \\
& & & & & x_6 & & & & & & & =1
\end{array}$$

$$x_{l_c} \in \{0,1\}, \quad (l_c \in \mathscr{L}_c) .$$

The minimal value $\mathrm{P}''_g = 4$ is obtained, using MINA program for the following vectors x^{*n}:

$$x^{*1} = [0, 0, 1, 0, 0, 1, 1, 0, 0, 0, 1, 0]^T ,$$
$$x^{*2} = [0, 0, 0, 1, 0, 1, 0, 1, 0, 1, 0, 0]^T ,$$
$$x^{*3} = [0, 1, 0, 0, 0, 1, 0, 1, 0, 0, 1, 0]^T ,$$
$$x^{*4} = [0, 0, 1, 1, 0, 1, 0, 0, 0, 0, 0, 1]^T .$$

Therefore,

$$\begin{aligned}
\mathfrak{D}_a^* &= \{\{D_3, D_6, D_7, D_{11}\}, \{D_4, D_6, D_8, D_{10}\}, \\
&\qquad \{D_2, D_6, D_8, D_{11}\}, \{D_3, D_4, D_6, D_{12}\}\} \\
&= \{\mathfrak{D}_a^7, \mathfrak{D}_a^8, \mathfrak{D}_a^9, \mathfrak{D}_a^{10}\}.
\end{aligned}$$

The pseudocode of MINA program is given in Appendix III. The program can be used only for determination of complete sets of signal groups with minimal number of elements.

3.3.3. Compatibility relation of signal groups

When determining the control (signal plan) for an intersection, essential information relates to the feasibility of simultaneous giving right-of-way to different signal groups, and to determination of signal group pairs, belonging to a complete set of signal groups, such that the components of the pair are not allowed to move simultaneously through the intersection. This information is contained in the compatibility relation of signal groups, as mentioned in Section 2.3. The compatibility relation is defined as follows: *Two signal groups are compatible if each traffic stream belonging to one signal group is compatible with each traffic stream belonging to the other signal group.*

The compatibility relation $C_g^m \subset \mathfrak{D}_a^m \times \mathfrak{D}_a^m$ can be defined by the following expression:

$$C_g^m = \{(D_p^m, D_q^m) \mid (\sigma_i \in D_p^m) \wedge (\sigma_j \in D_q^m) \Rightarrow (\sigma_i, \sigma_j) \in C\} \quad (m \in \mathcal{M}). \tag{3.27}$$

The compatibility relation is reflexive, i.e.,

$$D_p^m \, C_g^m \, D_p^m \qquad (D_p^m \in \mathfrak{D}_a^m),$$

and symmetric, i.e.,

$$D_p^m \, C_g^m \, D_q^m \Rightarrow \quad D_q^m \, C_g^m \, D_p^m, \quad (D_p^m, D_q^m \in \mathfrak{D}_a^m).$$

The graph

$$G_g^m = (\mathfrak{D}_a^m, C_g^m) \tag{3.28}$$

is the *graph of signal group compatibility of a complete set of signal groups* $\mathfrak{D}_a^m$. Graph G_g^m is a nonoriented graph, with a loop at each node.

Example 3.3

An intersection with seven traffic streams is presented in Fig. 3.5, together with the graph of compatibility of traffic streams, G_c, and the graph of signal groups compatibility, G_g^1, when the set of signal groups is:

$$\mathfrak{D}_a^1 = \{D_1^1, D_2^1, D_3^1, D_4^1, D_5^1\} = \\ = \{\{\sigma_2\}, \{\sigma_3\}, \{\sigma_6\}, \{\sigma_1, \sigma_4\}, \{\sigma_5, \sigma_7\}\}.$$

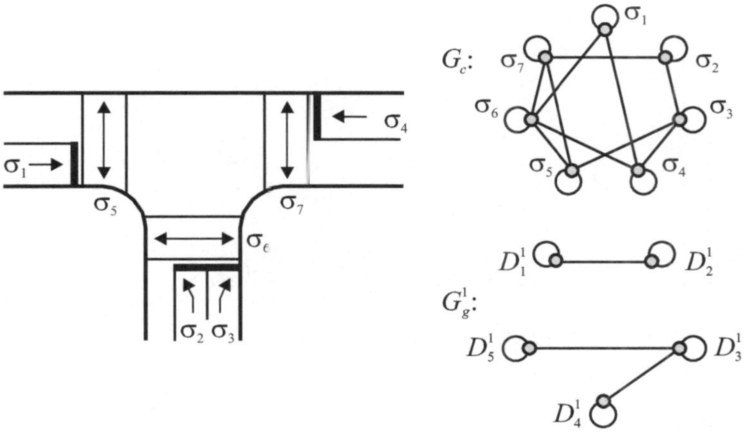

Figure 3.5

Example 3.4

Define and draw compatibility graphs for all complete sets of signal groups defined for the intersection presented in Fig. 3.4 in Example 3.1. The compatibility graph of traffic streams for this intersection is given in Fig. 2.5.

The intersection presented in Fig. 3.4 has six traffic streams, and the identity graph G_α. The set of all cliques, $\mathcal{D}$, of this graph is (Example 3.1):

$$\begin{aligned}\mathcal{D} = d'(G_\alpha) &= \{\{\sigma_1\},\{\sigma_2\},\{\sigma_3\},\{\sigma_4\},\{\sigma_5\},\{\sigma_6\},\{\sigma_1,\sigma_2\},\\ &\qquad \{\sigma_1,\sigma_3\},\{\sigma_1,\sigma_5\},\{\sigma_2,\sigma_5\},\{\sigma_4,\sigma_5\},\{\sigma_1,\sigma_2,\sigma_5\}\}\\ &= \{D_1, D_2, \ldots, D_{12}\}.\end{aligned}$$

The collection of all complete sets of signal groups, $\mathcal{D}_b$, is

$$\mathcal{D}_b = \{\mathcal{D}_a^1, \mathcal{D}_a^2, \ldots, \mathcal{D}_c^{10}\}.$$

The complete sets of signal groups are listed in Example 3.1. All relations C_ε^m, $(m \in \mathcal{M})$ are symmetric, and in defining these relations an element (D_j^m, D_i^m) is not included if element (D_i^m, D_j^m) exists.

The compatibility graphs that correspond to complete sets of signal groups are presented below, in Figures 3.6–3.15.

a) $G_g^1 = (\mathcal{D}_a^1, C_g^1)$

$$\begin{aligned}\mathcal{D}_a^1 &= \{D_1^1, D_2^1, D_3^1, D_4^1, D_5^1, D_6^1\} \\ &= \{D_1, D_2, D_3, D_4, D_5, D_6\} \\ &= \{\{\sigma_1\}, \{\sigma_2\}, \{\sigma_3\}, \{\sigma_4\}, \{\sigma_5\}, \{\sigma_6\}\}\end{aligned}$$

$$\begin{aligned}C_g^1 = \{&(D_1^1, D_1^1), (D_1^1, D_2^1), (D_1^1, D_3^1), (D_1^1, D_5^1), \\ &(D_2^1, D_2^1), (D_2^1, D_5^1), (D_2^1, D_6^1), (D_3^1, D_3^1), \\ &(D_4^1, D_4^1), (D_4^1, D_5^1), (D_5^1, D_5^1), (D_5^1, D_6^1), \\ &(D_6^1, D_6^1)\}\end{aligned}$$

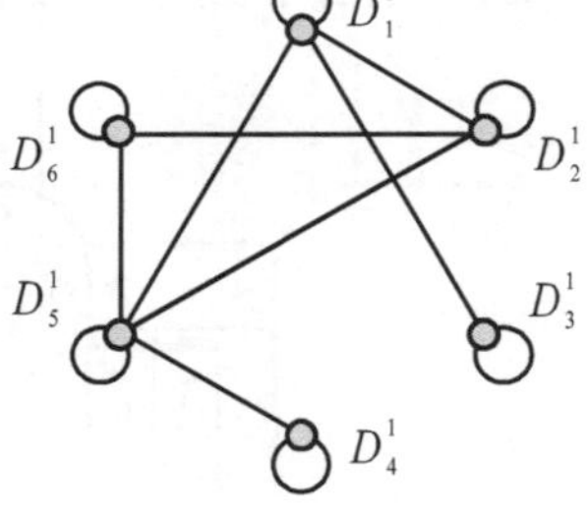

Figure 3.6

b) $G_g^2 = (\mathcal{D}_a^2, C_g^2)$

$$\begin{aligned}\mathcal{D}_a^2 &= \{D_1^2, D_2^2, D_3^2, D_4^2, D_5^2\} \\ &= \{D_3, D_4, D_5, D_6, D_7\} \\ &= \{\{\sigma_3\}, \{\sigma_4\}, \{\sigma_5\}, \{\sigma_6\}, \{\sigma_1, \sigma_2\}\}\end{aligned}$$

$$\begin{aligned}C_g^2 = \{&(D_1^2, D_1^2)\ (D_2^2, D_2^2), (D_2^2, D_3^2), (D_3^2, D_3^2), \\ &(D_3^2, D_4^2), (D_3^2, D_5^2), (D_4^2, D_4^2), (D_5^2, D_5^2)\}\end{aligned}$$

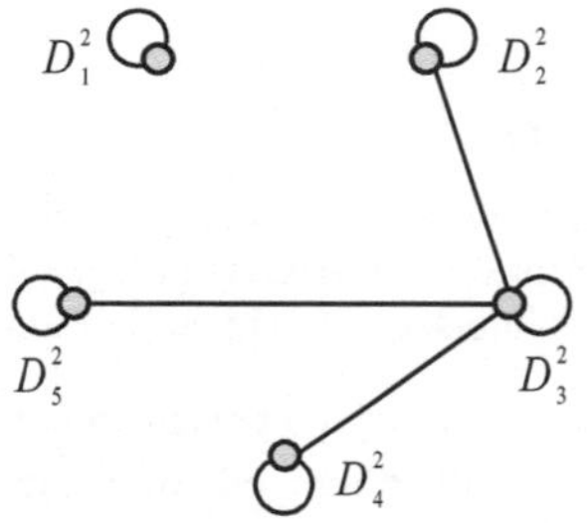

Figure 3.7

c) $G_g^3 = (\mathcal{D}_a^3, C_g^3)$

$$\begin{aligned}\mathcal{D}_a^3 &= \{D_1^3, D_2^3, D_3^3, D_4^3, D_5^3\} \\ &= \{D_2, D_4, D_5, D_6, D_8\} \\ &= \{\{\sigma_2\}, \{\sigma_4\}, \{\sigma_5\}, \{\sigma_6\}, \{\sigma_1, \sigma_3\}\}\end{aligned}$$

$$\begin{aligned}C_g^3 = \{&(D_1^3, D_1^3), (D_1^3, D_3^3), (D_1^3, D_4^3), \\ &(D_2^3, D_2^3), (D_2^3, D_3^3), (D_3^3, D_3^3), \\ &(D_3^3, D_4^3), (D_4^3, D_4^3), (D_5^3, D_5^3)\}\end{aligned}$$

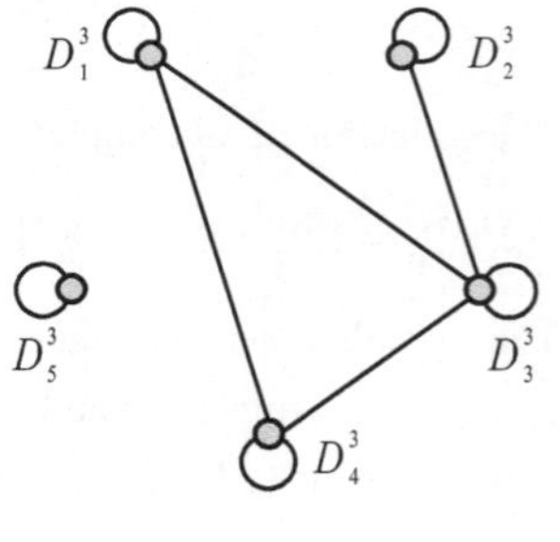

Figure 3.8

d) $G_g^4 = (\mathcal{D}_a^4, C_g^4)$

$$\mathcal{D}_a^4 = \{D_1^4, D_2^4, D_3^4, D_4^4, D_5^4\}$$
$$= \{D_2, D_3, D_4, D_6, D_9\}$$
$$= \{\{\sigma_2\}, \{\sigma_3\}, \{\sigma_4\}, \{\sigma_6\}, \{\sigma_1, \sigma_5\}\}$$

$$C_g^4 = \{(D_1^4, D_1^4), (D_1^4, D_4^4), (D_1^4, D_5^4), (D_2^4, D_2^4), (D_3^4, D_3^4), (D_4^4, D_4^4), (D_5^4, D_5^4)\}$$

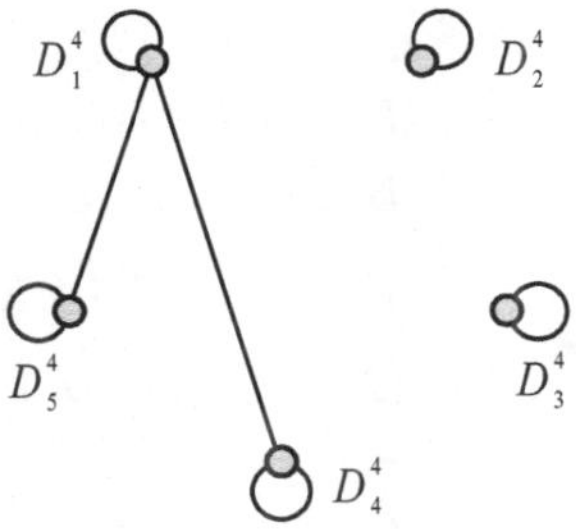

Figure 3.9

e) $G_g^5 = (\mathcal{D}_a^5, C_g^5)$

$$\mathcal{D}_a^5 = \{D_1^5, D_2^5, D_3^5, D_4^5, D_5^5\}$$
$$= \{D_1, D_3, D_4, D_6, D_{10}\}$$
$$= \{\{\sigma_1\}, \{\sigma_3\}, \{\sigma_4\}, \{\sigma_6\}, \{\sigma_2, \sigma_5\}\}$$

$$C_g^5 = \{(D_1^5, D_1^5), (D_1^5, D_2^5), (D_1^5, D_5^5), (D_2^5, D_2^5), (D_3^5, D_3^5), (D_4^5, D_4^5), (D_4^5, D_5^5), (D_5^5, D_5^5)\}$$

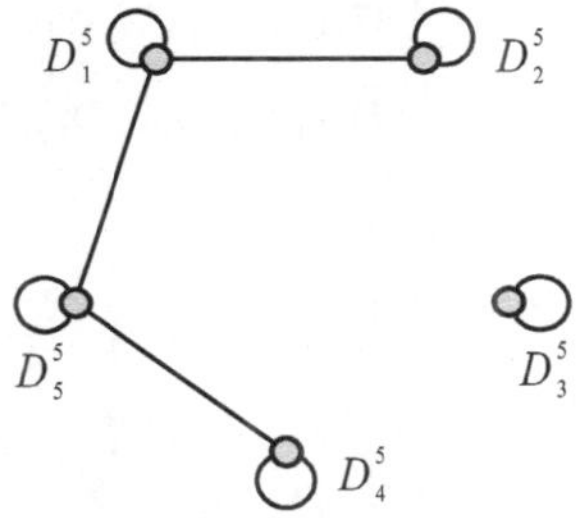

Figure 3.10

f) $G_g^6 = (\mathcal{D}_a^6, C_g^6)$

$$\mathcal{D}_a^6 = \{D_1^6, D_2^6, D_3^6, D_4^6, D_5^6\}$$
$$= \{D_1, D_2, D_3, D_6, D_{11}\}$$
$$= \{\{\sigma_1\}, \{\sigma_2\}, \{\sigma_3\}, \{\sigma_6\}, \{\sigma_4, \sigma_5\}\}$$

$$C_g^6 = \{(D_1^6, D_1^6), (D_1^6, D_2^6), (D_1^6, D_3^6), (D_2^6, D_2^6), (D_2^6, D_4^6), (D_3^6, D_3^6), (D_4^6, D_4^6), (D_5^6, D_5^6)\}$$

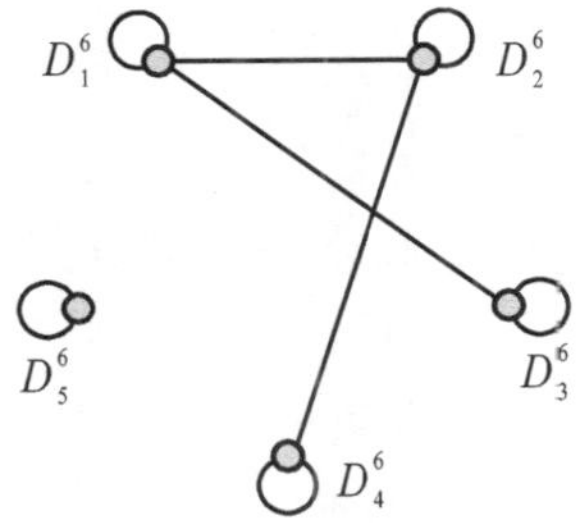

Figure 3.11

g) $G_g^7 = (\mathcal{D}_a^7, C_g^7)$

$$\mathcal{D}_a^7 = \{D_1^7, D_2^7, D_3^7, D_4^7\} = \{D_3, D_4, D_6, D_{12}\} = \{\{\sigma_3\},\{\sigma_4\},\{\sigma_6\},\{\sigma_1,\sigma_2,\sigma_5\}\}$$

$$C_g^7 = \{(D_1^7, D_1^7),(D_2^7, D_2^7),(D_3^7, D_3^7),(D_4^7, D_4^7)\}$$

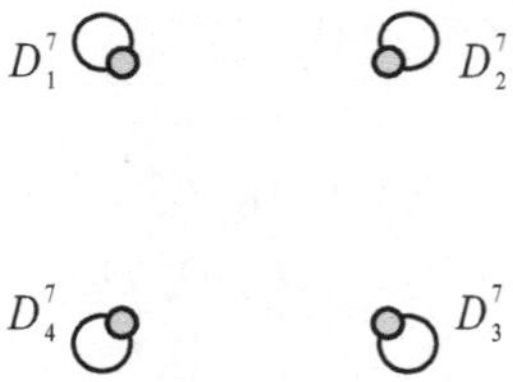

Figure 3.12

h) $G_g^8 = (\mathcal{D}_a^8, C_g^8)$

$$\mathcal{D}_a^8 = \{D_1^8, D_2^8, D_3^8, D_4^8\} = \{D_3, D_6, D_7, D_{11}\} = \{\{\sigma_3\},\{\sigma_6\},\{\sigma_1,\sigma_2\},\{\sigma_4,\sigma_5\}\}$$

$$C_g^8 = \{(D_1^8, D_1^8),(D_2^8, D_2^8),(D_3^8, D_3^8),(D_4^8, D_4^8)\}$$

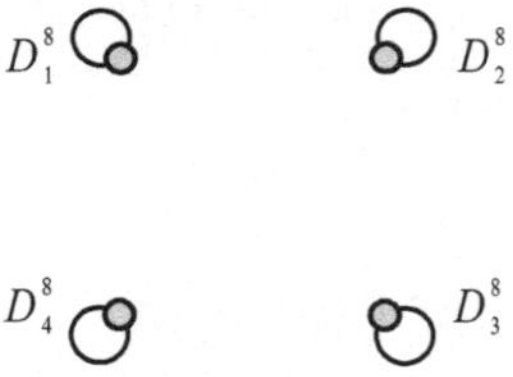

Figure 3.13

i) $G_g^9 = (\mathcal{D}_a^9, C_g^9)$

$$\mathcal{D}_a^9 = \{D_1^9, D_2^9, D_3^9, D_4^9\} = \{D_4, D_6, D_8, D_{10}\} = \{\{\sigma_4\},\{\sigma_6\},\{\sigma_1,\sigma_3\},\{\sigma_2,\sigma_5\}\}$$

$$C_g^9 = \{(D_1^9, D_1^9),(D_2^9, D_2^9),(D_2^9, D_4^9),(D_3^9, D_3^9),(D_4^9, D_4^9)\}$$

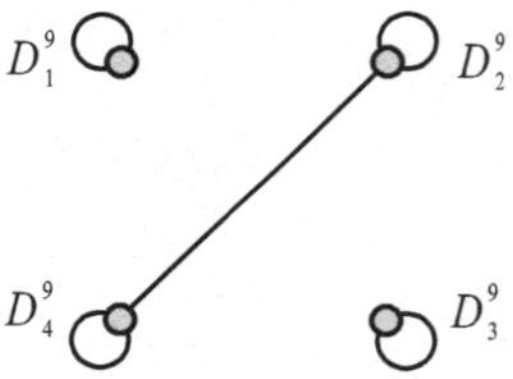

Figure 3.14

j) $G_g^{10} = (\mathcal{D}_a^{10}, C_g^{10})$

$$\mathcal{D}_a^{10} = \{D_1^{10}, D_2^{10}, D_3^{10}, D_4^{10}\} = \{D_2, D_6, D_8, D_{11}\} = \{\{\sigma_2\},\{\sigma_6\},\{\sigma_1,\sigma_3\},\{\sigma_4,\sigma_5\}\}$$

$$C_g^{10} = \{(D_1^{10}, D_1^{10}),(D_1^{10}, D_2^{10}),(D_2^{10}, D_1^{10}),(D_3^{10}, D_3^{10}),(D_4^{10}, D_4^{10})\}$$

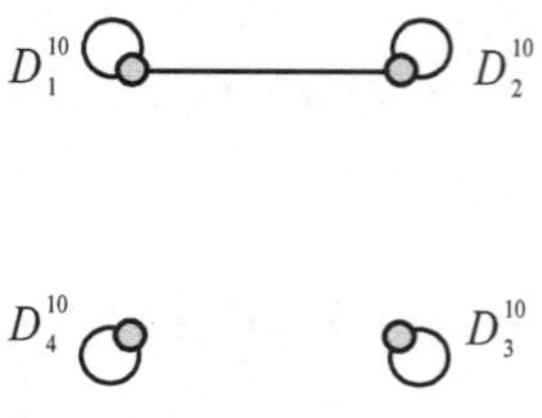

Figure 3.15

The collection of complete sets of signal groups has ten elements, i.e.,

$$\mathcal{D}_b = \{\mathcal{D}_a^1, \mathcal{D}_a^2, \mathcal{D}_a^3, \mathcal{D}_a^4, \mathcal{D}_a^5, \mathcal{D}_a^6, \mathcal{D}_a^7, \mathcal{D}_a^8, \mathcal{D}_a^9, \mathcal{D}_a^{10}\}.$$

3.3.4. Relation of partial ordering in collection $\mathfrak{D}_b$

The assignment of control variables to signal groups, i.e., the choice of the complete set of signal groups that will be used for traffic control, by all means affects performance indices of the intersection control process.

In order to make the best choice of the set of signal groups, it is necessary to introduce a measure of effectiveness that defines the quality of traffic control as a function of a complete set of signal groups, i.e., it is necessary to introduce the function

$$J_G : \mathfrak{D}_b \to \mathrm{R}\,, \tag{3.29}$$

where R is the set of real numbers.

Function J_G can represent the maximal capacity of an approach or the whole intersection, the minimal delay and number of stops, the maximal capacity factor, or any other function that can be used to access the quality of traffic control process.

Function J_G can represent even more complex performance index that includes some economic parameters, such as, for example, the cost of introducing traffic signals on the intersection, etc.

This function, in fact, introduces a linear ordering relation, R_l in collection $\mathfrak{D}_b$, such that:

$$J_G(\mathfrak{D}_a^m) \le J_G(\mathfrak{D}_a^q) \Rightarrow \mathfrak{D}_a^m \; R_l \; \mathfrak{D}_a^q\,, \tag{3.30}$$

i.e., the complete set of signal groups $\mathfrak{D}_a^m$ is better or equal to the complete set $\mathfrak{D}_a^q$ in respect to criterion J_G. In such a way, any two complete sets of signal groups can be compared, and the best complete set can be selected.

However, even without introducing relation R_l, there already exists the relation of partial ordering in collection $\mathfrak{D}_b$, which holds for all performance indices that are functions of traffic parameters.

A complete set of signal groups $\mathfrak{D}_a^m$ represents one partitioning of set $\mathscr{S}$, and collection $\mathfrak{D}_b$ represents collection of all complete sets of signal groups

$$\mathfrak{D}_b = \{\mathfrak{D}_a^1, \mathfrak{D}_a^2, \ldots, \mathfrak{D}_a^m, \ldots, \mathfrak{D}_a^{\mathrm{M}}\}\,.$$

In collection $\mathfrak{D}_b$ there exists a relation of partial ordering, or so-called *refinement relation* [51], R_p, defined as follows:

$$\begin{aligned} R_p = \{(\mathfrak{D}_a^r, \mathfrak{D}_a^q) \mid (D_s \in \mathfrak{D}_a^r, D_t \in \mathfrak{D}_a^q) \Rightarrow ((D_s = D_t) \vee (D_s \subset D_t)),\\ \mathfrak{D}_a^r \in \mathfrak{D}_b, \mathfrak{D}_a^q \in \mathfrak{D}_b, \mathfrak{D}_a^r \ne \mathfrak{D}_a^q\}. \end{aligned} \tag{3.31}$$

Taking into account the definition of relation R_p, i.e., the fact that $\mathfrak{D}_a^r R_p \mathfrak{D}_a^q$ if and only if each signal group in complete set of signal groups $\mathfrak{D}_a^r$ is a subset of a signal group that belongs to complete set $\mathfrak{D}_a^q$, there can be concluded that $\mathfrak{D}_a^r$ is a refinement of $\mathfrak{D}_a^q$.

In this case, the pair

$$\mathfrak{X} = (\mathfrak{D}_b, R_p) \tag{3.32}$$

represents a partially ordered set, and can be represented by a special graph that is called Hasse diagram (Appendix IV) [19], [25], [51]. A Hasse diagram is constructed in the following way. To each element of collection $\mathfrak{D}_a$ there is assigned a node in Hasse diagram. Nodes $\mathfrak{D}_a^r$ and $\mathfrak{D}_a^q$ are joined by an edge if $\mathfrak{D}_a^r R_p \mathfrak{D}_a^q$ and no $\mathfrak{D}_a^v$ exists such that $\mathfrak{D}_a^r R_p \mathfrak{D}_a^v$ and $\mathfrak{D}_a^v R_p \mathfrak{D}_a^q$. Hasse diagram is represented by a nonoriented graph, with the convention that when $\mathfrak{D}_a^r R_p \mathfrak{D}_a^q$, the node representing $\mathfrak{D}_a^r$ is drawn below the node representing $\mathfrak{D}_a^q$.

The number of complete sets of signal groups rapidly grows with the number of signal groups and thus, very often, the choice of the complete set of signal groups becomes a very complex problem. Using the refinement relation R_p can lead to simplification of this problem (see Part V).

Example 3.5

Define the relation of partial ordering (refinement) in collection $\mathfrak{D}_b$ determined in Example 3.1, and represent it by a Hasse diagram.

The collection $\mathfrak{D}_b$ (Example 3.1) has ten elements that represent complete sets of signal groups:

$$\mathfrak{D}_b = \{\mathfrak{D}_a^1, \mathfrak{D}_a^2, \ldots, \mathfrak{D}_a^{10}\}.$$

These elements represent the following collections:

$$\mathfrak{D}_a^1 = \{D_1, D_2, D_3, D_4, D_5, D_6\},\ \mathfrak{D}_a^2 = \{D_3, D_4, D_5, D_6, D_7\},$$

$$\mathfrak{D}_a^3 = \{D_2, D_4, D_5, D_6, D_8\},\ \mathfrak{D}_a^4 = \{D_2, D_3, D_4, D_6, D_9\},$$

$$\mathfrak{D}_a^5 = \{D_1, D_3, D_4, D_6, D_{10}\},\ \mathfrak{D}_a^6 = \{D_1, D_2, D_3, D_6, D_{11}\},$$

$$\mathfrak{D}_a^7 = \{D_3, D_4, D_6, D_{12}\},\ \mathfrak{D}_a^8 = \{D_3, D_6, D_7, D_{11}\},$$

$$\mathfrak{D}_a^9 = \{D_4, D_6, D_8, D_{10}\},\ \mathfrak{D}_a^{10} = \{D_2, D_6, D_8, D_{11}\}.$$

The collection of all signal groups, $\mathfrak{D}$, is

$$\mathfrak{D} = \{D_1, D_2, \ldots, D_{12}\}.$$

The signal groups are defined as follows:

$$D_1 = \{\sigma_1\},\ D_2 = \{\sigma_2\},\ D_3 = \{\sigma_3\},\ D_4 = \{\sigma_4\},\ D_5 = \{\sigma_5\},\ D_6 = \{\sigma_6\},$$

$$D_7 = \{\sigma_1, \sigma_2\},\ D_8 = \{\sigma_1, \sigma_3\},\ D_9 = \{\sigma_1, \sigma_5\},\ D_{10} = \{\sigma_2, \sigma_5\},$$

$$D_{11} = \{\sigma_4, \sigma_5\},$$

$$D_{12} = \{\sigma_1, \sigma_2, \sigma_5\}.$$

Relation R_p is expressed by the following set of pairs:

$$R_p = \{(\mathcal{D}_a^1, \mathcal{D}_a^2), (\mathcal{D}_a^1, \mathcal{D}_a^3), (\mathcal{D}_a^1, \mathcal{D}_a^4), (\mathcal{D}_a^1, \mathcal{D}_a^5), (\mathcal{D}_a^1, \mathcal{D}_a^6), (\mathcal{D}_a^1, \mathcal{D}_a^7),$$
$$(\mathcal{D}_a^1, \mathcal{D}_a^8), (\mathcal{D}_a^1, \mathcal{D}_a^9), (\mathcal{D}_a^1, \mathcal{D}_a^{10}), (\mathcal{D}_a^2, \mathcal{D}_a^7), (\mathcal{D}_a^2, \mathcal{D}_a^8), (\mathcal{D}_a^3, \mathcal{D}_a^9),$$
$$(\mathcal{D}_a^3, \mathcal{D}_a^{10}), (\mathcal{D}_a^4, \mathcal{D}_a^7), (\mathcal{D}_a^5, \mathcal{D}_a^7), (\mathcal{D}_a^5, \mathcal{D}_a^9), (\mathcal{D}_a^6, \mathcal{D}_a^8), (\mathcal{D}_a^6, \mathcal{D}_a^{10})\}.$$

There can be noted that the complete set of signal groups $\mathcal{D}_a^1$ is "better" than any other complete set (this holds for any performance index). Also, there can be noted, for example, that $\mathcal{D}_a^5$ is "better" than $\mathcal{D}_a^7$ and $\mathcal{D}_a^9$, and that it is "worse" than $\mathcal{D}_a^1$. However, nothing can be said about the relation of $\mathcal{D}_a^5$ to other complete sets of signal groups.

The Hasse diagram of the partially ordered set $\mathcal{D}_b$ is given in Fig. 3.16.

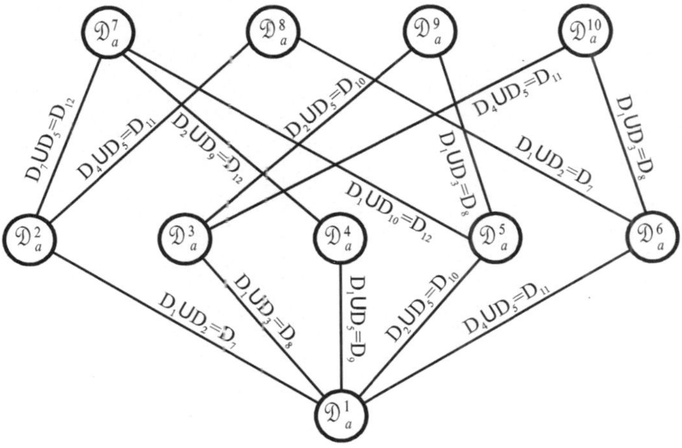

Figure 3.16

4. TRAFFIC CONTROL

Traffic process on a signalized intersection, as explained in Chapter 2, can be described by two functions: state transition function φ^1 and the reaction function, i.e., output function φ^2. In order to determine a "new" state and output, the uncontrolled inputs, i.e., arrival traffic flows (traffic streams) have to be known, as well as the controlled inputs, i.e., the control. The control is defined as a vector whose components are functions of time. These components control traffic streams by different signal indications. To different signal indications certain control variable values are assigned by a convention, as mentioned in Chapter 3.

A control variable is assigned to each signal group in the chosen complete set of signal groups.

In order to determine the control during an interval, values of control variables should be known at any instant of the interval. Therefore, the traffic control problem is the problem of determining time functions which, in fact, represent control variables in given interval.

In contemporary traffic control systems, control variables are periodic time functions. This means that it is necessary to determine values of these control variables only in the interval equal to the base period of these functions. Duration of this base period is called *cycle duration* or *cycle time*. It remains constant in time periods during which the stochastic variables, which represent arrival flows, can be regarded as stationary stochastic processes.

Real-time traffic control systems have the feature that data on current values of traffic flow parameters are included when determining values of control variables. The data are obtained using vehicle detectors, realized with various sensors (inductive, pneumatic, laser, etc.). These data are mostly used for modification of already determined values of control variables. Control variables in this case are also periodic function of time, determined on the basis of average values of traffic flow parameters, observed in longer time periods.

4.1. Control variables

A control variable is assigned to each signal group. Different control variable values correspond to different indications of traffic lights that control the signal group. Therefore, the set from which a control variable can "assume" its value must have the number of elements equal to the number of different light indications used to control the associated signal group.The set of values that can be assumed by variables assigned to vehicle signal groups will have four elements. Similarly, the set of values that can be assumed by variables assigned to pedestrian signal groups will have two elements, etc.

However, in all exact methods and programs for traffic control problem solving, the sequence of signal indications used to control vehicle signal groups, shown in Fig. 3.1, is substituted by the sequence, which contains only two indications: green and red. In this way, the number of elements of the set of available values of control variables assigned to vehicle signal groups is reduced to two.

This transformation is performed by substituting the real function that describes the queue discharging from a saturated intersection approach (the queue remains after the end of green indication) with a rectangular function, as shown in Fig. 4.1 [88], [89]. The area between the time axis and the volume function, which represents the number of vehicles that leave the approach during green indication, is the same in both cases. The edge of the rectangle lying on the ordinate axis represents the saturation flow volume, and the edge lying on the abscise axis is called the *effective green time*. The difference between the cycle time and effective green time is called the *effective red time*.

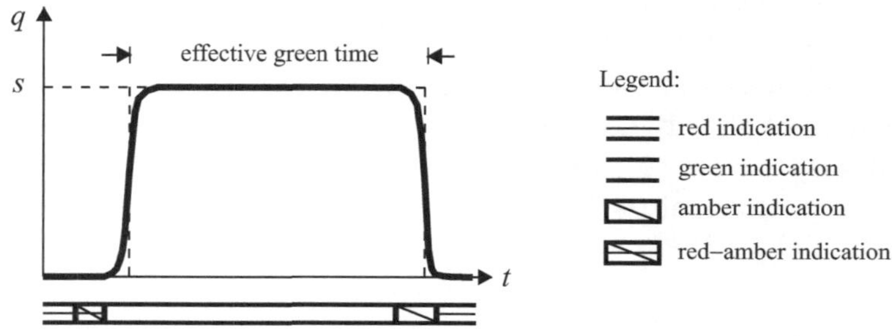

Figure 4.1

Therefore, the cycle time consists of the effective green time, during which the queue discharges with saturation flow volume, and the effective red time, during which the flow volume is equal to zero.

Relations between cycle time, displayed green, red, and amber times and effective green and red times are given in Appendix V.

For control of pedestrian and tram streams only two different traffic light indications are used, so that there is no need to transform them into effective values, as was the case with vehicle traffic streams.

Control variables are assigned to signal groups after the complete set of signal group is chosen, i.e., when one element is chosen from the collection of all complete sets of signal groups,

$$\mathfrak{D}_b = \{\mathfrak{D}_a^1, \mathfrak{D}_a^2, \ldots, \mathfrak{D}_a^m, \ldots, \mathfrak{D}_a^{\mathrm{M}}\}.$$

Let the chosen element be:

$$\mathfrak{D}_a^m = \{D_1^m, D_2^m, \ldots, D_p^m, \ldots, D_{\mathrm{P}_m}^m\}. \tag{4.1}$$

Since further discussion assumes the set $\mathfrak{D}_a^m$ is defined, index m can be neglected. Thus, the chosen complete set of signal groups can be represented as:

$$\mathfrak{D}_a = \{D_1', D_2', \ldots, D_p', \ldots, D_{\mathrm{P}}'\}. \tag{4.2}$$

The set of signal group indices in the chosen complete set of signal groups is:

$$\mathscr{P} = \{1, 2, \ldots, p, \ldots, \mathrm{P}\}. \tag{4.3}$$

Notation D_p' is introduced to distinguish $D_p' \in \mathfrak{D}_a$ from $D_{l_c} \in \mathfrak{D}$, when $l_c = p$.

The number of signal groups in the chosen complete set of signal groups is less than or equal to the number of traffic streams on the intersection, i.e., $\mathrm{P} \leq \mathrm{I}$.

Assignment of control variables to signal groups can be represented by a function, f_a, in the following way:

$$f_a : \mathfrak{D}_a \rightarrow \mathscr{U}',$$

where

$$\mathscr{U}' = \{u_1'(\cdot), u_2'(\cdot), \ldots, u_p'(\cdot), \ldots, u_{\mathrm{P}}'(\cdot)\}, \text{ and}$$

$$f_a(D_p') = u_p'(\cdot), \quad (p \in \mathscr{P}). \tag{4.4}$$

$u_p'(\cdot)$ is a periodic time function, with the property:

$$u_p'(t) = u_p'(t \bmod c), \quad (p \in \mathscr{P}).$$

Thus, function f_a assigns control variable $u_p'(\cdot)$ to signal group D_p' $(p \in \mathscr{P})$. Set $\mathscr{P}$ is, therefore, the index set for both signal groups and control variables.

To different signal indications certain control variable values are assigned by a convention, as mentioned in Chapter 3. Because after transformation into effective values only two different light indications exist, it can be assumed that control variable $u'_p(\cdot)$ can take either value 0 or 1. These values have the following meaning:

$$u'_p(t) = \begin{cases} 1, & \text{if signal group } D'_p \text{ has the right-of-way at time } t \\ & \text{(effective green indication of signals controlling } D'_p\text{)} \\ 0 & \text{if signal group } D'_p \text{ does not have the right-of-way at time } t \\ & \text{(effective red indication of signals controlling } D'_p\text{)} \end{cases} \tag{4.5}$$

During one control variable cycle, the right-of-way is once given and once taken from each signal group. Therefore, on the basis of data on effective green times (start time and end time), the cycle time, and the time the cycle starts, the value of control variable can be determined for any t, regardless whether it is greater or less than the cycle time, or $t < 0$ or $t > 0$.

The relation between start and end time of effective green for signal group D'_p and the time when cycle begins is illustrated by Fig. 4.2. Notations in this figure have the following meaning:

c – the cycle time,

t'_p – the time the effective green of signal groups D'_p starts, in regard to the beginning of the cycle,

t''_p – the time the effective green of signal groups D'_p ends, in regard to the beginning of the cycle.

The value of control variable $u'_p(\cdot)$ for any t can be determined using the following expression:

$$u'_p(t) = h((t'_p - t(\text{mod}\, c))(t''_p - t(\text{mod}\, c))\,(t'_p - t''_p))\,. \tag{4.6}$$

It can be noticed that values of control variable are determined using Heaviside function, which is defined as:

$$h(x) = \begin{cases} 1, & x \geq 0 \\ 0, & x < 0 \end{cases}. \tag{4.7}$$

Therefore, the control at time t can be described by the following vector:

$$u'(t) = [u'_1(t), u'_2(t), \ldots, u'_p(t), \ldots, u'_P(t)]^T, \tag{4.8}$$

where each variable $u'_p(\cdot)$ can get either the 0 or 1 value.

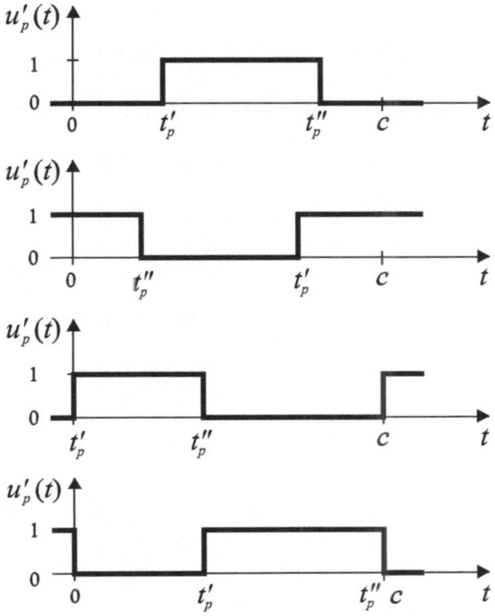

Figure 4.2

Interval $[t_0,t]$, for which the control $u_{[t_0,t]}$ has to be determined, is chosen so that the assumption on stationarity of the stochastic process represented by components of vector $q_{[t_0,t]} \in \mathscr{Q}_{[t_0,t]}$ is justified. The interval $[t_0,t] \subset \mathscr{T}$ and set $\mathscr{T}$ is defined in Section 2.1. This assumption is valid as long as stochastic process q is stationary. Interval $[t_0,t]$ can be considered as longer than 10 to 15 minutes [83].

Component $u_{[t_0,t]p}$ of control vector $u_{[t_0,t]}$ represents restriction of control $u'_p(\cdot) \in \mathscr{U}$ to interval $[t_0,t] \cap \mathscr{T}$. Function $u_{[t_0,t]}$ is a periodic vector function, with cycle duration c, and $t - t_0 >> c$.

In many countries, the maximal cycle time is limited to 2 minutes. Therefore, in order to determine $u_{[t_0,t]}$, which is a periodic function of time, it is enough to determine the control during the cycle time, i.e., the restriction of function $u_{[t_0,t]}$ to interval $[0,c] \cap [t_0,t]$.

The control vector, or control $u(\cdot)$ in further text, represents the restriction of function $u_{[t_0,t]}$ to interval $[0,c] \cap [t_0,t]$, and the following notations are used:

$u'(\cdot) \in \mathscr{U}'$, where $u'(\cdot) = \{u(t) \,|\, t \in \mathscr{T}\}$, $(\mathscr{T} = (-\infty, \infty))$

$u_{[t_0,t]} \in \mathscr{U}_{[t_0,t]}$, where $u_{[t_0,t]}$ is restriction of $u'(\cdot)$ to interval $[t_0,t] \cap \mathscr{T}$

$u(\cdot) \in \mathscr{U}$, where $u(\cdot)$ is restriction of $u_{[t_0,t]}$ to interval $[0,c] \cap [t_0,t]$, and

$u(t)$ is element of $u(\cdot)$ at time *t*.

The controls that satisfy certain conditions (Section 2.1) are elements of the set of feasible controls, $\mathscr{U}_f$.

4.2. The control—signal plan

Traffic control on a signalized intersection during a cycle is defined by the vector time function:

$$u(\cdot) = [u_1(\cdot), u_2(\cdot), \ldots, u_p(\cdot), \ldots, u_P(\cdot)]^T. \tag{4.9}$$

Components of this function are functions of time, in interval $[0,c]$. A common name for vector time function $u(\cdot)$ is *signal plan.*

As an example, the signal plan, i.e., functions $u_p(\cdot)$ presented in Fig. 4.3 are used for traffic control on the intersection presented in the same figure. The same figure contains, also, the graphic representation of the signal plan, which is common in traffic engineering practice. This representation contains the intervals of amber and red–amber indications, as well. The signal plan presented in Fig. 4.3 is:

$$u(\cdot) = [u_1(\cdot), u_2(\cdot), u_3(\cdot), u_4(\cdot)]^T.$$

Values of variables $u_1(\cdot), u_2(\cdot), u_3(\cdot), u_4(\cdot)$, for any *t* in interval $[0,c]$, can be determined using expression (4.6), with values of t'_p and t''_p known in interval $[0,c]$.

4.2.1. Control vectors—phases

In any signal plan there exist some intervals (see Fig. 4.3) in which no component changes its value. Therefore, a signal plan can be represented still in another way. Namely, a signal plan can be described by a sequence of control vectors, together with their durations. Each control vector represents the control in the interval in which no component changes its value. Thus, a signal plan can be described as:

$$u(\cdot)=[(\mathbf{u}^1,\tau^1)^T,(\mathbf{u}^2,\tau^2)^T,\ldots,(\mathbf{u}^k,\tau^k)^T,\ldots,(\mathbf{u}^K,\tau^K)^T], \tag{4.10}$$

where K is the number of control vectors in the signal plan.

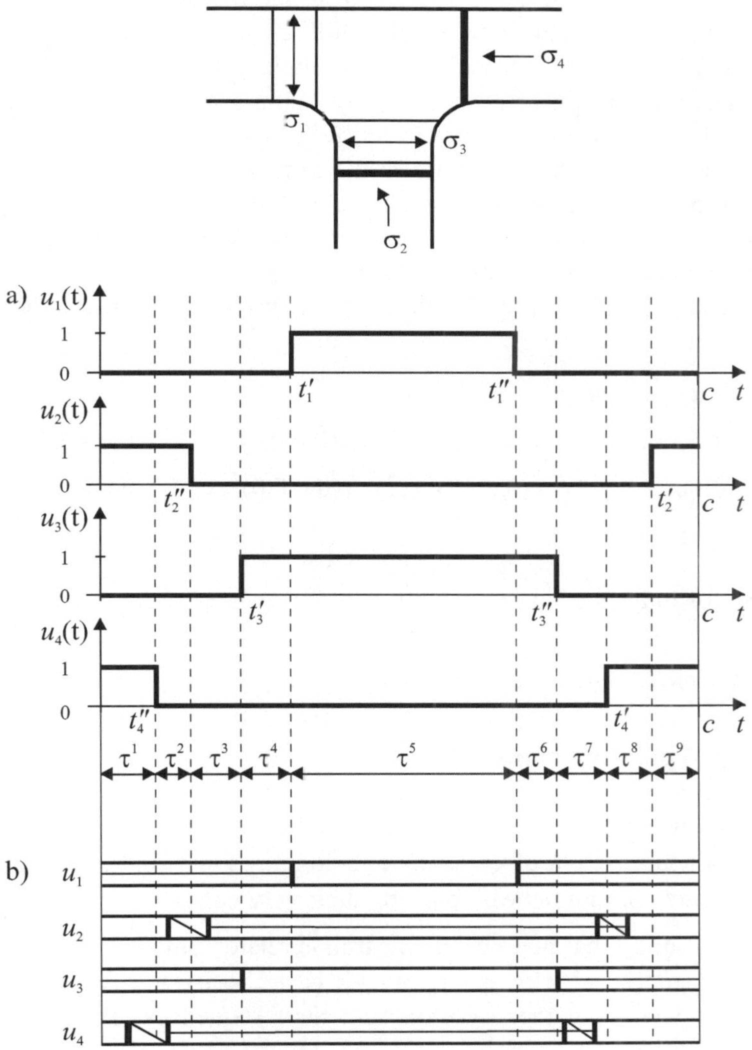

Figure 4.3

Control vectors $\mathbf{u}^1,\mathbf{u}^2,\ldots,\mathbf{u}^k,\ldots,\mathbf{u}^{\mathrm{K}}$ are often called *phases*. Both terms, control vector and phase, will be used in further discussions.

Signal plan determination now can be represented as the problem of finding the control vectors, their sequence, duration, and number.

The term *phase* is not uniquely defined in the literature. In accordance with the definition given above, here is the Stoffer's definition [77]: "... 'traffic signal phase' (or, shortly, 'phase') is the complete specification of signal indications which can appear simultaneously." Similar to this is the Akcelik's definition [2]. In British literature, Webster [89], Allsop [4], and other authors use the following definition: a phase is the sequence of signal indications used to control a signal group during the cycle. This definition represents the sequence of values of one control variable during the cycle. The definition used in this text is similar to notion of "stage" in British literature.

The signal plan from Fig. 4.3 now can be presented in the following way:

$$\begin{aligned} u(\cdot) &= [\ (\mathbf{u}^1,\tau^1)^T,(\mathbf{u}^2,\tau^2)^T,\ldots,(\mathbf{u}^9,\tau^9)^T\ ] \\ &= [((0,1,0,1)^T,\tau^1)^T,((0,1,0,0)^T,\tau^2)^T,((0,0,0,0)^T,\tau^3)^T,((0,0,1,0)^T,\tau^4)^T, \\ &\quad ((1,0,1,0)^T,\tau^5)^T,((0,0,1,0)^T,\tau^6)^T,((0,0,0,0)^T,\tau^7)^T,((0,0,0,1)^T,\tau^8)^T, \\ &\quad ((0101)^T,\tau^9)^T]. \end{aligned}$$

The set of indices of intervals with constant values of control variables is

$$\mathscr{K} = \{1,2,\ldots,k,\ldots,\mathrm{K}\}\,. \tag{4.11}$$

4.2.2. Signal plan structure

Variables $\mathbf{u}^k$ and τ^k are of a different nature. Variables $\mathbf{u}^k$ are vectors, whose components assume discrete values belonging to set $\{0,1\}$, while phase durations τ^k are continuous variables, by their very nature.

Bearing in mind that there exist constraints related only to variables $\mathbf{u}^k$ and their sequences, $\mathbf{u}^1,\mathbf{u}^2,\ldots,\mathbf{u}^k,\ldots,\mathbf{u}^{\mathrm{K}}$, it is necessary to give particular attention to selection of these vectors and their sequences. Therefore the notion of *signal plan structure* is introduced, which relates to these variables only [37]. The signal plan structure is defined as the sequence:

$$\mathbf{u} = [\mathbf{u}^1,\mathbf{u}^2,\ldots,\mathbf{u}^k,\ldots,\mathbf{u}^{\mathrm{K}}]\,. \tag{4.12}$$

Thus, changes in the number of phases, their sequence, or components of a phase, represent changes of the signal plan structure.

Related to signal plan structure is the classification of traffic controllers into so-called phase-based controllers and signal–group-based controllers. With phase-based controllers, the signal plan structure is generally fixed and cannot be changed by, say, some signal sent from a control center. Modification of signal plan structure can be done only by interventions in the controller. An example of such a controller is the electromechanical controller with a drum that rotates and periodically gives contacts for green and red indications of signal groups. Contemporary electronic traffic controllers, signal–group-based, do not have constraints related to signal plan structure. With them, the signal plan structure can be remotely changed, from a control center or automatically, on the base of data obtained from detectors located at intersection approaches.

The structure of the signal plan presented in Fig. 4.3 is:

$$\begin{aligned}\mathbf{u} &= [\mathbf{u}^1, \mathbf{u}^2, \ldots, \mathbf{u}^9] \\ &= [(0,1,0,1)^T, (0,1,0,0)^T, (0,0,0,0)^T, (0,0,1,0)^T, (1,0,1,0)^T, \\ &\quad (0,0,1,0)^T, (0,0,0,0)^T, (0,0,0,1)^T, (0,1,0,1)^T] \\ &= \left[\begin{bmatrix}0\\1\\0\\1\end{bmatrix}, \begin{bmatrix}0\\1\\0\\0\end{bmatrix}, \begin{bmatrix}0\\0\\0\\0\end{bmatrix}, \begin{bmatrix}0\\0\\1\\0\end{bmatrix}, \begin{bmatrix}1\\0\\1\\0\end{bmatrix}, \begin{bmatrix}0\\0\\1\\0\end{bmatrix}, \begin{bmatrix}0\\0\\0\\0\end{bmatrix}, \begin{bmatrix}0\\0\\0\\1\end{bmatrix}, \begin{bmatrix}0\\1\\0\\1\end{bmatrix}\right]\end{aligned}$$

or, in matrix form, when brackets are not used for individual vectors $\mathbf{u}^k$, $(k \in \mathscr{K})$:

$$\begin{array}{c} \quad\quad\quad\quad\quad \mathbf{u}^1\ \mathbf{u}^2\ \mathbf{u}^3\ \mathbf{u}^4\ \mathbf{u}^5\ \mathbf{u}^6\ \mathbf{u}^7\ \mathbf{u}^8\ \mathbf{u}^9 \\ \mathbf{u} = \begin{bmatrix}\mathbf{u}_1\\ \mathbf{u}_2\\ \mathbf{u}_3\\ \mathbf{u}_4\end{bmatrix} = \begin{bmatrix}0&0&0&0&1&0&0&0&0\\1&1&0&0&0&0&0&0&1\\0&0&0&1&1&1&0&0&0\\1&0&0&0&0&0&0&1&1\end{bmatrix}.\end{array}$$

If the vector of phase durations is introduced,

$$\tau = [\tau^1, \tau^2, \ldots, \tau^k, \ldots, \tau^K], \tag{4.13}$$

whose components are the durations of phases, $\tau^1, \tau^2, \ldots, \tau^K$, then the signal plan can be represented as the following pair:

$$u(\cdot) = \begin{bmatrix} \mathbf{u} \\ \tau \end{bmatrix} = (\mathbf{u}, \tau)^T = \begin{bmatrix} (\mathbf{u}^1, \mathbf{u}^2, \ldots, \mathbf{u}^k, \ldots, \mathbf{u}^{\mathrm{K}}) \\ (\tau^1, \tau^2, \ldots, \tau^{\mathrm{k}}, \ldots, \tau^{\mathrm{K}}) \end{bmatrix}, \tag{4.14}$$

where

$$\mathbf{u}^k = (u_1^k, u_2^k, \ldots, u_p^k, \ldots, u_{\mathrm{P}}^k)^T . \tag{4.15}$$

Control variable $u_p(\cdot)$, for signal group D_p', is then defined as:

$$u_p(\cdot) = \begin{bmatrix} (u_p^1, u_p^2, \ldots, u_p^k, \ldots, u_p^{\mathrm{K}}) \\ (\tau^1, \tau^2, \ldots, \tau^{\mathrm{k}}, \ldots, \tau^{\mathrm{K}}) \end{bmatrix} = \begin{bmatrix} \mathbf{u}_p \\ \tau \end{bmatrix}, \tag{4.16}$$

where

$$\mathbf{u}_p = (u_p^1, u_p^2, \ldots, u_p^k, \ldots, u_p^{\mathrm{K}}), \tag{4.17}$$

so that the signal plan structure can be expressed as:

$$\mathbf{u} = [\mathbf{u}_1, \mathbf{u}_2, \ldots, \mathbf{u}_p, \ldots, \mathbf{u}_{\mathrm{P}}]^T . \tag{4.18}$$

Therefore, $\mathbf{u}_p$ represents the sequence of values of control variable $u_p(\cdot)$ in the control vectors sequence.

Effective green time of signal group D_p' is determined by the following expression:

$$g_p = \sum_{k=1}^{\mathrm{K}} u_p^k \tau^k = \mathbf{u}_p \tau^T , \quad (p \in \mathscr{P}), \tag{4.19}$$

and effective red time is:

$$r_p = c - g_p = c - \sum_{k=1}^{\mathrm{K}} u_p^k \tau^k , \quad (p \in \mathscr{P}). \tag{4.20}$$

The set of signal groups to which control variables are assigned is:

$$\mathfrak{D}_a = \{D_1', D_2', \ldots, D_p', \ldots, D_{\mathrm{P}}'\} . \tag{4.21}$$

The first P' variables, where $\mathrm{P}' \leq \mathrm{P}$, are assigned to vehicle signal groups. Thus, the subset containing vehicle signal groups is:

$$\mathfrak{D}_a^v = \{D_1', D_2', \ldots, D_p', \ldots, D_{\mathrm{P}'}'\} . \tag{4.22}$$

The number of components, I', of the state vector is equal to the number of vehicle traffic streams. The number of vehicle signal groups is $\mathrm{P}' \leq \mathrm{I}'$. A vehicle signal group represents subsets of the set of vehicle traffic streams, $\mathscr{S}^1$, and can be represented as:

$$D'_p = \{\sigma_{p_1}, \mathrm{v}_{p_2}, ..., \sigma_{p_e}, ..., \sigma_{p_{\mathrm{E}(p)}}\}, \quad (p \in \mathscr{P}'), \tag{4.23}$$

where

$$\mathscr{P}' = \{1,2,...,\mathrm{P}'\}, \tag{4.24}$$

$$\sigma_{p_e} \in \mathscr{S}^1, \quad p \in \mathscr{P}', \quad e \in \{1,2,...,\mathrm{E}(p)\}$$

$$\mathrm{E}(p) = card\ D'_p.$$

There also holds:

$$\bigcup_{p=1}^{\mathrm{P}'} D'_p = \mathscr{S}^1 = \{\sigma_1, \sigma_2, ..., \sigma_i, ..., \sigma_{\mathrm{I}'}\}. \tag{4.25}$$

The set of indices of vehicle traffic streams is:

$$\mathscr{I}' = \{1,2,\ldots,i,\ldots,\mathrm{I}'\}.$$

5. QUEUES—STATE OF THE TRAFFIC PROCESS ON ISOLATED SIGNALIZED INTERSECTIONS

5.1. Definition of the state

In systems theory, [58], [90], [23], the state is defined as the minimal quantity of information about process history that is necessary to determine future output and state.

A component of the state of dynamic traffic process on an intersection is the length of vehicle queue on an approach to the intersection, as mentioned in Section 1.1. Thus, the state of the process is defined by a vector whose components are lengths of vehicle queues on all approaches. The choice of this variable for the state of the process is in accordance with the definition given above. This can be seen observing Fig. 1.1. For determination of volumes $q_3'(t)$ and $q_5'(t)$ at some time *t*, besides the values of input variables $q_3(t)$, $q_5(t)$, $u_3'(t)$, and $u_5'(t)$, it is necessary to know the values of queue lengths $w_3(t)$ and $w_5(t)$, i.e., the process state components. If there is no information on queue lengths, values of volumes $q_3'(t)$ or $q_5'(t)$ (output) cannot be determined, although values $q_3(t)$, $q_5(t)$, $u_3'(t)$, and $u_5'(t)$ are known. Namely, if queues exist, volumes $q_3'(t)$ or $q_5'(t)$ will be equal to saturation flow volumes when streams σ_3 or σ_5 have the right-of-way ($u_3'(t)=1$ or $u_5'(t)=1$). If there are no queues, output volumes $q_3'(t)$ and $q_5'(t)$ will be equal to input volumes $q_3(t)$ and $q_5(t)$.

To determine the state at some time *t*, it is necessary to know the state at some previous time t_0, and functions $q_{[t_0,t]}$ and $u_{[t_0,t]}$ in interval $[t_0,t]$.

The state at time *t* is represented by the following vector (Section 2.1):

$$w(t)=[w_1(t),w_2(t),\ldots,w_i(t),\ldots,w_{\mathrm{I}'}(t)]^T. \tag{5.1}$$

The set of traffic streams, $\mathcal{S}$, as pointed out in Section 3.2, has to be partitioned into subsets (3.5) in such a way that one subset contains only traffic streams of a particular type. These subsets form the following collection (3.5):

$$\overline{\mathcal{S}}=\{\mathcal{S}^1,\mathcal{S}^2,\ldots,\mathcal{S}^f,\ldots,\mathcal{S}^{\mathrm{F}}\}.$$

Obviously,

$$\bigcup_{f=1}^{F} \mathscr{S}^f = \mathscr{S}, \text{ and}$$

$$\mathscr{S}^a \cap \mathscr{S}^b = \varnothing \quad (\mathscr{S}^a, \mathscr{S}^b \in \bar{\mathscr{S}}).$$

The traffic streams having indices 1 to I' are vehicle streams (Section 2.3). If the subset of vehicle traffic streams is denoted by $\mathscr{S}^1$, then (4.25):

$$\mathscr{S}^1 = \{\sigma_1, \sigma_2, \ldots, \sigma_i, \ldots, \sigma_{\mathrm{I}'}\},$$

where $\mathrm{I}' \leq \mathrm{I}$ (I is the total number of traffic streams, and I' is the number of vehicle traffic streams). The number of state vector components is equal to the number of vehicle traffic streams.

In the introductory part of Chapter 1, it was stated that any dynamic process, and hence the traffic process on an intersection as well, can be mathematically described by two functions: the function of state transitions and the output function (reaction).

The state transition function can be represented in the following way (1.6):

$$\varphi^1 : \ \mathscr{W}_{t_0} \times (\mathscr{Q}_{[t_0,t]} \times \mathscr{U}_{[t_0,t]}) \to \mathscr{W}_t \ ,$$

where φ^1 is the function that determines state vector $w(t) \in \mathscr{W}_t$ at time t for known state $w(t_0) \in \mathscr{W}_{t_0}$ at time t_0 and known input volume functions $q_{[t_0,t]} \in \mathscr{Q}_{[t_0,t]}$ and control $u_{[t_0,t]} \in \mathscr{U}_{[t_0,t]}$ at any time in interval $[t_0,t]$. The number of vector $w(t)$ components is equal to the number of vehicle traffic streams.

Change of state in interval $[t_0,t]$—"evolution" of state over time—is represented by the following function:

$$\varphi^3 : \ \mathscr{W}_{t_0} \times (\mathscr{Q}_{[t_0,t]} \times \mathscr{U}_{[t_0,t]}) \to \mathscr{W}_{[t_0,t]}, \tag{5.2}$$

where

$$\varphi^3 = \{\varphi^1 \mid t_0, t \in \mathscr{T} \cap [t_0,t], t > t_0\}. \tag{5.3}$$

An element of set $\mathscr{W}_{[t_0,t]}$ is the function of time, $w_{[t_0,t]}$, in interval $[t_0,t]$.

There exist both deterministic and stochastic models of arrival flows (traffic streams), as pointed out in Section 2.2. In accordance with this, the queue *dynamics* can be treated either as deterministic or stochastic processes.

5.2. State transformation

According to expression (1.6), the state at time t is a function of the state at time t_0 and input in interval $[t_0,t]$. This is a general expression, which does not specify whether the elements of set $\mathcal{Q}_{[t_0,t]}$ are deterministic or stochastic processes. Volumes of arrival flows are, by their nature, stochastic processes, so that lengths of queues that form on intersection approaches are also stochastic processes. Therefore, the queuing process analysis should be performed by treating this process as a stochastic one.

The queuing process, however, can be regarded as a deterministic process —an approximation of the stochastic process. This approximation is often performed when stochastic processes are analyzed. In this case, it is assumed that the arrival flow volume is equal to an average value, which is particularly justifiable for higher volume values. Traffic flows can then be regarded as regular flows, similar to fluids [67].

One of the most important intersection performance indices is the delay, which is defined as the integral of queue length over time. The value of this integral can be determined as the sum of the delay resulting from average queue lengths, and the delay component that is the consequence of the stochastic nature of the process. Therefore, it is necessary to consider both stochastic and deterministic models of state transformation.

5.2.1. Deterministic model of state transformations (regular vehicle arrivals)

The discrete deterministic model of state transformation enables determination of state w^{l+1} at time $(l+1)\Delta t$ on the basis of state w^l at time $l\Delta t$, input volume q^l and control vector u^l, under the assumption that u^l and q^l do not change in interval Δt.

A component w_i^{l+1} of vector w^{l+1} depends on w_i^l, q_i^l, and u_i^l (with $\sigma_i \in D'_p$), i.e.,

$$\begin{aligned} w_i^{l+1} &= \varphi^1(w_i^l, q_i^l, u_p^l) \\ &= (w_i^l + (q_i^l - s_i u_p^l)\Delta t)\, h(w_i^l + (q_i^l - s_i u_p^l)\Delta t) \\ &= \max\{0, [w_i^l + (q_i^l - s_i u_p^l)\Delta t]\}, \quad (i \in \mathcal{J}') \end{aligned} \tag{5.4}$$

where $h(x)$ is the Heaviside function:

$$h(x) = \begin{cases} 0, & x < 0 \\ 1, & x \geq 0 \end{cases}.$$

If function $\bar{h}(x)$ is defined as:

$$\bar{h}(x) = \begin{cases} 0, & x < 0 \\ x, & x \geq 0 \end{cases},$$

then expression (5.4) can be written as:

$$w_i^{l-1} = \bar{h}(w_i^l + (q_i^l - s_i u_p^l)\Delta t), \quad (i \in \mathscr{Z}'). \tag{5.5}$$

The notations used in (5.4) and (5.5) have the following meanings:

w_i^l – the length of queue formed by stream σ_i in interval l,

q_i^l – the volume of stream σ_i in interval l,

s_i – saturation flow volume of stream σ_i,

u_p^l – the control of signal group D_p' in interval l, where $\sigma_i \in D_p'$.

Expression (5.5) enables determination of w_i^{l+1} even in the case the volumes q_i^l have different values in different intervals.

In the case of regular vehicle arrivals, which is a rough approximation of the real process, it is assumed that the volume q_i^l of traffic stream σ_i in interval l is deterministic, and same for any l, i.e., $q_i^l = q_i = \text{const}$ during a certain period.

If the queue length at the beginning of red indication is zero, and the beginning of red is at the same time the beginning of a cycle, i.e.,

$$w_{\beta i}(0) = w_{\beta i}^0 = 0 \text{ with } u_p^0 = 0,$$

then

$$w_{\beta i}^1 = q_i\Delta t \cdot h(q_i\Delta t) = \bar{h}(q_i\Delta t) = q_i\Delta t \quad (i \in \mathscr{Z}')$$

$$w_{\beta i}^2 = \bar{h}(w_{\beta i}^1 + q_i\Delta t) = q_i\Delta t + q_i\Delta t = 2q_i\Delta t \quad (i \in \mathscr{Z}')$$

$$\ldots$$

$$w_{\beta i}^l = l \cdot q_i\Delta t \quad (i \in \mathscr{Z}').$$

$w_{\beta i}$ is the queue that is formed under the constant volume q_i of stream σ_i $(i \in \mathscr{Z}')$.

If the effective red time of signal group D'_p is

$$r_p = l_p \Delta t \,,$$

then, at the end of red indication, i.e., at the beginning of green indication, the length of the queue formed by stream $\sigma_i \in D'_p$ is:

$$w_{\beta i}^{l_p} = l_p \, q_i \, \Delta t = r_p q_i \quad (i \in \mathcal{J}', p \in \mathcal{P}') \,. \tag{5.6}$$

The effective green time is determined as:

$$g_p = c - r_p \,, \text{ i.e.,}$$

$$g_p + r_p = c \,,$$

where c is the cycle time, and g_p and r_p are effective green and effective red times of signal group D'_p $(\sigma_i \in D'_p)$.

When the following condition is satisfied

$$q_i \, c < g_p \, s_i \,,$$

the queue will start decreasing as soon as green indication begins ($u_p^{l_p+1} = 1$), and it will *discharge* before the end of the cycle (Fig. 5.1). After the end of effective red, the queue changes in the following way:

$$\begin{aligned} w_{\beta i}^{l_p+1} &= (r_p q_i + (q_i - s_i)\Delta t) \, h(r_p q_i + (q_i - s_i)\Delta t) \\ &= r_p q_i + (q_i - s_i)\Delta t \\ &= (r_p + \Delta t) q_i - s_i \Delta t \end{aligned}$$

$$\begin{aligned} w_{\beta i}^{l_p+2} &= (w_{\beta i}^{l_p+1} + (q_i - s_i)\Delta t \; h(w_{\beta i}^{l_p+1} + (q_i - s_i)\Delta t) \\ &= (r_p + 2\Delta t) q_i - 2 s_i \Delta t \end{aligned}$$

...

$$w_{\beta i}^{l_p+\gamma} = (r_p + \gamma \, \Delta t) q_i - \gamma \, s_i \Delta t \,. \tag{5.7}$$

Expression (5.7) defines queue lengths for $\gamma = 1, 2, \ldots$, as long as $w_{\beta i}^{l_p+\gamma}$ is positive. If γ_p^i is the number of Δt intervals from the end of effective red time of signal group D'_p until the queue formed by stream $\sigma_i \in D'_p$ is discharged, then the following equality holds (Fig. 5.1):

$$(r_p + \gamma_p^i \Delta t) \, q_i - \gamma_p^i \Delta t \, s_i = 0$$

and

$$\gamma_p^i \Delta t = \frac{r_p q_i}{s_i - q_i} = \bar{\gamma}_p^i .$$

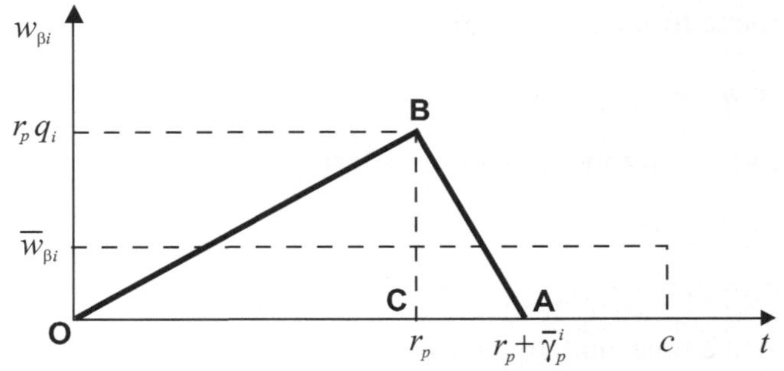

Figure 5.1

When flow intensity $\rho_i = \frac{c q_i}{g_p s_i}$ is less than 1, ($\sigma_i \in D'_p$), the value of average queue length can be determined as follows. The area of triangle OAB in Fig. 5.1 (queue length integral) represents time losses, i.e., the total delay of vehicles on approach T_i used by stream σ_i. The same total delay is obtained as the product of average queue length $\bar{w}_{\beta i}$ and cycle time c.

Thus, the average queue length $\bar{w}_{\beta i}$ can be expressed as:

$$\bar{w}_{\beta i} = \frac{1}{2c}(r_p + \bar{\gamma}_p^i) r_p q_i = \frac{r_p q_i}{2c}\left(r_p + \frac{r_p q_i}{s_i - q_i} \right)$$

$$\bar{w}_{\beta i} = \frac{r_p^2 q_i}{2c}\left(\frac{s_i}{s_i - q_i} \right) = \frac{r_p^2 q_i}{2c\left(1 - \frac{q_i}{s_i}\right)} \tag{5.8}$$

$$= \frac{c q_i (1 - \lambda_p)^2}{2(1 - \theta_i)}, \quad (i \in \mathscr{I}', p \in \mathscr{P}')$$

where

$$\lambda_p = \frac{g_p}{c}, \ \theta_i = \frac{q_i}{s_i},$$

with $\sigma_i \in D'_p$ and $\sigma_i = \sigma_{p_e} \in D'_p$.

If $\rho_i = 1, \quad (i \in \mathscr{I}')$, i.e., $\overline{\gamma}^i_p = g_p$, then the average queue length has the value:

$$\overline{w}_{\beta i} = \frac{q_i r_p}{2} = \frac{q_i(c - g_p)}{2} = \frac{q_i c(1 - \lambda_p)}{2}.$$

Since $q_i c = g_p s_i$, then

$$q_i = \frac{g_p s_i}{c} = \lambda_p s_i,$$

and the average queue length becomes:

$$\overline{w}_{\beta i} = \frac{c s_i}{2} \lambda_p (1 - \lambda_p). \tag{5.9}$$

The maximal queue length is attained at the end of effective red time, and it is (5.6):

$$w^{l_p}_{\beta i} = l_p q_i \Delta t = r_p q_i \quad (i \in \mathscr{I}', p \in \mathscr{P}').$$

5.2.2. Stochastic queuing models

The queuing model, which describes the real process more precisely than the deterministic model, assumes that elements of set $\mathscr{Q}_{[t_0,t]}$ are vectors whose elements are stochastic processes. In this case, elements of set $\mathscr{W}_t$ are vectors whose components are random variables.

Transformation of state on approach T_i can be described by the following expression:

$$w^{l+1}_i = f_v(q^l_i, w^l_i, u^l_i) + \xi_i(q^l_i, w^l_i, u^l_i), \tag{5.10}$$
$$(i \in \mathscr{I}', \; l = 1,2,\ldots,l_p, \; \sigma_i \in D'_p),$$

where f_v is the conditional average value w^{l+1}_i, with given w^l_i, and ξ_i the random variable with average value equal to zero.

Also, $l_p = \dfrac{r_p}{\Delta t}$.

Considering the state definition, expression (5.10) can be used to describe state transformation if the conditional probability distribution of variable w^{l+1}_i, with w^l_i given, does not depend on w^s_i, where $s < l$.

The process described by expression (5.10) is in this case Marcovian.

The states at the beginning and at the end of green signal indications are particularly important when constraints and optimization criterion are considered in traffic control problems statements. Since queuing is a stochastic process, determination of average queue length is also very significant.

Queues are formed at the stop line of an approach due to the following reasons [83]:

- Alteration of green and red signal indications, which results in queue formation even when vehicle arrivals are regular and the volume less than capacity.
- Stochastic nature of arrival flows.
- Congestion that arises when the queue is not discharged until the end of green signal indication. This is also a consequence of the stochastic nature of arrival flows. Congestion can appear, in come cycles, even when the average arrival flow volume is less than capacity, i.e., regardless of the average volume value.

Each of the queue components resulting from these three reasons can be considered by itself.

There exist many stochastic models of arrival flows, the main difference between them being the adopted probability distribution of the number of vehicles arriving on an intersection approach during a unit of time. Some of these models are described in Section 2.2. Also, there exist several stochastic models of queuing process.

In signal plan determination, average queue lengths are used most often, and in some traffic control problems it is necessary to find the average maximal queue length as well. These average values are determined using various expressions, depending on the probability distribution ξ_i used in defining the stochastic queuing process. These expressions can be used for determination of signal plan that will be applied in time periods for which it can be assumed that the process is stationary, i.e., that the average volumes are constant. Such periods last approximately 10 to 15 minutes or longer [17], [49].

Mathematical expectation of queue length w_i formed by traffic stream σ_i is, thus, equal to the sum of mathematical expectations of the components listed above, i.e.,

$$\begin{aligned} \mathrm{M}w_i &= \mathrm{M}w_{\beta i} + \mathrm{M}w_{\eta i} \\ &= \mathrm{M}w_{\beta i} + \mathrm{M}w'_{\eta i} + \mathrm{M}w''_{\eta i} \quad (i \in \mathscr{I}') \end{aligned}$$

where:

$w_{\beta i}$ – the regular queue component, resulting from average volume q_i,

$w'_{\eta i}$ – the component resulting from the stochastic nature of arrival streams, which exists even when the queue is discharged until the end of green indication,

$w''_{\eta i}$ – the average queue length at the end of effective green time.

The expressions for mathematical expectation of queue length presented here are the expressions proposed by Newell [46] and Webster [89]. The expression proposed by Newell is significant because it is of a general nature, derived in a pure theoretical manner, without specification of the probability distribution of the number of vehicles arriving to an intersection approach during a unit of time. It is also significant because it can be used to calculate the length of each of three components of the queue mentioned above. Webster's expression is derived in a more empirical manner, but it is very often used due to its simplicity. However, similar results are obtained when using both methods [46].

a) Mathematical expectation of queue length according to Newell

Results obtained with Newell expression are a good match of real processes when $s_i g_p >> 1$ and $q_i c >> 1$. Satisfactory results are obtained if $s_i g_p > 10$ [46].

According to Newell [46], mathematical expectation of queue length formed by stream σ_i is given by the following expression:

$$\begin{aligned} \mathrm{M}w_i &= \mathrm{M}w_{\beta i} + \mathrm{M}w'_{\eta i} + \mathrm{M}w''_{\eta i} \\ &= \frac{cq_i(1-\lambda_p)^2}{2(1-\theta_i)} + \frac{q_i I_i(1-\lambda_p)}{2s_i(1-\theta_i)^2} + \frac{q_i I_i H(\mu_i)}{2s_i(1-\theta_i)}, \end{aligned} \tag{5.15}$$

where:

c – the cycle time

q_i – the average volume of stream σ_i

$\lambda_p = \dfrac{g_p}{c}$ – the ratio between the effective green time of signal group D'_p and cycle time ($\sigma_i \in D'_p$)

$\theta_i = \dfrac{q_i}{s_i}$, $(i \in \mathscr{J}')$

s_i – the saturation flow volume of stream σ_i, $(i \in \mathscr{J}')$

$I_i = I_{Ai} + I_{Di}$, $(i \in \mathscr{J}')$

$I_{Ai} = \dfrac{DA_i(\bar{t}_i^0)}{\mathrm{M}A_i(\bar{t}_i^0)}$ – variation coefficient of random variable $A_i(\bar{t}_i^0) = q_i \bar{t}_i^0$

$I_{Di} = \dfrac{DB_i(\bar{t}_i^0 - r_p)}{\mathrm{M}B_i(\bar{t}_i^0 - r_p)}$ – variation coefficient of random variable $B_i(\bar{t}_i^0 - r_p) = s_i(\bar{t}_i^0 - r_p)$

$\mathrm{M}A_i(t) = q_i t$ – mathematical expectation of the number of vehicles of stream σ_i arriving on approach T_i during interval $[0,t]$

$\mathrm{M}A_i(\bar{t}_i^0) = q_i \bar{t}_i^0$

$\mathrm{M}B_i(\bar{t}_i^0 - r_p) = s_i(\bar{t}_i^0 - r_p)$ – mathematical expectation of the number of vehicles of stream σ_i leaving intersection in interval $[r_p, \bar{t}_i^0]$

$DA_i(\bar{t}_i^0)$ – dispersion of random variable $A_i(\bar{t}_i^0)$

$DB_i(\bar{t}_i^0 - r_p)$ – dispersion of random variable $B_i(\bar{t}_i^0 - r_p)$

$$\bar{t}_i^0 = \frac{s_i r_p}{s_i - q_i}$$

$$\mu_i = \frac{\mathrm{M}[A_i(c) - B_i(g_p)]}{\sqrt{I_i s_i g_p}} = \frac{q_i c - s_i g_p}{\sqrt{I_i s_i g_p}}$$

$$H(\mu_i) = \frac{2\mu_i^2}{\pi} \int_0^{\pi/2} \frac{tg^2\alpha}{e^{\frac{\mu_i^2}{2\cos^2\alpha} - I_i}} \cdot d\alpha$$

b) Mathematical expectation of queue length according to Webster

Webster’s expression is not entirely theoretic, and it is based, partially, on Pollaczek–Khintchin formula [48], [86]. Webster's expression can be used to calculate the average queue length for Poisson arrival flow, when service distribution is known, and flow intensity ρ_i is less than 1.

According to Webster [89], mathematical expectation of queue length formed by stream σ_i, in time periods during which it can be assumed that the arrival volume is a stationary random process, is given by the following expression:

$$Mw_i = \frac{q_i c(1-\lambda_p)^2}{2(1-\theta_i)} + \frac{\rho_i^2}{2(1-\rho_i)} - 0.65(q_i c)^{1/3} \rho_i^{(2+5\lambda_p)} . \tag{5.12}$$

Notations in expression (5.12) have the same meanings as in (5.11).

The first term in expression (5.12) represents the queue component, which results from uniform vehicle arrivals, with volume equal to the average volume q_i.

The second term represents, in fact, Pollaczek–Khintchin formula for the case the service time is deterministic, i.e., vehicles leave the queue in constant intervals.

The third term represents an empiric correction, and it is obtained by computer simulations.

The fact that the value of the third correction term lies in the range 5% to 15% of Mw_i value gives the possibility to use a simplified form of the expression for mathematical expectation of queue:

$$Mw_i \approx 0.9\left(\frac{q_i c(1-\lambda_p)^2}{2(1-\theta_i)} + \frac{\rho_i^2}{2(1-\rho_i)}\right), \quad (i \in \mathscr{T}') . \tag{5.13}$$

This simplified form is used in many practical applications and gives very good results for any volume value.

c) Mathematical expectation of the number of vehicles in a queue at the end of green signal indication

Mathematical expectation of the number of vehicles in a queue at the end of green signal indication varies with volume changes. Good results can be obtained if the period in which volume changes is divided into intervals having duration 10 to 15 minutes, so that it can be assumed for each interval that volumes are approximately constant during that interval [5], [83]. Several mathematical models have been defined [49], [17] for determination of the number of vehicles in a queue at time $t \in [t_0, t_1]$, with constant volume value in interval $[t_0, t_1]$.

Mathematical expectation w_{gi}^t of the number of vehicles in queue formed by vehicles of stream σ_i, at the end of effective green time $t = \mathrm{k}c$ ($\mathrm{k} = 1,2,\ldots$), can be determined using expression [83]:

$$\overline{w}_{gi}^t = (\sqrt{B_i^2 + 4A_i C} - B_i)/2A_i , (i \in \mathscr{T}') \tag{5.14}$$

where:

$$A_i = \lambda_p s_i t$$

$$B_i = (1-\rho_i)(\lambda_p s_i t)^2 + 1.2 q_i t + w_{gi}^0 (1.2 - \lambda_p s_i t)$$

$$C_i = 0.6(2w_{gi}^0 + q_i t)^2$$

$$\rho_i = \frac{q_i c}{g_p s_i}, \ (\sigma_i \in D'_p)$$

w_{gi}^0 – the number of vehicles in the queue formed by stream σ_i at time $t = 0$ (the start of red time).

6. THE OUTPUT FUNCTION

As pointed out in Section 2.1, among variables that characterize output flows, i.e., components of vector $y(\cdot) \in \mathscr{Y}$, the variables representing flow volumes, which are components of vector $y(\cdot) \in \mathscr{Y}$, will be used in mathematical model of traffic process. The output function (reaction) in this case, as already stated in Section 2.1, can be described in the following way (1.8):

$$\varphi^2 : \quad \mathscr{W}_t \times (\mathscr{Q}_t \times \mathscr{U}'_t) \to \mathscr{Q}^e_t .$$

The output of traffic process on an intersection is represented by vehicle flows departing from the intersection. When determining these flows, the following should be considered:

- Arrival flows are transformed by traffic signals.
- Departure flows are formed from one or more transformed arrival flows.

As an example, Fig. 1.1 shows, as functions of time, the arrival volumes of streams σ_3 and σ_5, their volumes $q'_3(t)$ and $q'_5(t)$ after control is applied, and volume $q^e_2(t)$ of the output flow $y_2(t)$, which is composed from volumes $q'_3(t)$ and $q'_5(t)$. Function φ^2 represents a composition of two functions, φ^2_{at} and φ^2_{bt}, i.e.,

$$\varphi^2 = \varphi^2_{at} \circ \varphi^2_{bt} , \qquad (6.1)$$

where φ^2_{at} defines transformation of arrival flows by control at time *t*, and function φ^2_{bt} represents formation of output flows from transformed input flows.

The transformation of input flows to output ones, as already mentioned, will be considered through transformation of their volumes. Evidently, dimensions of input and output vectors are not the same. If the vector of input volumes is

$$q(\cdot) = [q_1(\cdot), q_2(\cdot), \ldots, q_i(\cdot), \ldots, q_{I'}(\cdot)]^T ,$$

and the vector of volumes transformed by control is

$$q'(\cdot)=[q'_1(\cdot),q'_2(\cdot),...,q'_i(\cdot),...,q'_{I'}(\cdot)]^T,$$

then function φ^2_{at}, in discrete form, can be expressed as follows:

$$q'^l=\varphi^2_a(w^l,u^l_p,q^l), \qquad l\in\{1,2,...,l_e\}=\mathscr{L}_e,$$

where $l_e=\dfrac{c}{\Delta t}$.

For each component q'^l_i of vector q'^l there holds:

$$q'^l_i=u^l_p(q^l_i+(s_i-q^l_i)h(w^l_i)), \quad (i\in\mathscr{I}',l\in\mathscr{L}_e) \tag{6.2}$$

$(\sigma_i\in D'_p)$.

If traffic stream volumes are constant in the period under observation, then:

$$q'^l=\varphi^2_a(w^l,u^l_p,q), \qquad (l\in\mathscr{L}_e).$$

In this case, components q'^l_i are determined according to the following expression:

$$q'^l_i=u^l_p[q_i+(s_i-q_i)\,h(w^l_i)], \qquad (i\in\mathscr{I}',l\in\mathscr{L}_e). \tag{6.3}$$

Transformed volumes in each interval l represent components of vector

$$q'^l=[q'^l_1,q'^l_2,...,q'^l_i,...,q'^l_{I'}]^T, \quad (l\in\mathscr{L}_e). \tag{6.4}$$

Forming of the vector of output volumes in interval l,

$$q^{el}=[q^{el}_1,q^{el}_2,...,q^{el}_h,...,q^{el}_H]^T,$$

can now be represented as follows [55]:

$$q^{el}=\varphi^2_b(q'^l)=Aq'^l, \quad (l\in\mathscr{L}_e), \tag{6.5}$$

where A is the output matrix, containing information on participation of traffic streams in output flows. Function φ^2_b remains the same for every l.

An element a_{hi} of matrix $A=[a_{hi}]_{H\times I'}$ represents the fraction of volume q'^l_i of traffic stream σ_i, which takes part in forming volume q^{el}_h of output flow y_h. The value of any element is $a_{hi}\in[0,1]$, $(i\in\mathscr{I}',h\in\mathscr{H})$, where

$$\mathscr{H}=\{1,2,...,h,...,H\},$$

and H is the number of vehicle output flows (the number of output vector components) from the intersection.

Values of matrix A elements are determined in the following way:

$$a_{hi} = \begin{cases} 1 \text{ , if whole traffic stream } \sigma_i \text{ takes part in forming} \\ \quad \text{output flow } y_h \\ a'_h \text{ , if a partial stream of traffic stream } \sigma_i \text{ takes part} \\ \quad \text{in forming output flow } y_h \\ 0 \text{ , if traffic stream } \sigma_i \text{ does not take part in forming} \\ \quad \text{output flow } y_h \end{cases} .$$

Here, $a'_{hi} < 1$ represents the fraction of volume q_i that takes part in forming volume q_h^e.

On the basis of expressions (6.2) and (6.3) it can be seen that the volume leaving a stop line is equal to the arrival volume (if no queue exists) or to the saturation flow volume, s_i (if queue exists) (Fig. 1.1). The saturation flow volume is the queue discharge rate (Fig. AV.1 in Appendix V). The saturation flow volume has an approximately constant value that is attained after the acceleration of vehicles leaving the stop line is performed at the beginning of green time, i.e., when vehicles crossing the stop line do not accelerate any more. The value of saturation flow volume is influenced by many factors [88], [14], such as approach width, number of lanes, grade, flow composition, turnings, pavement condition, etc. The value of saturation flow may be different in peak and off-peak periods. The way traffic is controlled on an intersection can also influence saturation flow values. Namely, if conflicting traffic streams are allowed to move simultaneously through an intersection (e.g., when a vehicle stream intersects a pedestrian stream or other vehicle stream), then the saturation flow of the priority stream does not change, but the saturation flow of the other stream decreases.

Expressions giving the relation between approach width and saturation flow value were experimentally determined by Branston [15] and Kimber and Semmens [50]. A typical saturation flow volume value is 1800 PCU/h per lane.

Saturation flow volumes can also be determined by measurements. One of frequently used methods is the method designed by TRRL [TRRL, Road Note No. 34].

Example 6.1

Determine output flow volumes for the intersection given in Fig. 6.1, together with its output matrix A, using expression (6.5).

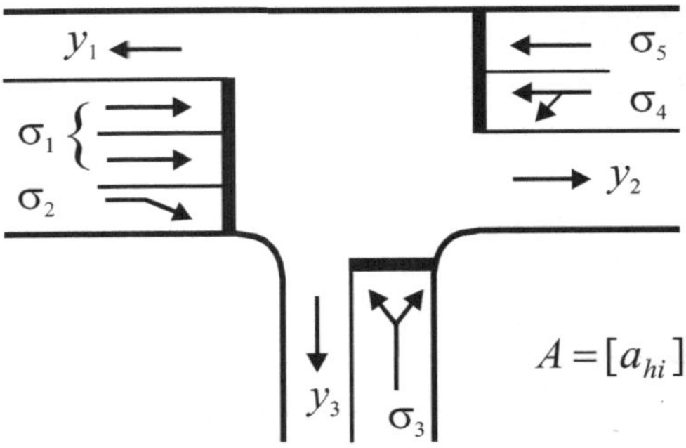

$$A = [a_{hi}]_{3x5} = \begin{bmatrix} 0 & 0 & 0.6 & 0.8 & 1 \\ 1 & 0 & 0.4 & 0 & 0 \\ 0 & 1 & 0 & 0.2 & 0 \end{bmatrix}$$

Figure 6.1

$$q^{el} = Aq'^{l} = \begin{bmatrix} 0 & 0 & 0.6 & 0.8 & 1 \\ 1 & 0 & 0.4 & 0 & 0 \\ 0 & 1 & 0 & 0.2 & 0 \end{bmatrix} \cdot \begin{bmatrix} q_1'^{l} \\ q_2'^{l} \\ q_3'^{l} \\ q_4'^{l} \\ q_5'^{l} \end{bmatrix} = \begin{bmatrix} 0{,}6q_3'^{l} + 0.8q_4'^{l} + q_5'^{l} \\ q_1'^{l} + 0.4q_3'^{l} \\ q_2'^{l} + 0.2q_4'^{l} \end{bmatrix} = \begin{bmatrix} q_1^{el} \\ q_2^{el} \\ q_3^{el} \end{bmatrix}$$

Therefore:

$$q_1^{el} = 0.6q_3'^{l} + 0.8q_4'^{l} + q_5'^{l}$$
$$q_2^{el} = q_1'^{l} + 0.4q_3'^{l}$$
$$q_3^{el} = q_2'^{l} + 0.2q_4'^{l}$$

Output variables are not important in solving traffic control problems for isolated intersections. However, in solving traffic control problems for networks of signalized intersections, output variables from intersections represent so-called "platoons," and their transformation when traveling between intersections is the most significant component of models used in solving these problems [72].

Part II

CONTROL PROBLEM STATEMENT

Solutions of traffic control problems are today based on application of modern mathematical optimization methods and digital computers. For efficient application of the mentioned means, it is necessary to give exact formulations of traffic control problems and develop algorithms for their solution. According to the general systems theory [58], a control problem can be formulated as a satisfaction problem or as an optimization problem.

In this book the traffic control problem will be formulated and solved as an optimal control problem.

7. GENERAL STATEMENT OF TRAFFIC CONTROL PROBLEM (SIGNAL PLAN CHOICE)

The optimal control problem is the problem of choosing control $u(\cdot)$ from the set of feasible controls $\mathscr{U}_f \subset \mathscr{U}$, which gives the optimal value of the chosen optimization criterion J_c. Therefore, it is necessary to:

a) Define the set whose elements are the controls that can be chosen, i.e., define the set of feasible controls $\mathscr{U}_f \subset \mathscr{U}$.

b) Define the measure that can be used for comparing effects of any two controls, and for the choice of the best control because set $\mathscr{U}_f$ is unordered. In the optimal control problems, this measure is termed *the optimality criterion, the goal function, the performance index, or the objective function.*

The performance index is defined by the following mapping:

$$J_c'' : \mathscr{X}_{[t_0,t]} \times \mathscr{W}_{[t_0,t]} \to \mathrm{R}', \tag{7.1}$$

i.e., the performance index is the function of mapping an ordered pair $(x_{[t_0,t]}, w_{[t_0,t]})$ to an element of the linearly ordered set R'. Most frequently R' is the set of real numbers, i.e., $\mathrm{R}' = \mathrm{R}$. The optimal value of the performance index is determined using values of input and state in interval $[t_0,t]$. Therefore, this value represents some "integral" measure of control quality on the intersection in $[t_0,t]$ interval, rather than a measure related to a specific time.

For isolated signalized intersections $\mathscr{X}_{[t_0,t]} = \mathscr{Q}_{[t_0,t]} \times \mathscr{U}_{[t_0,t]}$, so that the performance index J_c'' can be presented by the expression:

$$J_c'' : \mathscr{Q}_{[t_0,t]} \times \mathscr{U}_{[t_0,t]} \times \mathscr{W}_{[t_0,t]} \to \mathrm{R}. \tag{7.2}$$

Bearing in mind that $\mathscr{W}_{[t_0,t]}$ is defined by (5.2):

$$\varphi^3_{[t_0,t]} : \mathscr{W}_{t_0} \times (\mathscr{Q}_{[t_0,t]} \times \mathscr{U}_{[t_0,t]}) \to \mathscr{W}_{[t_0,t]}$$

and that the initial state $w(t_0)$ is known, as well as elements of $\mathscr{Q}_{[t_0,t]}$, whose components are traffic streams volumes in interval $[t_0,t]$, the performance index can be presented by the following mapping:

$$J'_c : \mathcal{U}_{[t_0,t]} \to \mathrm{R} . \tag{7.3}$$

The assumption is that traffic streams' volumes are either constants or can be described as stationary stochastic processes in $[t_0,t]$ interval.

Elements of $\mathcal{U}_{[t_0,t]}$ are periodical, vector time functions, and because of that, it is sufficient to determine their values, $u(\cdot)$, at each instant of the basic period, $[0,c]$. Using these values, it is possible to determine values of each component of vector $u_{[t_0,t]}$ at any instant $t' \in [t_0,t]$ (4.6). Function $u(\cdot)$ is a restriction of $u_{[t_0,t]}$ to $[0,c] \cap [t_0,t]$.

If the set of feasible controls is $\mathcal{U}_f$, and

$$u(\cdot) \in \mathcal{U}_f ,$$

then the control quality can be determined using the restriction $u(\cdot)$ of control $u_{[t_0,t]}$ to interval $[0,c] \cap [t_0,t]$. Thus, the performance index can be described as the mapping:

$$J_c : \mathcal{U}_f \to \mathrm{R} . \tag{7.4}$$

Since the set of real numbers, R, is linearly ordered by $\leq$ relation, introduction of the J_c function makes it possible to introduce the linear order relation in the set $\mathcal{U}_f$, as well. The order relation, R_π, is introduced in $\mathcal{U}_f$ by the convention that $u^1(\cdot)$ is better, or at least as good as $u^2(\cdot)$, i.e., $u^1(\cdot) R_\pi u^2(\cdot)$ if and only if $J_c(u^1(\cdot)) \leq J_c(u^2(\cdot))$, i.e.,

$$J_c(u^1(\cdot)) \leq J_c(u^2(\cdot)) \Rightarrow u^1(\cdot) R_\pi u^2(\cdot) . \tag{7.5}$$

The problem of the choice of optimal control can now be stated as follows: The set of feasible controls, $\mathcal{U}_f$, is given, and the performance index is:

$$J_c : \mathcal{U}_f \to \mathrm{R} .$$

Let

$$\inf_{u(\cdot) \in \mathcal{U}_f} J_c(u(\cdot) = J_c^* . \tag{7.6}$$

Determine the set of optimal controls:

$$\mathcal{U}^* = \{ u^*(\cdot) \mid u^*(\cdot) \in \mathcal{U}_f,\ J_c(u^*(\cdot)) = J_c^* \} . \tag{7.7}$$

Here it is considered that the best control is the control mapped to the minimal performance index value.

There may be some cases, of course, with $\mathcal{U}^* = \varnothing$, i.e., no feasible solution can be found ($\mathcal{U}^* = \varnothing \quad \Rightarrow \quad \mathcal{U}_f = \varnothing$).

Solving optimal control problem always implies solving some minimization or maximization problem. A maximization problem can always be transformed into a minimization problem, and vice versa. This is possible because:

$$\max_{u(\cdot)\in\mathcal{U}_f} J_c(u(\cdot)) = -\min_{u(\cdot)\in\mathcal{U}_f} \left(-J_c(u(\cdot))\right).$$

The fact stated by (7.7) can also be expressed as follows:

$$J_c(u^*(\cdot)) = \inf\{ J_c(u(\cdot)) \mid (u(\cdot)) \in \mathcal{U}_f \} = \inf_{u(\cdot)\in\mathcal{U}_f} J_c(u(\cdot)) \,. \tag{7.8}$$

Obviously, the problem of the choice of the optimal control can be defined by the pair $(\mathcal{U}_f, J_c)$ and the statement whether the performance index shall be minimized or maximized.

8. THE SET OF FEASIBLE CONTROLS (SIGNAL PLANS)

The formulation of the optimal control problem $(\mathcal{U}_f, J_c)$ implies defining the feasible control set $\mathcal{U}_f$, whose element $u(\cdot) \in \mathcal{U}_f$ is defined by an ordered pair as follows:

$$u(\cdot) = \begin{pmatrix} \mathbf{u} \\ \tau \end{pmatrix} = (\mathbf{u}, \tau)^T,$$

where

$$\mathbf{u} = [\mathbf{u}^1, \mathbf{u}^2, \ldots, \mathbf{u}^k, \ldots, \mathbf{u}^{\mathrm{K}}]$$

is the control structure (signal plan structure) (4.12). A control structure component

$$\mathbf{u}^k = [u_1^k, u_2^k, \ldots, u_p^k, \ldots, u_{\mathrm{P}}^k]^T \quad (k \in \mathcal{K})$$

is a vector (4.15), whose components assume values from set $\{0,1\}$, i.e.,

$$u_p^k \in \{0,1\} \quad (k \in \mathcal{K}, p \in \mathcal{P}),$$

where $\mathcal{P}$ is the set of signal group indices in the complete set of signal groups. The complete set of signal groups has to be chosen before start of the problem solving.

The second element of the pair (4.14), τ, represents the cycle time split, i.e.,

$$\tau = [\tau^1, \tau^2, \ldots, \tau^k, \ldots, \tau^{\mathrm{K}}].$$

The terms *control* and *signal plan* are used as synonyms. Component $\mathbf{u}^k$ is termed the *control vector* or *phase*.

Control is also described as the vector time function:

$$u(\cdot) = [u_1(\cdot), u_2(\cdot), \ldots, u_p(\cdot), \ldots, u_{\mathrm{P}}(\cdot)]^T,$$

where one vector component, $u_p(\cdot)$, represents the control variable assigned to signal group D'_p. Values of this variable during a cycle time are defined by the sequence:

$$\mathbf{u}_p = [u_p^1, u_p^2, \ldots, u_p^k, \ldots, u_p^{\mathrm{K}}], \quad u_p^k \in \{0,1\}, (p \in \mathcal{P}),$$

i.e., by the values of variables u_p^k for $k = 1,2,\ldots,\mathrm{K}$. The duration of each component u_p^k of that sequence is determined by the value of component τ^k of vector τ with the same index, *k*. Hence, the control variable $u_p(\cdot)$ can be defined by the pair (4.16):

$$u_p(\cdot) = (\mathbf{u}_p, \tau)^T, \quad (p \in \mathscr{P}).$$

The problems of optimal signal plan choice, $(\mathscr{U}_f, J_c)$, can be formulated as mathematical optimization problems. The set of feasible controls, $\mathscr{U}_f$, is determined by the constraints that have to be satisfied by chosen controls. In some problems, constraints on state variables of the process are also present.

The set of feasible controls, $\mathscr{U}_f$, depends also on the problem being solved, and it is not completely the same in all problems of the choice of the optimal control. Namely, there exist constraints on control that have to be satisfied regardless of the problem considered, and there exist constraints that are present in some problems and not in others. For instance, in the problem of choosing the signal plan that minimizes the capacity, the sum of phase durations has to be equal to a given cycle time, whereas in the problem of cycle time maximization this constraint doesn't exist.

8.1. The constraints that define the set of feasible controls

Bearing in mind the elements of the control (signal plan) included in the constraints, one can note that there are groups of constraints related to particular components of control variables, i.e., the constraints related to control components assigned to signal groups, constraints related to phases (their sequence and structure), and constraints related to phase durations (cycle time allocation to phases).

In order to formulate the constraints related to phases, their sequence, and structure, it is necessary to determine the relations that exist in the set of control variables. On the basis of these relations, it is possible to determine the control variables that can simultaneously assume value 1, which is necessary for determination of the set of feasible phases. For finding feasible phase sequences, it is necessary to know whether a signal group can gain the right-of-way as soon as another group has lost it, or a time delay is needed for intersection clearance, i.e., the intergreen time has to be greater than zero.

The control variables have to satisfy the following conditions:

- Each control variable (the component assigned to one signal group) has to satisfy the condition that the signal group can gain the right-of-way only once in the cycle.
- The duration of green indication (green time) for each signal group has to be longer than the predefined minimum green time.
- The duration of green time for each vehicle signal group has to satisfy the capacity constraint, i.e., its value has to be large enough to accommodate all vehicles that arrive during a cycle to leave the intersection during the same cycle. This constraint, of course, makes sense only in the case when no approach is oversaturated.

The definition of the set of feasible phases, i.e., the set of control vectors giving the right-of-way to more than one signal group, is based on the information on the pairs of signal groups that can simultaneously gain the right-of-way. This information can be obtained from the signal group compatibility relation or compatibility graph (Subsection 3.3.3). All feasible phases can be determined using this relation.

The signal plan structure, i.e., the phase sequence, has to be determined bearing in mind that when a signal group loses its right-of-way, an incompatible signal group usually cannot immediately gain the right-of-way. Some time has to pass (intergreen time) before the incompatible signal group gains the right-of-way. Thus, in the signal plan structure, a phase can be followed only by particular feasible phases.

Time constraints refer to phase durations and the cycle duration. The sum of phase durations has to be less than or equal to the determined cycle time. The cycle time has to be less than or equal to a predefined maximal value (usually set to 120 s, and only exceptionally longer).

Constraints related to traffic process states, i.e., vehicle queue lengths on some approaches, appear in some control problems. Such constraints are usually transformed to constraints on duration of the red signal indication – red time constraints. Namely, it is possible, for a given average flow volume, to determine the maximal red time so that the queue length doesn't exceed some prescribed value.

The feasible set of signal plans can be, thus, defined by the following constraints:

a) *Control variable constraints*

1. The constraints of one green interval in the cycle for each signal group – Each signal group must get the right-of-way once and only once during the cycle.

2. Minimal green times constraints – The duration of green indication of traffic lights—green time allocated to a signal group has to be longer, or at least equal to a given minimum green time value defined for each signal group.
3. Maximal red time constraints – The duration of red indication of traffic lights—red time allocated to a signal group has to be shorter than a given maximal red time value defined for some or each signal group.
4. Capacity constraints (flow balance constraints) – The green time allocated to a vehicle signal group has to be long enough to enable all vehicles arriving during a cycle to intersection approaches belonging to that signal group to leave the intersection during the green time.

b) Constraints on the composition of control vectors—phases

In some intervals during the cycle, the right-of-way can be simultaneously given only to compatible signal groups (Subsection 3.3.3).

c) Constraints on control vector sequence and on signal plan structure

1. Minimal intergreen constraints – The duration of phases positioned between the phase that takes off the right-of-way to some signal groups, and the phase that gives the right-of-way to some incompatible signal groups, has to be longer than the specified minimal intergreen times.
2. Phase sequence constraints – For each feasible phase, a subset of the set of feasible phases is defined, containing the phases that can be chosen as next in the signal plan structure. This means that each phase in the signal plan structure has to belong to the subset of possible followers defined for the preceding phase in the structure.

d) Time constraints

1. The sum of phase durations has to be equal to the cycle time.
2. The sum of phase durations has to be equal or less than a given maximal cycle time value.

These constraints are present in most problems of optimal signal plan choice. Some of the constraints exist in almost all problems, whereas some constraints appear in some problems and not in others. Constraints a.1, a.2, b, and c have to be satisfied by any signal plan.

8.2. Constraints on control variables

As mentioned in the previous section, there exists a set of constraints that have to be satisfied by control variables, i.e., the variables that control signal groups. The assignment of control variables to signal groups can be done only after the choice of the complete set of signal groups. Because of that, all constraints presented in this section are related to one, chosen, complete set of signal groups.

This section presents mathematical expressions for control variable constraints. Control variable is defined by expression (4.16):

$$u_p(\cdot) = (\mathbf{u}_p, \tau)^T = \begin{bmatrix} \mathbf{u}_p \\ \tau \end{bmatrix} = \begin{bmatrix} (u_p^1, u_p^2, \ldots, u_p^k, \ldots, u_p^K) \\ (\tau^1, \tau^2, \ldots, \tau^k, \ldots, \tau^K) \end{bmatrix}, \quad (p \in \mathscr{P}).$$

Thus, control variable constraints are functions of vectors $\mathbf{u}_p$ and τ.

8.2.1. The constraint of one interval of green indication during the cycle

This constraint, mentioned as a.1, has to be satisfied by each control variable. If this condition is satisfied, the time interval between two subsequent starts of green interval for a signal group is equal to the cycle time. Thus, the phases giving the right-of-way to one signal group have to be consecutive.

This constraint is common in existing types of traffic control by fixed signal plans, and drivers and pedestrians have gotten accustomed to it. Also, time losses generally increase if the number of intervals of green indication for a signal group is greater than one. Omitting this constraint leads to an extension of the set of feasible signal plans, and because of that, in some cases the optimal signal plan can have more than one interval of green indication for a signal group.

The analytical expression of this constraint has to be valid for any position of green interval in the cycle relative to the beginning of the cycle. Figure 4.2 presents all possible positions of green interval (more precisely, effective green interval). In formulating the analytical expression, the fact is used that in this case the number of changes of control variable values (from 0 to 1, and from 1 to 0) for each signal group during one cycle has to be equal to 2 (see Fig. 4.2).

The analytical expression of this constraint for control variable $u_p(\cdot) = (\mathbf{u}_p, \tau)^T$, where $\mathbf{u}_p = [u_p^1, u_p^2, \ldots, u_p^k, \ldots, u_p^K]$, $u_p^k \in \{0,1\}$, $(p \in \mathscr{P})$ can be formulated as follows:

$$\sum_{k=1}^{K}(u_p^k + u_p^{k(\text{mod}\,K)+1})(\text{mod}\,2) = 2, \quad (p \in \mathscr{P}), \tag{8.1}$$

where $\mathscr{P} = \{1,2,\ldots,P\}$ is the set of signal group indices.

Example 8.1

Check whether the signal plan presented in Fig. 8.1 satisfies constraints (8.1).

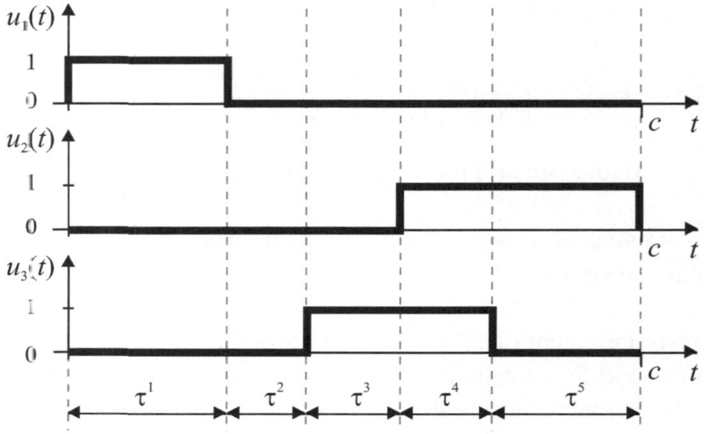

Figure 8.1

The signal plan structure, $\mathbf{u}$, is:

$$\mathbf{u} = [\mathbf{u}^1, \mathbf{u}^2, \mathbf{u}^3, \mathbf{u}^4, \mathbf{u}^5] = \begin{bmatrix} 1 & 0 & 0 & 0 & 0 \\ 0 & 0 & 0 & 1 & 1 \\ 0 & 0 & 1 & 1 & 0 \end{bmatrix}.$$

The constraints of one green interval in a cycle are:

For $p = 1$: $\sum_{k=1}^{5}(u_1^k + u_1^{k(\text{mod}\,5)+1})(\text{mod}\,2) = (1+0)(\text{mod}\,2) + (0+0)(\text{mod}\,2) +$

$$(0+0)(\text{mod}\,2) + (0+0)(\text{mod}\,2) + (0+1)(\text{mod}\,2) = 2$$

For $p = 2$: $\sum_{k=1}^{5}(u_2^k + u_2^{k(\text{mod}\,5)+1})(\text{mod}\,2) = (0+0)(\text{mod}\,2) + (0+0)(\text{mod}\,2) +$

$$(0+1)(\text{mod}\,2) + (1+1)(\text{mod}\,2) + (1+0)(\text{mod}\,2) = 2$$

For $p = 3: \sum_{k=1}^{5}(u_3^k + u_3^{k(\text{mod}5)+1})(\text{mod}2) = (0+0)(\text{mod}2) + (0+1)(\text{mod}2) + (1+1)(\text{mod}2) + (1+0)(\text{mod}2) + (0+0)(\text{mod}2) = 2$

Obviously, the constraints of one green interval in a cycle for each signal group are satisfied by the signal plan in this example.

8.2.2. Constraints of minimal duration of green indication intervals

Sometimes it happens that the calculated durations of green indications (displayed green time) are very short, just a couple of seconds. Usually, such short green times are not permitted, and because of that, for each control variable, the *minimal displayed green time* is defined.

There are many reasons for introducing *minimal displayed green times*. Some of the reasons are psychological—drivers, not accustomed to very short displayed green time, could assume a traffic light mistake and stop respecting it; also, noting that the green time is short, drivers could accelerate in order to pass through the intersection so that the safety would be significantly reduced, etc. Technical reasons are also present, particularly with vehicle-actuated signals, where the data obtained from vehicle detectors are used for generation of control. In some of these systems, the green indication is given to a signal group only if particular detectors are "actuated." The possibility exists that some vehicle, because of a very short displayed green time, stays trapped between the detector loop and stop line, and won't get the right-of-way unless another vehicle arrives behind it. To prevent such situations, a minimal displayed green time is periodically assigned to each control variable (in each cycle) regardless of detector actuation.

Minimal values of minimal displayed green times are recommended by technical standards in many countries. In Germany, for instance, the following values are fixed as lower limits for minimal displayed green times [69]:

- Vehicle signal group control variables: (5–10) s
- Pedestrian signal groups control variables: 5 s
- Tram signal groups control variables: 5 s

Feasible signal plans have to satisfy the constraints of minimal displayed green times. These times are component of the vector

$$G_m = (G_{m1}, G_{m2}, \ldots, G_{mp}, \ldots, G_{mP}) . \tag{8.2}$$

Minimal displayed green time, G_{mp}, for control variable controlling signal group D'_p ($p \in \{1,2,\ldots,P\}$), is determined using the minimal displayed green times for each traffic stream, σ_i, that belongs to signal group D'_p. If minimal displayed green time for σ_i ($i \in \mathscr{I}$) is G_i^m, then the minimal displayed green time, G_{mp}, for control variable that controls signal group D'_p is given by the following expression:

$$G_{mp} = \max\{ G_i^m \mid \sigma_i \in D'_p \} = G_{i0}^m, \quad (p \in \mathscr{P}), \tag{8.3}$$

i.e.,

$$G_{i0}^m \geq G_i^m, \quad (\sigma_i \in D'_p).$$

Likewise, the minimal effective green times for signal groups, g_{mp}, ($p \in \mathscr{P}$) are given by the expression:

$$g_{mp} = \max\{ g_i^m \mid \sigma_i \in D'_p \}, \quad (p \in \{1,2,\ldots \mathrm{P}\}). \tag{8.4}$$

In the computing process of signal plan choice the effective values of green and red time are used.

Minimal effective green times for vehicle signal groups are calculated in the same way as other effective green times for vehicle signal groups (Appendix V), i.e.,

$$g_{mp} = G_{mp} + a_p - (l'_p + l''_p) = G_{mp} + a_p - l_p, \quad (p \in \mathscr{P}) \tag{8.5}$$

where a_p and l_p are usually the standard values, 3 s and 2 s, respectively.

The minimal effective green times for pedestrian and tram traffic streams are usually of the same duration as green signal indications, i.e., displayed green times.

Minimal effective green times are components of vector g_m, i.e.,

$$g_m = (g_{m1}, g_{m2}, \ldots, g_{mp}, \ldots, g_{m\mathrm{P}}). \tag{8.6}$$

The effective green time for a signal group has to be longer than the minimal effective green time for that group. This means that the interval in which the respective control variable assumes value 1 has to be longer than the minimal effective green time, i.e.,

$$\mathbf{u}_p \tau^T = \sum_{k=1}^{\mathrm{K}} u_p^k \tau_p^k \geq g_{mp}, \quad (p \in \mathscr{P}). \tag{8.7}$$

8.2.3. The constraints of maximal red times

In some traffic control problems there exist constraints related to states, i.e., to queues. These constraints, expressed as maximal feasible values of queue lengths, can be transformed into constraints of maximal displayed red times. Since a queue forms during red signal indication, it is obvious that longer red time leads to longer queues. Thus, the maximal duration of red signal indication, i.e., the *maximal displayed red time*, corresponds to the maximal feasible queue length.

In one signal group there can be several traffic streams with maximal displayed red time constraints. Therefore, it is necessary to determine values of maximal red times for each signal group that contains traffic streams with constrained queues. These values are determined as follows (Appendix V):

$$r_{Mq} = \min\{ r_j^M \mid \sigma_j \in D'_q \}, \quad (q \in \mathscr{P}^M), \tag{8.8}$$

where:

r_{Mq} – the maximal effective red time for signal group D'_q,

r_j^M – the maximal effective red time,

$\mathscr{P}^M \subseteq \mathscr{P}'$ – the index set of vehicle signal groups with maximal effective red time constraints. In each element of $\mathscr{P}^M$ there is one or more traffic streams with such type of constraints.

Analytical expressions of the maximal effective red constraints are given by the following inequalities:

$$\sum_{k=1}^{K} (1 - u_q^k) \cdot \tau^k \le r_{Mq}, \quad (q \in \mathscr{P}^M). \tag{8.9}$$

8.2.4. The flow balance (capacity) constraints

These constraints are formulated for two cases:

- Saturation flow is constant during the cycle.
- Saturation flow can have two values during the cycle, which is the case when "filtering" one traffic stream "through" another is allowed.

a) The capacity constraint when saturation flow is constant during the cycle

All vehicles of an arrival traffic stream, σ_i, with the average flow volume, q_i, coming to the intersection during one cycle, can leave the intersection in time not longer than the cycle time if the effective green time, g_p, $(\sigma_i \in D'_p)$, satisfies the following inequality:

$$q_i c \le g_p s_i \, , \quad (i \in \mathcal{J}') \, ,$$

i.e.,

$$q_i \le \frac{g_p s_i}{c} \, , \quad (i \in \mathcal{J}'). \tag{8.10}$$

The condition (8.10) can be presented in the form:

$$\frac{q_i c}{g_p s_i} \le 1 \, , \quad (i \in \mathcal{J}', p \in \mathcal{P}'). \tag{8.11}$$

The arrival flow volume equal to $\frac{g_p s_i}{c}$ is termed the *theoretical capacity* of traffic stream σ_i. However, when the arrival volume has this value, queue lengths formed during red signal indication will be extremely long. Because of that, the *practical capacity*, which is equal to $\rho_i \frac{g_p s_i}{c}$, is used in practice.

The condition (8.11) can now be expressed as:

$$\rho_i = \frac{q_i}{\frac{s_i g_p}{c}} = \frac{q_i c}{s_i g_p} \le 1 \, , \quad (i \in \mathcal{J}', p \in \mathcal{P}'). \tag{8.12}$$

The ratio defined by ρ_i is called the *saturation degree* of traffic stream σ_i (or the traffic intensity, in the queuing theory terminology). The value of the maximal acceptable saturation degree, $\overline{\rho}_i$, is less than 1, meaning that for each traffic stream the condition $\rho_i \le \overline{\rho}_i$ has to be satisfied. This degree, $\overline{\rho}_i$, usually represents an estimation of traffic engineers and can be different for different intersection approaches. Most commonly the value of 0.9 is used, as suggested by Webster and Cobbe [88]. If it is necessary to prevent formation of longer queues on an approach, the value of $\overline{\rho}_i$ has to be lower than 0.9; if longer queues can be tolerated, then the value can be slightly greater than 0.9.

To each traffic stream enough effective green time has to be allocated to ensure that the practical capacity is greater than the average arrival flow volume. The necessary green times for traffic streams belonging to one signal group need not be the same. The effective green time of a signal group has to be greater than the effective green time necessary for any traffic stream belonging to this signal group. Analytic expression of this constraint is:

$$\mathbf{u}_p \tau^T = \sum_{k=1}^{K} u_p^k \tau^k \geq \gamma_p \, , \quad (p \in \mathscr{P}') \, , \tag{8.13}$$

where:

$$\gamma_p = \max_i \left\{ \frac{c q_i}{\bar{\rho}_i s_i} \mid \sigma_i \in D'_p \right\}, \quad (p \in \mathscr{P}'), \tag{8.14}$$

$\bar{\rho}_i$ – the maximal acceptable saturation degree for traffic stream σ_i,

$\mathscr{P}'$ – the index set of vehicle signal groups set.

b) The capacity constraints when filtering is allowed

When a pair of traffic streams, (σ_a, σ_b), is at the same time element of both, the conflictness relation, C_1, and the compatibility relation, C, i.e., $(\sigma_a, \sigma_b) \in C_1$ and $(\sigma_a, \sigma_b) \in C$, then a feasible signal plan can contain phases by which simultaneous right-of-way is given to both traffic streams.

The phases (control vectors) with $u_a^k = u_b^k = 1$ belong to the set of feasible phases. One traffic stream, in this case, "filters" through the other traffic stream, under priority rules. When left-turning vehicles, i.e., the *opposed turning traffic*, filter (by the right hand rule) through the traffic stream that passes straight through the intersection, i.e., the *opposing traffic*, the opposing traffic stream has the priority. The vehicles in the opposed traffic stream, in this case, have so-called *conditional* right-of-way. Vehicle traffic streams can be filtered through pedestrian traffic streams, as well (e.g., the right-turning vehicles that have right-of-way at the same time as the pedestrian stream through which they filter).

Filtering is possible if gaps between vehicles in the opposing stream have acceptable duration, and if there is enough space on the intersection to accommodate turning vehicles waiting for acceptable gaps.

In formulating capacity constraints, a.4, saturation flow volumes, $s_i, (i \in \mathscr{J}')$ are constant during the cycle.

In the case when filtering is permitted, the saturation flow volume of the opposed traffic stream has one value in the case when both opposed and opposing traffic streams get the right-of-way simultaneously, and another, different value when its movement is "protected," i.e., its right-of-way is not completely simultaneous with the right-of-way of the opposing traffic stream (e.g., late start or early stop). This means that the value of saturation flow volume of the opposed stream, s_b^k, in some interval *k*, depends on the control vector in that interval, $\mathbf{u}^k$, i.e.:

$$s_b^k = s_b^k(\mathbf{u}^k). \tag{8.15}$$

For instance, if traffic streams σ_4 and σ_5 in Fig. 8.2 are compatible, namely $(\sigma_4,\sigma_5) \in C$, and if one control variable is assigned to each traffic stream, then the saturation flow volume in interval *k*, s_4^k, when control vector $(0,0,1,1,1,0)^T$ is applied, giving the right-of-way to both streams, is not the same as the saturation flow value in interval *r*, s_4^r, when control vector $(0,0,1,1,0,0)^T$ is applied, giving the right-of-way only to stream σ_5 and not to stream σ_4. This can be described by the following inequality:

$$s_4^k(\mathbf{u}^k) < s_4^r(\mathbf{u}^r), \text{ i.e.,}$$

$$s_4^k((0,0,1,1,1,0)^T) < s_4^r((0,0,1,1,0,0)^T).$$

By permitting filtering, it is possible to improve some intersection performance indices.

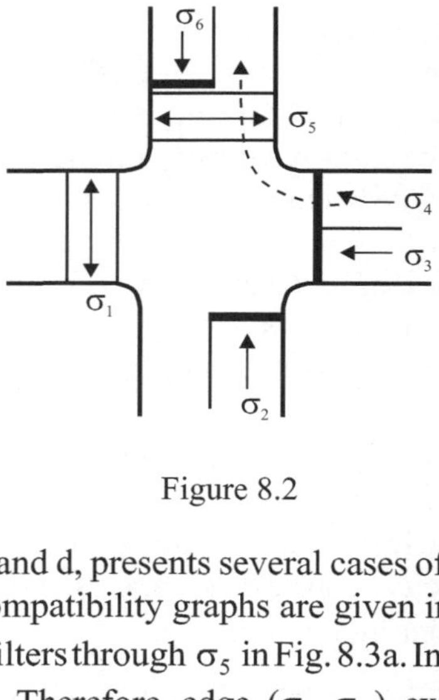

Figure 8.2

Figure 8.3, a, b, c, and d, presents several cases of permitted filtering. The conflictness and the compatibility graphs are given in the same figure.

Traffic stream σ_2 filters through σ_5 in Fig. 8.3a. In this case, $(\sigma_2,\sigma_5) \in C_1$, and also $(\sigma_2,\sigma_5) \in C$. Therefore, edge (σ_2,σ_5) exists in the compatibility graph G_c, as well as in the conflictness graph, G_k.

In Fig. 8.3b the left-turning partial stream of σ_2 filters through vehicle stream σ_5, and in Fig. 8.3c a partial stream of σ_2 filters through vehicle stream σ_4. Fig. 8.3d illustrates filtering of partial stream of σ_1 through pedestrian traffic stream σ_2.

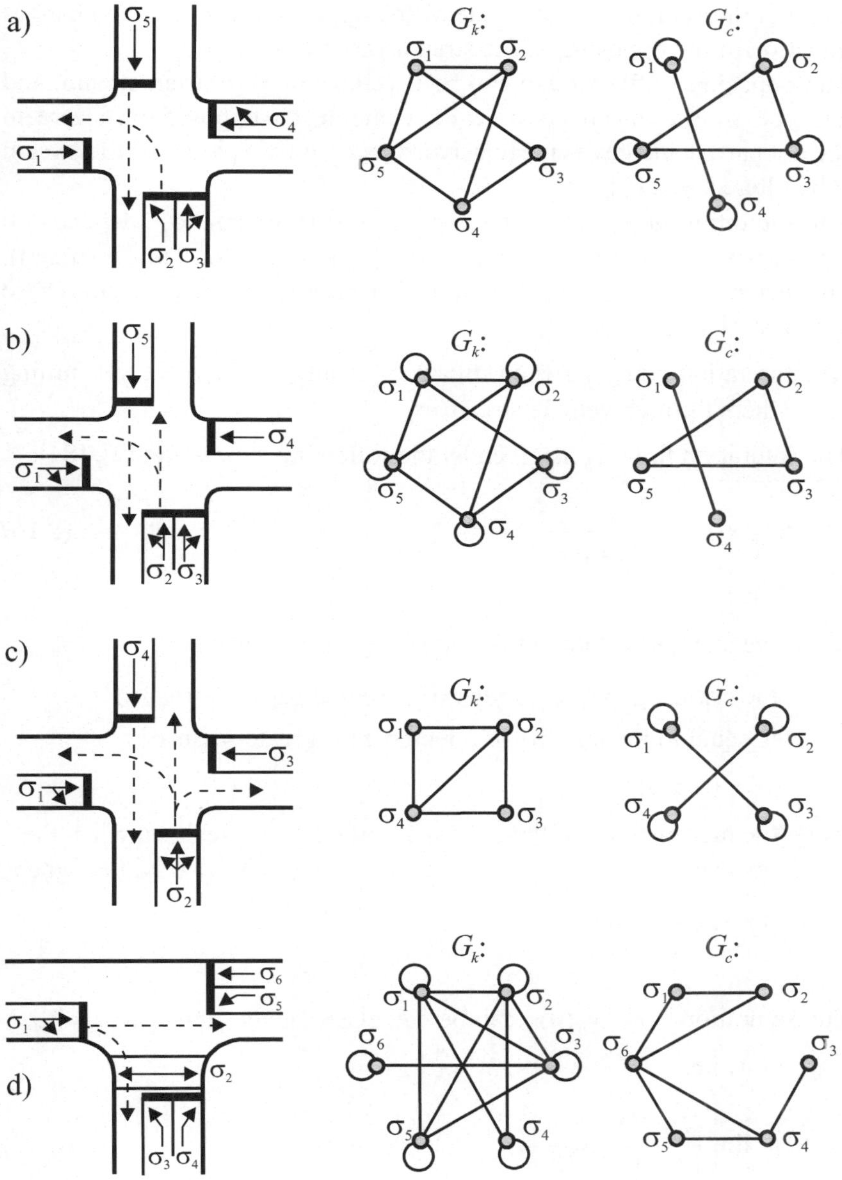

Figure 8.3

During filtering of one traffic stream through another, the saturation flow value of the priority traffic stream (the opposing traffic stream) does not change, and the saturation flow value of the opposed traffic stream changes as a function of the opposing traffic stream volume [45], [42].

The opposing traffic stream can be a vehicle or pedestrian stream, and various types of filtering are possible. For example, the opposed traffic stream may be not partial and have an exclusive lane; or, only a partial traffic stream could be filtered, etc. [6].

The saturation flow, s_b, of the opposed traffic stream, σ_b, depends on the type of the opposing traffic stream, σ_a (a pedestrian or vehicle stream), and on the opposing traffic stream volume in case it is a vehicle stream. These two cases are discussed below.

b.1) The saturation flow, s_b, when stream σ_b, using an exclusive left-turning lane, filters through vehicle stream σ_a

The saturation flow, s_b ,is given by the following expression [2], [84]:

$$s_b = s_b(q_a) = \frac{q_a e^{-\alpha' q_a}}{1 - e^{-\beta_b q_a}}, \tag{8.16}$$

where:

- q_a – the average volume of the opposing traffic stream σ_a,
- α' – the critical gap (the number of accepted gaps less than α' is equal to the number of rejected gaps greater than α'), $\alpha' = (4.5–5)\,\mathrm{s}$
- β_b – the minimal gap of the opposed traffic stream, realized when the opposing traffic stream does not have the right-of-way, i.e., when $s_b = s_b(0)$, and $\beta_b = \dfrac{1}{s_b(0)}$.

The saturation flow, $s_b(0)$, can be calculated as the limit value $s_b(q_a)$ when $q_a \to 0$, i.e.,

$$s_b(0) = \lim_{q_a \to 0} \frac{q_a e^{-\alpha' q_a}}{1 - e^{-\beta_b q_a}} = \frac{1}{\beta_b}. \tag{8.17}$$

The $s_b(q_a)$ function is presented in Fig. 8.4.

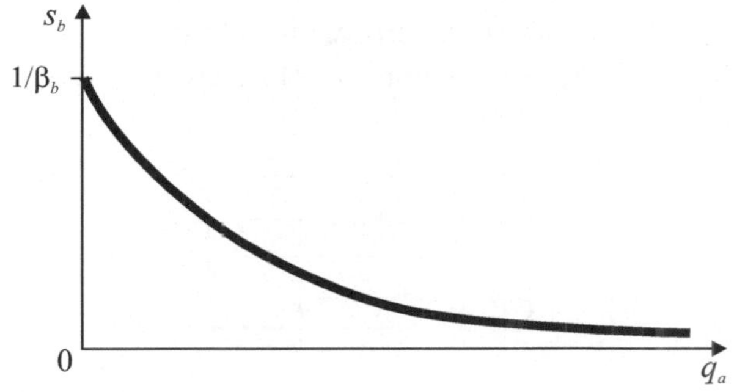

Figure 8.4

In any case, when filtering is permitted, there exists an interval when two conflicting and compatible vehicle traffic streams get the right-of-way. They have a part of the interval with simultaneous right-of-way, but the opposing traffic stream can get the right-of-way before the opposed one, or the opposed stream can get the right-of-way before the opposing stream. If the opposed traffic stream is the first to get the right-of-way, its saturation flow will fall down to zero when the opposing traffic stream gets the simultaneous right-of-way. This happens because no vehicle from the opposed stream can filter through the opposing stream when its volume is equal to the saturation flow volume. The conditions necessary for filtering establish when the volume of the opposing stream becomes less than saturation flow volume. From that point onward, the value of the opposed stream saturation flow becomes different than zero. This means that when the opposed stream has the right-of-way two intervals can be noted in which the opposed flow volume is greater than zero, separated by an interval in which this volume equals zero. Some authors [2], therefore, recommend two right-of-way intervals in a cycle for the opposed traffic stream, i.e., the intervals in which the volume of the stream can be different than zero.

In the case when first the opposing stream gets the right-of-way, and only after its queue discharges the simultaneous right-of-way with an opposed traffic stream begins, then the opposed traffic stream volume will be equal to its saturation flow, which depends on the flow of the opposing stream. If, after the simultaneous right-of-way, the opposing stream is stopped, the saturation flow volume of the opposed traffic stream will be equal to the saturation flow volume of the "protected" stream if a queue exists.

The volumes of the opposing stream, σ_a, and the opposed stream, σ_b, are presented in Fig. 8.5 and Fig. 8.6 for these two cases. In the case presented in Fig. 8.6, the volume of the opposed stream, σ_b, during its green indication

($u_b^k = 1$) is constantly greater than zero, while in the other case (Fig. 8.5) there exists an interval when its volume, $q_b(t)$, is equal to zero.

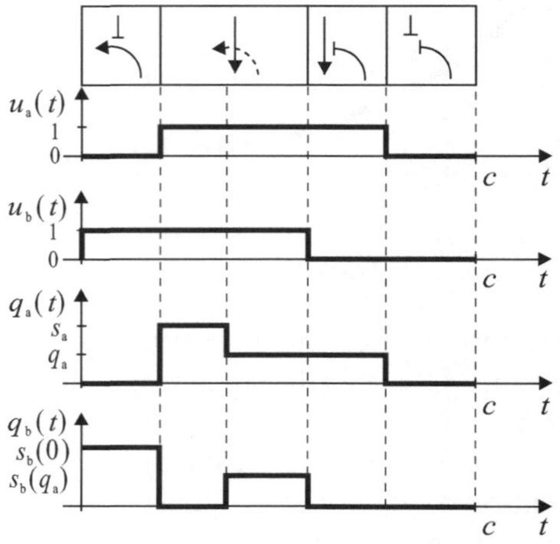

Figure 8.5

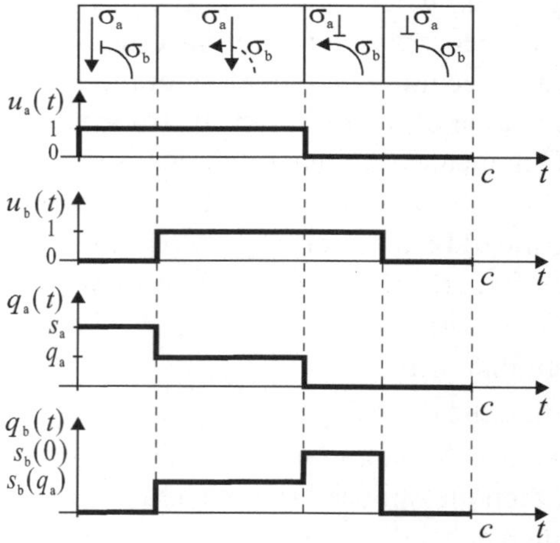

Figure 8.6

The control presented in Fig. 8.6 has more advantages than the control presented in Fig. 8.5. Therefore, here are presented constraints for this case.

The constraint for the opposing stream, σ_a:

I. The flow balance constraint for σ_a:

$$\sum_{k=1}^{K}(s_a u_a^k(1-u_b^k)+q_a u_a^k u_b^k)\,\tau^k \geq cq_a,\quad (a,b\in\mathcal{J}') \tag{8.18}$$

II. The condition of discharging the queue of stream σ_a before giving the simultaneous right-of-way to σ_a and σ_b:

$$\begin{aligned}&\sum_{k=1}^{K}q_a(1-u_a^k)\,\tau^k-(s_a-q_a)\sum_{k=1}^{K}u_a^k(1-u_b^k)\,\tau^k\\&=\sum_{k=1}^{K}((1-u_a^k u_b^k)q_a-u_a^k(1-u_b^k)s_a)\tau^k\geq 0\end{aligned} \tag{8.19}$$

The constraints related to the opposed traffic stream, σ_b:

$$\sum_{k=1}^{K}(u_a^k u_b^k s_b(q_a)+u_b^k(1-u_a^k)s_b(0))\,\tau^k \geq q_b c\,,$$

and when expression (8.16) for $s_b(q_a)$ is included, the constraint gets the following form:

$$\sum_{k=1}^{K}\left(u_a^k u_b^k\frac{q_a e^{-\alpha' q_a}}{1-e^{-\beta_b q_a}}-u_b^k(1-u_a^k)s_b(0)\right)\tau^k\geq q_b c\,. \tag{8.20}$$

It should be noted that opposed and opposing traffic streams can have one interval of the simultaneous right-of-way, but if they don't have simultaneous right-of-way, i.e., when they separately get the right-of-way, then the interval between the end of the right-of-way for one traffic stream and start of another stream's right-of-way should be longer then a predefined intergreen time.

b.2) The saturation flow, s_b, in the case when opposed traffic stream, σ_b, filters through a pedestrian traffic stream, σ_a

When a turning traffic stream, σ_b, filters through a pedestrian (opposing) traffic stream, σ_a, then the dependence of saturation flow, $s_b(q_a)$, on the pedestrian traffic volume, q_a, according to the results of Hoppa and Krystek [42], is given by the expression:

$$s_b(q_a)=s_b(200)(k_1-k_2 q_a)\,, \tag{8.21}$$

where:

$s_b(200)$ – saturation flow of stream σ_b if the volume of the pedestrian traffic stream σ_a is 200 ped/h,

$k_1 = 1.03$.

$k_2 = 0.00015$ h/ped.

The expression (8.21) is valid if the pedestrian volume, q_a, belongs to interval (200–1.500) ped/h. If q_a is less than 200 ped/h, its influence to σ_b is not significant, and in that case $s_b \geq 0.78\, s_b(0)$. If $s_a > 1.500$ ped/h, this influence becomes significant and $s_b = 0.21\, s_b(0)$.

8.3. The set of feasible control vectors

The information contained in the signal group compatibility relation, C_g, and compatibility graph of signal groups, G_g, (Subsection 3.3.3), can be used for determination of the set of feasible control vectors (phases).

Relation C_g contains information on pairs of signal groups that can simultaneously get the right-of-way.

The control vector – phase in interval k is:

$$\mathbf{u}^k = [u_1^k, u_2^k, \ldots, u_p^k, \ldots, u_q^k, \ldots, u_{\mathrm{P}}^k]^T, \quad (k \in \mathcal{K}),$$

where:

$$u_p^k \in \{0,1\}, \quad (k \in \mathcal{K}, p \in \mathcal{P}).$$

Introducing the notation $B = \{0,1\}$, the expression can be written as:

$$\mathbf{u}^k \in B^{\mathrm{P}}, \quad (k \in \mathcal{K}),$$

where:

$$B^{\mathrm{P}} = \underbrace{B \times B \times \ldots \times B}_{\text{P groups}}, \tag{8.22}$$

i.e., $\mathbf{u}^k$ is a vector whose components are equal to 0 or 1.

Two components, u_p^k and u_q^k, of this vector can have value 1 only if their respective signal groups, D'_p and D'_q, are compatible, i.e.,

$$(u_p^k \cdot u_q^k = 1) \Rightarrow (D'_p, D'_q) \in C_g. \tag{8.23}$$

However, one phase can give the right-of-way to more than two signal groups, or only to one or no signal group. It is necessary, of course, that the signal groups getting the simultaneous right-of-way are mutually compatible.

For determining all feasible phases, it is necessary to form all subsets of the chosen complete set of signal groups, $\mathfrak{D}_a$, with the property that each pair of signal groups from one subset belongs to the compatibility relation C_g. The subsets having this property are in fact cliques of graph $G_g = (\mathfrak{D}_a, C_g) = (\mathfrak{D}_a, \Gamma_g)$.

In order to determine the set of feasible phases, it is necessary to find all the cliques of graph G_g, and assign to each clique the phase that gives the right-of-way to signal groups that are members of the clique. In the set of feasible phases one more element shall be added: the phase whose components all have value 0 (all signal indications are red).

Subset $\overline{\mathfrak{D}}_a^r$ of the complete set of signal groups, $\mathfrak{D}_a$, $\overline{\mathfrak{D}}_a^r \subset \mathfrak{D}_a$ is a clique of graph G_g if the following relation holds:

$$(D'_p \in \overline{\mathfrak{D}}_a^r,\ D'_q \in \overline{\mathfrak{D}}_a^r) \ \Rightarrow\ D'_q \in \Gamma_g D'_p,\ \ (p, q \in \mathscr{P}) \,. \qquad (8.24)$$

The procedure of clique determination is described in Subsection 3.3.1, where the determination of signal group set is presented. The set of all cliques of graph $G_g = (\mathfrak{D}_a, C_g) = (\mathfrak{D}_a, \Gamma_g)$, where Γ_g is the mapping:

$$\Gamma_g : \mathfrak{D}_a \to \mathcal{P}(\mathfrak{D}_a),$$

which is determined by function $d'(G_g)$. $\mathcal{P}(\mathfrak{D}_a)$ is partitive set of set $\mathfrak{D}_a$.

Mapping $d'(G_g)$ is defined by the expression (Subsection 3.3.1):

$$\begin{aligned} d'(G_g) = \overline{\mathfrak{D}}_a = \{\overline{\mathfrak{D}}_a^r \mid ((D'_p \in \overline{\mathfrak{D}}_a^r) \wedge (D'_q \in \overline{\mathfrak{D}}_a^r)) \\ \Rightarrow (D'_q \in \Gamma_g D'_p)\ (p, q \in \mathscr{P})\} \\ = \{\overline{\mathfrak{D}}_a^1, \overline{\mathfrak{D}}_a^2, \ldots, \overline{\mathfrak{D}}_a^r, \ldots, \overline{\mathfrak{D}}_a^{\overline{\mathrm{R}}}\} \end{aligned} \qquad (8.25)$$

where $\overline{\mathrm{R}}$ is the number of graph G_g cliques, and $\mathscr{R}$ is the index set of the cliques, i.e., $\mathscr{R} = \{1, 2, \ldots, r, \ldots, \overline{\mathrm{R}}\}$.

Mapping d' determines the unique set of all cliques $\overline{\mathfrak{D}}_a$ of nonoriented graph G_g. CLIQ program [34] is developed for realization of this function. The pseudocode of CLIQ program is given in Appendix III.

To each clique $\overline{\mathfrak{D}}_a^r \in \overline{\mathfrak{D}}_a$, $(r \in \mathscr{R})$, there corresponds one feasible phase $\mathbf{u}(r)$, whose components, $u_p(r)$, are determined by the following expression:

$$u_p(r) = \begin{cases} 1, \text{ if } D'_p \in \overline{\mathfrak{D}}_a^r \\ 0, \text{ if } D'_p \notin \overline{\mathfrak{D}}_a^r \end{cases}, \quad (p \in \mathscr{P}, r \in \mathscr{R}). \tag{8.26}$$

It means that the number of feasible phases is $\overline{\mathrm{R}}+1$, i.e., $\overline{\mathrm{R}}$ phases that correspond to graph G_g cliques, and one phase whose components are all equal to 0 ("all-red" phase).

All feasible phases, $\mathbf{u}(1),\mathbf{u}(2),\ldots,\mathbf{u}(r),\ldots,\mathbf{u}(\overline{\mathrm{R}})$, are elements of set $\mathbf{U}_f$, i.e.,

$$\mathbf{U}_f = \{\mathbf{u}(1),\mathbf{u}(2),\ldots,\mathbf{u}(r),\ldots,\mathbf{u}(\overline{\mathrm{R}})\} \cup \{(0,0,\ldots,0)\}.$$

Each phase $\mathbf{u}^k = [u_1^k, u_2^k, \ldots, u_p^k, \ldots, u_{\mathrm{P}}^k]^T$ in interval k belongs to the set of feasible phases:

$$\mathbf{u}^k \in \mathbf{U}_f, \quad (k \in \mathscr{K}). \tag{8.27}$$

If the sequence

$$\mathbf{u} = [\mathbf{u}^1, \mathbf{u}^2, \ldots, \mathbf{u}^k, \ldots, \mathbf{u}^{\mathrm{K}}]$$

represents the structure of a feasible signal plan, then elements of the sequence (i.e., phases) in each interval $k \in \mathscr{K}$, have to be chosen from the set of feasible phases $\mathbf{U}_f$.

Example 8.2

For the intersection presented in Fig. 3.4, determine the set of feasible control vectors (phases) for two complete sets of signal groups:

a) $\mathfrak{D}_a^1 = \{D_1^1, D_2^1, D_3^1, D_4^1, D_5^1, D_6^1\}$, where

$D_1^1 = \{\sigma_1\}$, $D_2^1 = \{\sigma_2\}$, $D_3^1 = \{\sigma_3\}$, $D_4^1 = \{\sigma_4\}$, $D_5^1 = \{\sigma_5\}$, $D_6^1 = \{\sigma_6\}$,

and

b) $\mathfrak{D}_a^3 = \{D_1^3, D_2^3, D_3^3, D_4^3, D_5^3\}$, where

$D_1^3 = \{\sigma_2\}$, $D_2^3 = \{\sigma_4\}$, $D_3^3 = \{\sigma_5\}$, $D_4^3 = \{\sigma_6\}$, $D_5^3 = \{\sigma_1, \sigma_3\}$.

a) Fig. 8.7 presents the intersection, the signal group compatibility graph, G_g^1, and subgraphs of G_g^1 that are complete graphs. Node sets of these subgraphs are cliques of the compatibility graph. The cliques are elements of set $\mathcal{D}_a^1$, and were obtained using CLIQ program.

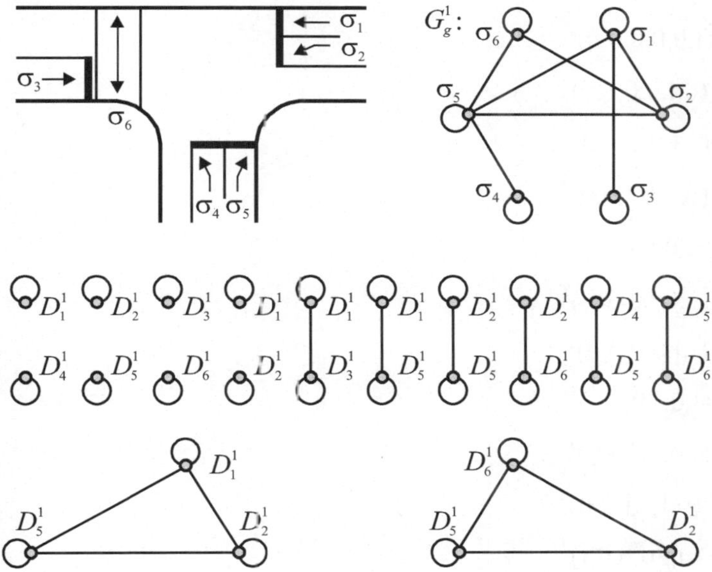

Figure 8.7

The complete set of signal groups, $\mathcal{D}_a^1$, in this case is:

$$\mathcal{D}_a^1 = \{D_1^1, D_2^1, D_3^1, D_4^1, D_5^1, D_6^1\}\,.$$

The set of cliques is:

$$\overline{\mathcal{D}}_{a1} = \{\overline{\mathcal{D}}_a^1, \overline{\mathcal{D}}_a^2, \ldots, \overline{\mathcal{D}}_a^{15}\},$$

where:

$$\overline{\mathcal{D}}_a^1 = \{D_1^1\},\ \overline{\mathcal{D}}_a^2 = \{D_2^1\},\ \overline{\mathcal{D}}_a^3 = \{D_3^1\},\ \overline{\mathcal{D}}_a^4 = \{D_4^1\},\ \overline{\mathcal{D}}_a^5 = \{D_5^1\},$$

$$\overline{\mathcal{D}}_a^6 = \{D_6^1\},\ \overline{\mathcal{D}}_a^7 = \{D_1^1, D_2^1\},\ \overline{\mathcal{D}}_a^8 = \{D_1^1, D_3^1\},\ \overline{\mathcal{D}}_a^9 = \{D_1^1, D_5^1\},$$

$$\overline{\mathcal{D}}_a^{10} = \{D_2^1, D_5^1\},\ \overline{\mathcal{D}}_a^{11} = \{D_2^1, D_6^1\},\ \overline{\mathcal{D}}_a^{12} = \{D_4^1, D_5^1\},\ \overline{\mathcal{D}}_a^{13} = \{D_5^1, D_6^1\},$$

$$\overline{\mathcal{D}}_a^{14} = \{D_1^1, D_2^1, D_5^1\},\ \overline{\mathcal{D}}_a^{15} = \{D_2^1, D_5^1, D_6^1\}\,.$$

The feasible phases, which correspond to cliques, are vectors with six components because the complete set of signal groups $\mathcal{D}_a^1$ has six elements, i.e.,

$$\mathbf{u}(r) = [u_1(r), u_2(r), \ldots, u_p(r), \ldots, u_\mathrm{P}(r)]^T \ , \quad (\mathrm{P} = 6, r \in \mathscr{R}) \ .$$

The phases that correspond to the cliques are:

$$\mathbf{u}(1) = (1,0,0,0,0,0)^T$$

$$\mathbf{u}(2) = (0,1,0,0,0,0)^T$$

$$\mathbf{u}(3) = (0,0,1,0,0,0)^T$$

$$\mathbf{u}(4) = (0,0,0,1,0,0)^T$$

$$\mathbf{u}(5) = (0,0,0,0,1,0)^T$$

$$\mathbf{u}(6) = (0,0,0,0,0,1)^T$$

$$\mathbf{u}(7) = (1,1,0,0,0,0)^T$$

$$\mathbf{u}(8) = (1,0,1,0,0,0)^T$$

$$\mathbf{u}(9) = (1,0,0,0,1,0)^T$$

$$\mathbf{u}(10) = (0,1,0,0,1,0)^T$$

$$\mathbf{u}(11) = (0,1,0,0,0,1)^T$$

$$\mathbf{u}(12) = (0,0,0,1,1,0)^T$$

$$\mathbf{u}(13) = (0,0,0,0,1,1)^T$$

$$\mathbf{u}(14) = (1,1,0,0,1,0)^T$$

$$\mathbf{u}(15) = (0,1,0,0,1,1)^T$$

The number of all cliques is $\overline{\mathrm{R}} = 15$, and thus the number of all feasible phases is $card\ \mathbf{U}_f = 16$.

The set of feasible phases is:

$$\begin{aligned}\mathbf{U}_f = \{&(0,0,0,0,0,0)^T, (1,0,0,0,0,0)^T, (0,1,0,0,0,0)^T, (0,0,1,0,0,0)^T, \\ &(0,0,0,1,0,0)^T, (0,0,0,0,1,0)^T, (0,0,0,0,0,1)^T, (1,1,0,0,0,0)^T, \\ &(1,0,1,0,0,0)^T, (1,0,0,0,1,0)^T, (0,1,0,0,1,0)^T, (0,1,0,0,0,1)^T, \\ &(0,0,0,1,1,0)^T, (0,0,0,0,1,1)^T, (1,1,0,0,1,0)^T, (0,1,0,0,1,1)^T\}.\end{aligned}$$

b) The complete set of signal groups in this case is

$$\mathfrak{D}_a^3 = \{D_1^3, D_2^3, D_3^3, D_4^3, D_5^3\}.$$

Fig. 8.8 presents the signal group compatibility graph, $G_g^3 = (\mathfrak{D}_a^3, \Gamma_c^3)$, and subgraphs of G_g^3 that are complete graphs. Node sets of these subgraphs are cliques of the compatibility graph. The cliques were obtained using CLIQ program.

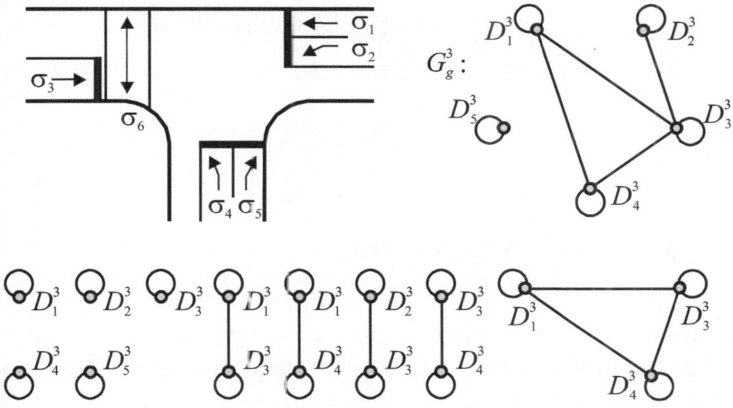

Figure 8.8

The set of all cliques, $\overline{\mathfrak{D}}_{a3}$, is defined by the following expression:

$$\overline{\mathfrak{D}}_{a3} = \{\{D_1^3\}, \{D_2^3\}, \{D_3^3\}, \{D_4^3\}, \{D_5^3\}, \{D_1^3, D_3^3\}, \{D_1^3, D_4^3\}, \{D_2^3, D_3^3\}, \{D_3^3, D_4^3\}, \{D_1^3, D_3^3, D_4^3\}\}.$$

The phases that correspond to these cliques are:

$$\mathbf{u}(1) = (1,0,0,0,0)^T$$

$$\mathbf{u}(2) = (0,1,0,0,0)^T$$

$$\mathbf{u}(3) = (0,0,1,0,0)^T$$

$$\mathbf{u}(4) = (0,0,0,1,0)^T$$

$$\mathbf{u}(5) = (0,0,0,0,1)^T$$

$$\mathbf{u}(6) = (1,0,1,0,0)^T$$

$$\mathbf{u}(7) = (1,0,0,1,0)^T$$

$$\mathbf{u}(8) = (0,1,1,0,0)^T$$
$$\mathbf{u}(9) = (0,0,1,1,0)^T$$
$$\mathbf{u}(10) = (1,0,1,1,0)^T .$$

The set of feasible phases is:

$$\mathbf{U}_f = \{(0,0,0,0,0)^T,(1,0,0,0,0)^T,(0,1,0,0,0)^T,(0,0,1,0,0)^T,$$
$$(0,0,0,1,0)^T,(0,0,0,0,1)^T,(1,0,1,0,0)^T,(1,0,0,1,0)^T,$$
$$(0,1,1,0,0)^T,(0,0,1,1,0)^T,(1,0,1,1,0)^T\}.$$

The number of feasible phases is $card\ \mathbf{U}_f = \overline{\mathrm{R}} + 1 = 10 + 1 = 11$.

8.4. The intergreen time constraints

In the process of signal plan design, it is necessary to respect the constraints that have to be satisfied by feasible sequences of control vectors. These constraints are the minimal intergreen times constraints, and constraints on phase sequences, i.e., the constraints related to signal plan structure.

The set of feasible phases is determined on the basis of signal groups compatibility relation, C_g, which contains information on pairs of signal groups that can simultaneously get the right-of-way. The pairs of incompatible traffic streams in the complete set of signal groups, $\mathcal{D}_a$, i.e., the elements of relation:

$$\overline{C}_g = (\mathcal{D}_a \times \mathcal{D}_a) \setminus C_g \tag{8.28}$$

comprise signal groups that must not simultaneously get the right-of-way. Moreover, if

$$(D'_p, D'_q) \in \overline{C}_g ,$$

these two signal groups not only cannot have a simultaneous right-of-way, but some time has to elapse from the end of the right-of-way for signal group D'_p until the start of the right-of-way for signal group D'_q. This time shall be greater than so-called minimal intergreen time in order to avoid conflicts of traffic participants whose movement is controlled by control variables $u_p(\cdot)$ and $u_q(\cdot)$.

The stated requirement can be expressed in the following way: In the phase sequence, representing a feasible signal plan structure between phase $\mathbf{u}^k$ with $u_p^k = 1$ and $u_p^{k+1} = 0$, and phase $\mathbf{u}^{k+r}$ with $u_q^{k+r} = 1$ and $u_q^{k+r-1} = 0$ (where $r-1$ is the number of phases between the last phase giving the right-of-way to signal group D'_p and the first phase giving the right-of-way to signal group D'_q) there have to exist phases such that their entire duration is greater than the minimal intergreen time defined for the pair (D'_p, D'_q).

For determination of these phase duration constraints, it is necessary to determine minimal intergreen times for all pairs of incompatible signal groups, based on minimal intergreen times for all pairs of incompatible traffic streams, which have to be calculated in advance.

With regard to the fact that the real sequence of signal indications (in the majority of cases: green – amber – red – red–amber) is transformed to effective green and effective red indications, it is necessary to determine minimal intergreen time between the end of effective green time of the signal group losing the right-of-way and the beginning of effective green time of the signal group gaining the right-of-way, i.e., the *minimal effective intergreen time* (m.e.i.t.) (Appendix V).

During the minimal effective intergreen time the volume of both traffic streams, the traffic stream losing the right-of-way, and the stream gaining the right-of-way is equal to zero if m.e.i.t. has a positive value.

8.4.1. Minimal intergreen times for pairs of traffic streams

a) Minimal intergreen times for pairs of incompatible traffic streams

To each pair of incompatible traffic streams, (σ_i, σ_j), which is a member of relation

$$\overline{C} = (\mathcal{S} \times \mathcal{S}) \setminus C,$$

one number, $\overline{z}_{ij}$ – the minimal intergreen time, is assigned by mapping:

$$\overline{Z} : \overline{C} \to R, \tag{8.29}$$

where R is the set of real numbers.

For calculation of minimal effective intergreen times, it is necessary first to determine minimal intergreen times for real sequences of signal indications.

In the case when σ_i and σ_j are vehicle streams, the minimal intergreen time is calculated by the following expression (Fig. 8.9):

$$\overline{z}_{ij} = t_{ij}^r - t_{ij}^e + t^p + t^g, \quad ((\sigma_i, \sigma_j) \in \overline{C}), \tag{8.30}$$

where:

- t_{ij}^{r} – the time necessary for the last vehicle of σ_i (the stream losing the right-of-way) to pass the distance between the stop line and the end of the conflict area with σ_j,
- t_{ij}^{e} – the time necessary for the first vehicle of σ_j (the stream gaining the right-of-way) to arrive at the beginning of the conflict area with σ_i,
- t^{p} – vehicle travel time through the conflict area,
- t^{g} – the part of the amber time used by vehicles of stream σ_i,
- $\bar{z}_{ij}$ – the minimal intergreen time.

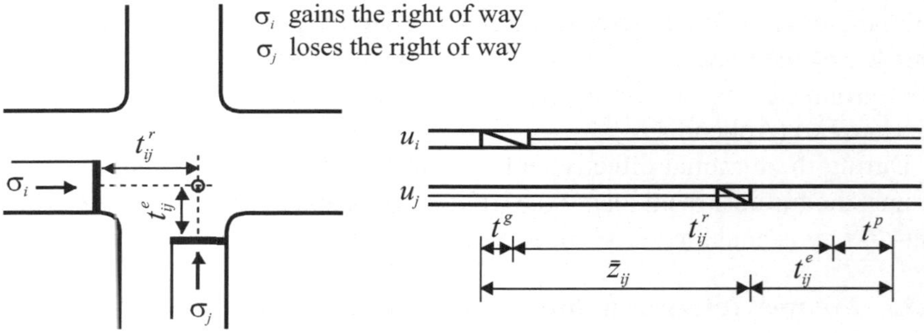

Figure 8.9

The formula for minimal green time calculation (8.30) is based on the assumption that vehicles leaving the conflict area travel with the lowest speed, while the vehicles approaching the area travel with the highest speed, under existing conditions. It is usually assumed that the speed of the vehicles leaving the conflict area is in the range of (25–30) km/h and the speed of vehicles arriving to the conflict area is in the range of (40–50) km/h [69]. The pedestrian speed is usually 1.2 m/s to 1.5 m/s.

The minimal effective green time, in this case, can be obtained by the expression (Fig. AV.3 in Appendix V):

$$z'_{ij} = \bar{z}_{ij} - a + l\,, \tag{8.31}$$

where:

a – the duration of amber indication,

$l = l' + l''$ – the lost time.

Expression (8.31) holds when a pair of incompatible vehicle streams is considered. Expressions for determination of minimal effective intergreen times for other cases are given in Appendix V.

As already mentioned, minimal effective intergreen times for pairs of incompatible traffic streams are determined by the mapping:

$$Z' : \overline{Z} \to R. \tag{8.32}$$

Determination of Z' elements, $z'_{ij} = Z'(\overline{z}_{ij})$, depends on the type of traffic streams in (σ_i, σ_j) pair (both vehicle streams, or one vehicle and the other pedestrian, etc.).

The values of minimal effective intergreen times between incompatible signal groups belong to the set of real numbers and usually are expressed as integer number of seconds. These values can be positive, negative, or zero. For example, by observing expression (8.31) it is obvious that if the intersection geometry is such that

$$t_{ij}^e \geq t_{ij}^r + t^p + t^g - a + l,$$

then $z'_{ij} \leq 0$.

In some countries there exist recommendations that minimal intergreen times should not be shorter than a prescribed value, $\overline{z}_{\min}$ (in some countries 4 s). Minimal intergreen times are then determined by the following expression:

$$z'_{ij} = Z'(\max\{\overline{z}_{\min}, \overline{z}_{ij}\}), \quad (i, j \in \mathcal{J}). \tag{8.33}$$

In this case, minimal intergreen times can have positive values only.

b) Intergreen times for pairs of compatible traffic streams

For effective intergreen times related to pairs of compatible traffic streams there are no constraints on their minimal values. In this case, the intergreen times have to satisfy the following constraint:

$$z'_{ij} \in [0, c], \quad ((\sigma_i, \sigma_j) \in C), \tag{8.34}$$

i.e., the minimal effective intergreen time for a pair of compatible traffic streams can assume any value in $[0, c]$ interval.

c) The matrix of minimal effective intergreen times

The minimal effective intergreen times for all pairs of incompatible traffic streams are defined by the function:

$$Z' : \mathscr{S} \times \mathscr{S} \to R, \tag{8.35}$$

i.e., $Z' = \{z'_{ij} \mid (\sigma_i, \sigma_j) \in \mathscr{S} \times \mathscr{S}\}$.

It means that the values of the function Z' are arranged in a matrix:

$$Z' = [z'_{ij}]_{I \times I},$$

whose elements are calculated as follows:

$$z'_{ij} = \begin{cases} Z'(\bar{z}_{ij}), (\sigma_i, \sigma_j) \in \overline{C} \\ \quad 0 \quad , (\sigma_i, \sigma_j) \in C \end{cases}, \quad (i, j \in \mathscr{I}). \tag{8.36}$$

Fig. 8.10 presents an intersection and the related matrix of minimal intergreen times, Z', and Fig. 8.11 presents another intersection with matrix Z' containing elements with negative values (z'_{14} and z'_{32}).

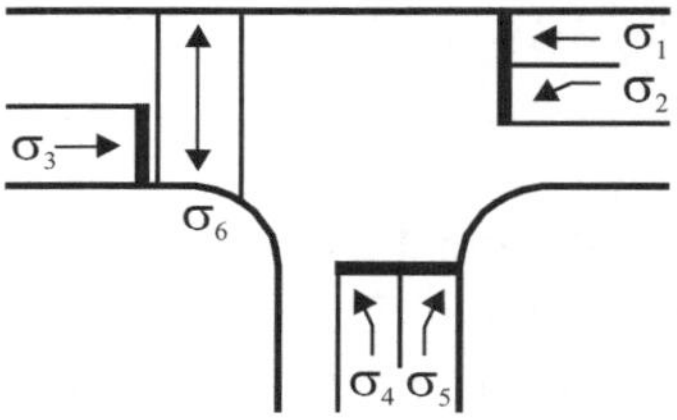

$$Z' = \begin{bmatrix} 0 & 0 & 0 & 4 & 0 & 4 \\ 0 & 0 & 3 & 5 & 0 & 0 \\ 0 & 3 & 0 & 3 & 6 & 2 \\ 2 & 1 & 2 & 0 & 0 & 2 \\ 0 & 0 & 1 & 0 & 0 & 0 \\ 8 & 0 & 4 & 8 & 0 & 0 \end{bmatrix}$$

Figure 8.10

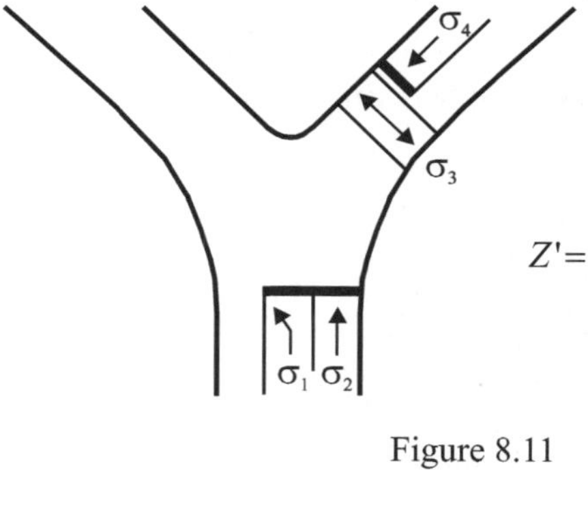

$$Z' = \begin{bmatrix} 0 & 0 & 0 & -2 \\ 0 & 0 & 9 & 0 \\ 0 & -2 & 0 & 6 \\ 10 & 0 & 1 & 0 \end{bmatrix}$$

Figure 8.11

8.4.2. Minimal effective intergreen times for signal group pairs

The matrix of effective intergreen times for all traffic stream pairs is determined using expression (8.36) and the formulas given in Appendix V. Since control variables are assigned to signal groups, the matrix of effective intergreen times related to signal group pairs has to be determined. These intergreen times are defined by the function:

$$Z : \mathfrak{D}_a \times \mathfrak{D}_a \to R , \tag{8.37}$$

where $\mathfrak{D}_a$ is the chosen complete set of signal groups, i.e.,

$$\mathfrak{D}_a = \{ D'_1, D'_2, \ldots, D'_p, \ldots, D'_P \} .$$

Function Z is defined as follows:

$$Z = \{ z_{pq} \mid (D'_p, D'_q) \in \mathfrak{D}_a \times \mathfrak{D}_a \} = [z_{pq}]_{P \times P} , \tag{8.38}$$

where

$$z_{pq} = \begin{cases} \max\{ z'_{ij} \mid \sigma_i \in D'_p, \sigma_j \in D'_q, (D'_p, D'_q) \in (\mathfrak{D}_a \times \mathfrak{D}_a) \setminus C_g \} \\ 0, \quad (D'_p, D'_q) \in C_g . \end{cases} \tag{8.39}$$

Example 8.3

Determine the matrix of minimal effective intergreen times for signal group pairs for the intersection presented in Fig. 8.10. The chosen complete set of signal groups is:

$$\mathfrak{D}_a = \{ D'_1, D'_2, D'_3, D'_4 \} ,$$

where

$$D'_1 = \{\sigma_4\} , \ D'_2 = \{\sigma_6\} , \ D'_3 = \{\sigma_1, \sigma_3\} , \ D'_4 = \{\sigma_2, \sigma_5\} .$$

The compatibility relation is given by the following expression:

$$C_g = \{ (D'_1, D'_1), (D'_2, D'_2), (D'_3, D'_3), (D'_4, D'_4), (D'_2, D'_4), \ldots, (D'_4, D'_2) \} .$$

The compatibility graph, $G_g = (\mathfrak{D}_a, C_g)$, is shown in Fig. 3.14.

Minimal effective intergreen times, z'_{ij}, for each pair of traffic streams are elements of Z' matrix presented in Fig. 8.10. Applying expression (8.39), the following values of minimal effective intergreen times for pairs of signal groups are obtained:

$$z_{12} = \max\{ z'_{ij} \mid \sigma_i \in \{\sigma_4\}, \sigma_j \in \{\sigma_6\} \} = \max\{ z'_{46} \} = z'_{46} = 2$$

$$z_{13} = \max\{ z'_{ij} \mid \sigma_i \in \{\sigma_4\}, \sigma_j \in \{\sigma_1, \sigma_3\} \} = \max\{2,2\} = 2$$

$$z_{14} = \max\{ z'_{ij} \mid \sigma_i \in \{\sigma_4\}, \sigma_j \in \{\sigma_2, \sigma_5\} \} = \max\{1,0\} = 1$$

$$z_{21} = \max\{z'_{ij} \mid \sigma_i \in \{\sigma_6\}, \sigma_j \in \{\sigma_4\}\} = \max\{8\} = 8$$

$$z_{23} = \max\{z'_{ij} \mid \sigma_i \in \{\sigma_6\}, \sigma_j \in \{\sigma_1, \sigma_3\}\} = \max\{8,4\} = 8$$

$$z_{24} = 0, \quad (D'_2, D'_4) \in C_g$$

$$z_{31} = \max\{z'_{ij} \mid \sigma_i \in \{\sigma_1, \sigma_3\}, \sigma_j \in \{\sigma_4\}\} = \max\{4,4\} = 4$$

$$z_{32} = \max\{z'_{ij} \mid \sigma_i \in \{\sigma_1, \sigma_3\}, \sigma_j \in \{\sigma_6\}\} = \max\{4,2\} = 4$$

$$z_{34} = \max\{z'_{ij} \mid \sigma_i \in \{\sigma_1, \sigma_3\}, \sigma_j \in \{\sigma_2, \sigma_5\}\} = \max\{0,0,3,6\} = 6$$

$$z_{41} = \max\{z'_{ij} \mid \sigma_i \in \{\sigma_2, \sigma_5\}, \sigma_j \in \{\sigma_4\}\} = \max\{5,0\} = 5$$

$$z_{42} = 0, \quad (D'_2, D'_4) \in C_g$$

$$z_{43} = \max\{z'_{ij} \mid \sigma_i \in \{\sigma_2, \sigma_5\}, \sigma_j \in \{\sigma_1, \sigma_3\}\} = \max\{0,3,0,1\} = 3$$

Hence, the matrix of minimal effective intergreen times for pairs of signal groups is:

$$Z = [z_{pq}]_{4\times 4} = \begin{bmatrix} - & 2 & 2 & 1 \\ 8 & - & 8 & 0 \\ 3 & 4 & - & 6 \\ 5 & 0 & 3 & - \end{bmatrix}.$$

8.4.3. The extension of the set of feasible phases

The fact that minimal effective intergreen times can assume negative values leads to the necessity of extending the set of feasible phases.

The definition of feasible phases and the procedure for determining the set of feasible phases are presented in Section 8.3. The relation of signal group compatibility is used in determining the set of feasible phases because this relation contains all signal group pairs with the property that the right-of-way can simultaneously be given to both members of the pair. This means that control variables $u_p(\cdot)$ and $u_q(\cdot)$ can simultaneously have the value 1 if $(D'_p, D'_q) \in C_g$.

However, if $z_{pq} \le 0$, an interval exists in which variables $u_p(\cdot)$ and $u_q(\cdot)$ can simultaneously have value 1 although $(D'_p, D'_q) \notin C_g$ (Fig. AV.6). The duration of this interval shall be less than or equal to the absolute value of the minimal effective intergreen time. The constraint that prevents this interval from being longer than $|z_{pq}|$ can be formulated as follows:

$$\sum_{k=1}^{\mathrm{K}} (u_p^k \cdot u_q^k)\tau^k \le |z_{pq}|, \quad (p, q \in \mathscr{P}). \tag{8.40}$$

Besides this time constraint, there exists a structural constraint that results from the fact that interval k with $u_p^k \cdot u_q^k = 1$ can exist only if in the previous interval $u_p^{k-1} = 1$ and $u_q^{k-1} = 0$, and in the subsequent interval $u_p^{k+1} = 0$ and $u_q^{k+1} = 1$. If the sequence is reversed, the interval k such that $u_p^k \cdot u_q^k = 1$ cannot exist.

The set of pairs (D'_p, D'_q) with the property $z_{pq} < 0$ represents the relation

$$C'_g = \{(D'_p, D'_q) \mid z_{pq} < 0,\ D'_p, D'_q \in \mathfrak{D}_a\}. \tag{8.41}$$

In this case, the set of feasible phases, $\mathbf{U}_f$, has to be extended by set $\mathbf{U}'_f$, which contains the phases with $u_p(s) = u_q(s) = 1$ and $z_{pq} < 0$ (s here denotes the index of a phase in the set of feasible phases, $\mathbf{U}'_f$).

Set $\mathbf{U}'_f$ can be defined as follows:

$$\mathbf{U}'_f = \{\mathbf{u}(s) \mid u_p(s) = u_q(s) = 1;\ z_{pq} < 0, s \in \{\overline{\mathrm{R}}+2, \ldots, \overline{\mathrm{R}}+\mathrm{N}\}\}, \tag{8.42}$$

because

$$card\,\mathbf{U}_f = \overline{\mathrm{R}} + 1,$$

where

$$\mathbf{u}(s) = [u_1(s), u_2(s), \ldots, u_p(s), \ldots, u_q(s), \ldots, u_{\mathrm{P}}(s)]^T,$$

and $\mathrm{N}-1$ is the number of phases with $u_p(s) = u_q(s) = 1$.

The extended set of feasible phases, $\mathbf{U}''_f$, is the union of sets $\mathbf{U}_f$ and $\mathbf{U}'_f$, i.e.,

$$\mathbf{U}''_f = \mathbf{U}_f \cup \mathbf{U}'_f. \tag{8.43}$$

Set $\mathfrak{D}_a$ and relation C'_g define the graph

$$G'_g = (\mathfrak{D}_a, C'_g). \tag{8.44}$$

Graph G''_g is obtained from graphs G_g and G'_g:

$$G''_g = (\mathfrak{D}_a, C_g \cup C'_g) = (\mathfrak{D}_a, C''_g) = (\mathfrak{D}_a, \Gamma''_g). \tag{8.45}$$

Relations C_g, C'_g, and C''_g are reflexive and symmetric, so that graphs G_g, G'_g, and G''_g are nonoriented graphs with a loop in each node.

All feasible phases are defined by cliques of graph G''_g, in the same way as when no negative minimal effective intergreen times exist (Section 8.3).

Example 8.4

Determine relations C_g, C'_g, and C''_g, graphs G_g, G'_g and G''_g, and the set of feasible phases, $\mathbf{U}''_f$, for the intersection presented in Fig. 8.11.

The set of signal groups is

$$\mathcal{D}_a = \{D'_1, D'_2, D'_3, D'_4\},$$

where

$$D'_1 = \{\sigma_1\}, D'_2 = \{\sigma_2\}, D'_3 = \{\sigma_3\}, D'_4 = \{\sigma_4\}.$$

The matrix of minimal effective intergreen times is

$$Z = \begin{bmatrix} - & 0 & 0 & -2 \\ 0 & - & 9 & 0 \\ 0 & -2 & - & 6 \\ 10 & 0 & 1 & - \end{bmatrix}.$$

The matrix, A_c, of the compatibility relation, C_g, is:

$$A_c = \begin{bmatrix} - & 1 & 1 & 0 \\ 1 & - & 0 & 1 \\ 1 & 0 & - & 0 \\ 0 & 1 & 0 & - \end{bmatrix}.$$

Graph $G_g = (\mathcal{D}_a, C_g)$ is shown in Fig. 8.12.

Elements a_{c14} and a_{c32} in matrix A_c have 0 value, representing that pairs (D'_1, D'_4) and (D'_3, D'_2) are not pairs of compatible signal groups. The minimal effective intergreen times corresponding to these pairs are negative, so that the control variables $u_1(\cdot)$ and $u_4(\cdot)$, as well as $u_3(\cdot)$ and $u_2(\cdot)$, can simultaneously provide green indications of 2 seconds duration $(z_{14} = z_{32} = -2)$.

Relation C'_g is defined by the following set of ordered signal group pairs:

$$C'_g = \{(D'_1, D'_4), (D'_3, D'_2)\}.$$

Relation C''_g is presented in Fig. 8.12 by A''_c matrix and graphs G_g, G'_g, and G''_g. In order to determine all elements of set $\mathbf{U}''_f$, i.e., all feasible phases, the cliques of graph G''_g have to be determined. The cliques of this graph are the sets of nodes of all complete subgraphs of graph G''_g, as presented in Fig. 8.13.

The set of all cliques, $\overline{\mathcal{D}}_a$, is:

$$\overline{\mathcal{D}}_a = \{\{D'_1\}, \{D'_2\}, \{D'_3\}, \{D'_4\}, \{D'_1, D'_2\}, \{D'_1, D'_3\}, \{D'_1, D'_4\}, \{D'_2, D'_3\}, \{D'_2, D'_4\}, \{D'_1, D'_2, D'_3\}, \{D'_1, D'_2, D'_4\}\}.$$

Set $\overline{\mathfrak{D}}_a$ is obtained using CLIQ program.

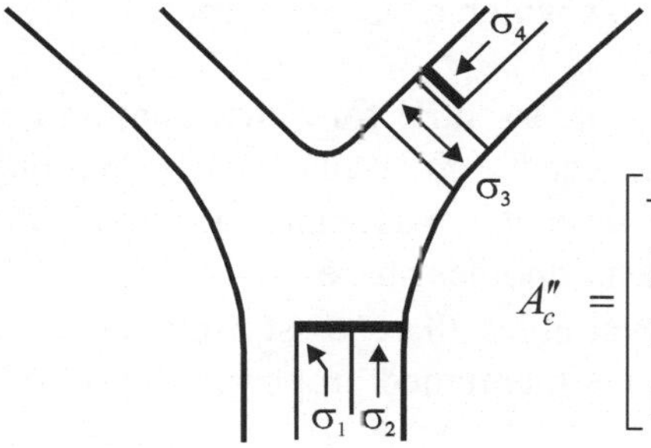

$$A''_c = \begin{bmatrix} - & 1 & 1 & 1 \\ 1 & - & 1 & 1 \\ 1 & 1 & - & 0 \\ 1 & 1 & 0 & - \end{bmatrix}$$

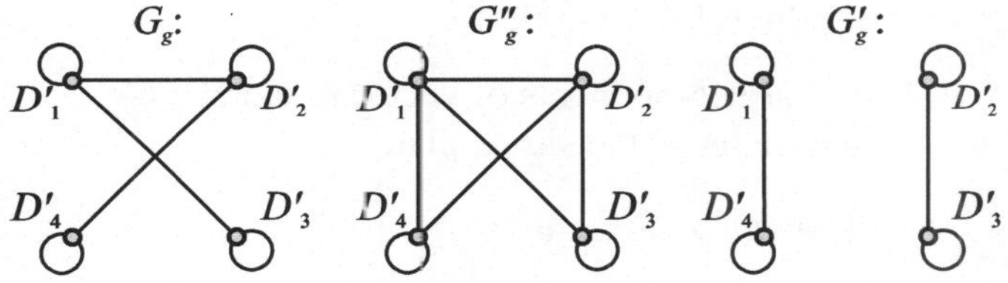

Figure 8.12

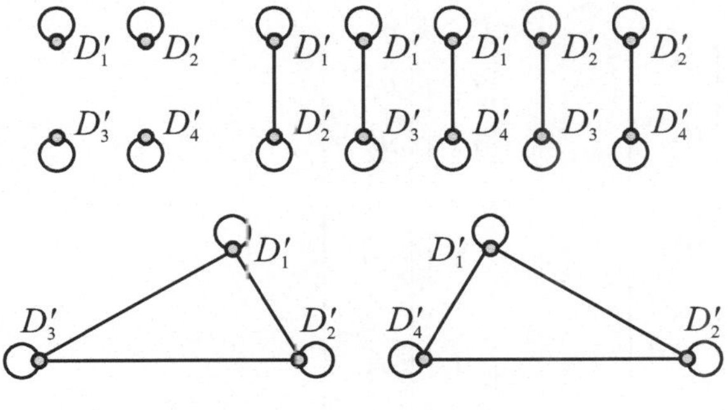

Figure 8.13

Thus, the complete set of signal groups is

$$\mathfrak{D}_a = \{\{\sigma_1\}, \{\sigma_2\}, \{\sigma_3\}, \{\sigma_4\}\} = \{D'_1, D'_2, D'_3, D'_4\},$$

and the set of feasible phases is:

$$\mathbf{U}''_f = \{(1,0,0,0)^T, (0,1,0,0)^T, (0,0,1,0)^T, (0,0,0,1)^T, (1,1,0,0)^T, (1,0,1,0)^T,$$
$$(1,0,0,1)^T, (0,1,1,0)^T, (0,0,1,1)^T, (1,1,1,0)^T, (1,0,1,1)^T, (0,0,0,0)^T\}.$$

8.5. The relation of green indications succession

When determining the sequence of phases that make a feasible signal plan structure, it is necessary to know which phases can immediately succeed a particular phase.

The information about m.e.i.t. is essential for determination of the set of phases that can immediately succeed a given phase. The data on m.e.i.t. define whether the effective green time of a phase can start immediately after the end of the effective green time of another phase.

Thus, in the set of control variables there exists relation R_n, which contains the pairs of phases with the mentioned property. The set of control variables is:

$$\mathcal{U}_n = \{u_1(\cdot), u_2(\cdot), \ldots, u_p(\cdot), \ldots, u_{\mathrm{P}}(\cdot)\}\,. \tag{8.46}$$

The elements of this set are components of vector function $u(\cdot)$.

A signal plan structure, **u**, of the signal plan

$$u(\cdot) = \begin{bmatrix} \mathbf{u} \\ \tau \end{bmatrix} = \begin{bmatrix} (\mathbf{u}^1, \mathbf{u}^2, \ldots, \mathbf{u}^k, \ldots, \mathbf{u}^{\mathrm{K}}) \\ (\tau^1, \tau^2, \ldots, \tau^k, \ldots, \tau^{\mathrm{K}}) \end{bmatrix}$$

can include the sequence $\ldots \mathbf{u}^k, \mathbf{u}^{k+1}, \ldots$, with $u_p^k = 1 \wedge u_p^{k+1} = 0$, and $u_p^k = 1 \wedge u_p^{k+1} = 1$, i.e.,

$$\ldots, \begin{bmatrix} u_1^k \\ u_2^k \\ \vdots \\ u_p^k \\ \vdots \\ u_q^k \\ \vdots \\ u_{\mathrm{P}}^k \end{bmatrix}, \begin{bmatrix} u_1^{k+1} \\ u_2^{k+1} \\ \vdots \\ u_p^{k+1} \\ \vdots \\ u_q^{k+1} \\ \vdots \\ u_{\mathrm{P}}^{k+1} \end{bmatrix}, \ldots = \ldots, \begin{bmatrix} u_1^k \\ u_2^k \\ \vdots \\ 1 \\ \vdots \\ 0 \\ \vdots \\ u_{\mathrm{P}}^k \end{bmatrix}, \begin{bmatrix} u_1^{k+1} \\ u_2^{k+1} \\ \vdots \\ 0 \\ \vdots \\ 1 \\ \vdots \\ u_{\mathrm{P}}^{k+1} \end{bmatrix}, \ldots$$

only if the control variables $u_p(\cdot)$ and $u_q(\cdot)$ are in relation R_n, i.e.,

$$(u_p(\cdot), u_q(\cdot)) \in R_n\,, \quad (p \neq q, \quad p, q \in \mathcal{P})\,.$$

An ordered pair $(u_p(\cdot), u_q(\cdot))$ is an element of relation R_n in the following cases:

α) Signal groups D'_p and D'_q are compatible, i.e.,

$$(D'_p, D'_q) \in C_g\,, \quad (p,q \in \mathscr{P})\,.$$

β) Signal groups D'_p and D'_q are incompatible, and m.e.i.t. is zero, i.e.,

$$((D'_p, D'_q) \notin C_g) \wedge (z_{pq} = 0)\ , \quad (p,q \in \mathscr{P})\,.$$

γ) Signal groups D'_p and D'_q are incompatible, and m.e.i.t. is negative, i.e.,

$$((D'_p, D'_q) \notin C_g) \wedge (z_{pq} < 0)\ , \quad (p,q \in \mathscr{P})\,.$$

Therefore, relation R_n is defined by the following expression:

$$R_n = \{(u_p(\cdot), u_q(\cdot)) \mid (((D'_p, D'_q) \in C_g) \vee ((D'_p, D'_q) \notin C_g) \wedge (z_{pq} \le 0)),\ p,q \in \mathscr{P},\ \ p \ne q\}. \tag{8.47}$$

By this relation the graph of green indications succession, G_n, is defined:

$$G_n = (\mathscr{U}_n, R_n)\,. \tag{8.48}$$

In the case when no pairs exist that satisfy conditions β and γ, the graph of green indications succession can be obtained from the graph of signal groups compatibility, G_g. In this case the following expression is valid:

$$((D'_p, D'_q) \in C_g) \Rightarrow (((u_p(\cdot), u_q(\cdot)) \in R_n) \wedge ((u_q(\cdot), u_p(\cdot)) \in R_n))\,. \tag{8.49}$$

The nodes of graph G_g represent signal groups, while the nodes of graph G_n represent control variables assigned to these signal groups. Graph G_n can be obtained by substituting each nonoriented edge of graph G_g with two oppositely oriented edges, i.e., by presenting graph G_g as a digraph. Each node that in G_g represents a signal group, $D'_p \in \mathfrak{D}_a$, in G_n will represent the control variable, $u_p(\cdot)$, assigned to that group.

Example 8.5

α) Determine the relation of green indication succession, R_n, and graph $G_n = (\mathscr{U}_n, R_n)$ for the intersection presented in Fig. 8.10. The complete set of signal groups is:

$$\mathfrak{D}_a = \{D'_1, D'_2, D'_3, D'_4, D'_5\}\,,$$

and the signal groups represent the following subsets of traffic streams set $\mathscr{S}$:

$$D'_1 = \{\sigma_2\}\,,\ D'_2 = \{\sigma_4\}\,,\ D'_3 = \{\sigma_5\}\,,\ D'_4 = \{\sigma_6\}\,,\ D'_5 = \{\sigma_1, \sigma_3\}\ .$$

The matrix of m.e.i.t. has no negative elements in this example.

The signal group compatibility graph, $G_g = (\mathfrak{D}_a, \Gamma_g)$, and the graph of green indication succession, $G_n = (\mathscr{U}_n, R_n)$, are presented in Fig. 8.14.

Relation R_n is:

$$R_n = \{(u_1(\cdot),u_3(\cdot)),(u_3(\cdot),u_1(\cdot)),(u_1(\cdot),u_4(\cdot)),(u_4(\cdot),u_1(\cdot)),\\ (u_2(\cdot),u_3(\cdot)),(u_3(\cdot),u_2(\cdot)),(u_3(\cdot),u_4(\cdot)),(u_4(\cdot),u_3(\cdot))\}.$$

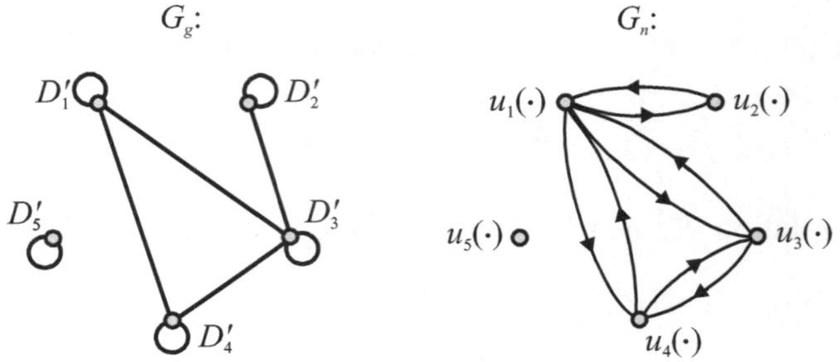

Figure 8.14

β) Determine the relation of green indication succession, R_n, and graph $G_n = (\mathscr{U}_n, R_n)$ for the intersection presented in Fig. 8.15. The matrix of m.e.i.t., Z, and graphs G'_k and G_c are given in the same figure.

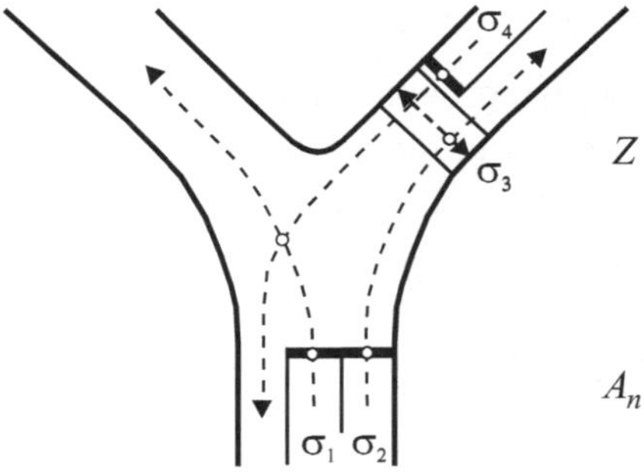

$$Z = \begin{bmatrix} 0 & 0 & 0 & -2 \\ 0 & 0 & 9 & 0 \\ 0 & -2 & 0 & 6 \\ 10 & 0 & 1 & 0 \end{bmatrix}$$

$$A_n = \begin{bmatrix} - & 1 & 1 & 1 \\ 1 & - & 0 & 1 \\ 1 & 1 & - & 0 \\ 0 & 1 & 0 & - \end{bmatrix}$$

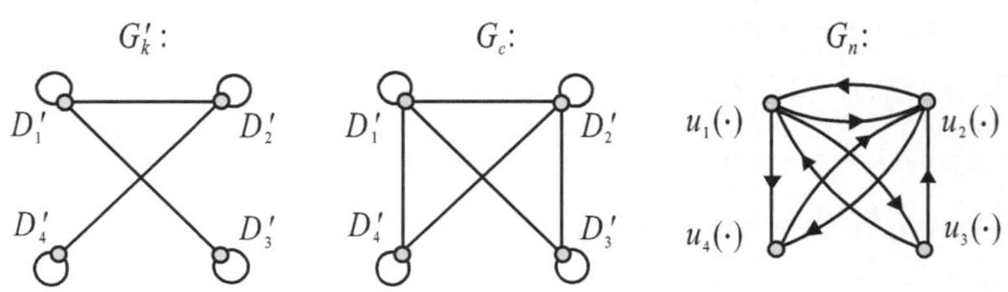

Figure 8.15

In the matrix of m.e.i.t., Z, there exist two negative elements.

Relation R_n is determined by the following set of ordered pairs:

$$R_n = \{(u_1(\cdot),u_2(\cdot)),(u_2(\cdot),u_1(\cdot)),(u_1(\cdot),u_3(\cdot)),(u_3(\cdot),u_1(\cdot)),\\ (u_2(\cdot),u_4(\cdot)),(u_4(\cdot),u_2(\cdot)),(u_1(\cdot),u_4(\cdot)),(u_3(\cdot),u_2(\cdot))\}.$$

The graph of green indications succession, $G_n = (\mathscr{U}_n, R_n)$, is given in Fig. 8.14. Digraph G_n does not present a nonoriented graph, as was the case in the previous example, because relation R_n is not symmetric.

8.6. The relation and graph of phase transitions

The signal group compatibility relation, the m.e.i.t. function, and the relation of green indications succession supply all information necessary for determination of the relation and graph of phase transitions.

8.6.1. The phase transitions relation

In the procedure of a feasible signal plan determination, phases are chosen from the set $\mathbf{U}_f$, and these phases are ordered in a sequence that makes a signal plan structure. The fact that a phase can be followed only by certain phases from $\mathbf{U}_f$ has to be taken into account when making this sequence.

If to a signal group D'_p the right-of-way is given by phase $\mathbf{u}^k$, then the succeeding phase in a feasible signal plan structure, $\mathbf{u}^{k+1}$, cannot be a phase by which the right-of-way begins for a signal group that is controlled by the control variable that is not in the relation of interval succession with control variables giving the right-of-way to some signal groups in the preceding phase. Namely, the right-of-way to D'_p cannot be stopped in phase $\mathbf{u}^k$ (it means, $u_p^k = 1$, $u_p^{k+1} = 0$) and given to some other signal group, D'_q, in phase $\mathbf{u}^{k+1}$ (it means $u_q^k = 0$, $u_q^{k+1} = 1$) if $(u_p(\cdot),u_q(\cdot)) \notin R_n$. Therefore, for each phase there exists a subset of set $\mathbf{U}_f$, comprising the phases that can immediately follow the given phase.

It is necessary to determine the conditions that have to be satisfied by a phase $\mathbf{u}(b) \in \mathbf{U}_f$ so that this phase can immediately follow a given phase $\mathbf{u}(a) \in \mathbf{U}_f$ (a and b are here the index numbers of elements in $\mathbf{U}_f$).

If $\mathbf{u}(a)$ and $\mathbf{u}(b)$ are included in a feasible signal plan structure, in the form $[\ldots,\mathbf{u}(a),\mathbf{u}(b),\ldots]=[\ldots,\mathbf{u}^k,\mathbf{u}^{k+1},\ldots]$, it is necessary to determine the conditions that $\mathbf{u}(b)$ has to satisfy so that it can be included in the signal plan structure as $\mathbf{u}^{k+1}$.

For determination of these conditions, it is suitable to consider two cases:

$$u_p(a)\,u_q(b)=0 \;, \quad (p,q\in\mathscr{P};\ \mathbf{u}(a),\mathbf{u}(b)\in\mathbf{U}_f),\text{ and}$$

$$u_p(a)\,u_q(b)=1 \;, \quad (p,q\in\mathscr{P};\ \mathbf{u}(a),\mathbf{u}(b)\in\mathbf{U}_f).$$

Case I: $u_p(a)\,u_q(b)=0$

This case arises when $u_p(a)\neq u_q(b)$ or $u_p(a)=u_q(b)=0$. Obviously, if $u_p(a)=0$ (red indication of the signals controlling signal group $D'_p\in\mathfrak{D}_a$), the control variables of the next phase, $\mathbf{u}(b)$, can assume any value (0 or 1), i.e., $u_q(b)=0$ or $u_q(b)=1$ $(q\in\mathscr{P})$.

If $u_p(a)=1$, then in the next phase, $\mathbf{u}(b)$, any control variable can have the value 0, i.e., $u_q(b)=0$ $(q\in\mathscr{P})$.

The value $u_p(a)=0$ can extend to the next phase, $\mathbf{u}(b)$. It is always possible to have $u_p(a)=u_q(b)=0$.

Hence, in determining whether phase $\mathbf{u}(b)$ can immediately follow phase $\mathbf{u}(a)$, the condition of succession is always satisfied for pairs $(u_p(a),u_q(b))$ if $u_p(a)\,u_q(b)=0$. Therefore, it is necessary to analyze only the following case:

Case II: $u_p(b)\,u_q(a)=1, \quad (p,q\in\mathscr{P})$

In this case, phase $\mathbf{u}(b)$ can immediately follow $\mathbf{u}(a)$ if at least one of the following conditions is satisfied:

II.1 $(u_p(b)\,u_q(a)=1)\wedge(p=q)\,,\,(p,q\in\mathscr{P})$

It is obvious that the green indication controlling one signal group can always extend to the next phase, $\mathbf{u}(b)$.

II.2 $(u_p(b)\,u_q(a)=1)\wedge((u_p(\cdot),u_q(\cdot))\in R_n), \quad (p\neq q,\ p,q\in\mathscr{P})$

This condition states the fact that control variable $u_p(b)$ in phase $\mathbf{u}(b)$ can have value 1 if the pair $(u_p(\cdot),u_q(\cdot))$ is the element of the relation of green intervals succession, R_n.

II.3 $(u_p(b)\,u_q(a)=1)\wedge(u_q(a)\,u_p(b)=1)\wedge(z_{pq}<0)$, $(p,q\in\mathscr{P})$

When negative m.e.i.t. exist, during this interval the control can be composed of two or more successive phases. This possibility is not included in previous cases.

Therefore, a phase $\mathbf{u}(b)$ can follow phase $\mathbf{u}(a)$ in a feasible signal plan structure, $\mathbf{u}=[\mathbf{u}^1,\mathbf{u}^2,\ldots,\mathbf{u}^k,\mathbf{u}^{k+1},\ldots,\mathbf{u}^K]$, i.e., a part of the structure can be:

$$[\ldots,\mathbf{u}^k,\mathbf{u}^{k+1},\ldots]=[\ldots,\mathbf{u}(a),\mathbf{u}(b),\ldots]$$

if the pairs of successive phases, ($\mathbf{u}(a),\mathbf{u}(b)$) satisfy the listed conditions.

The *phase transition relation*, R_s, thus, represents the set of phase pairs defined by the following expression:

$$\begin{aligned}R_s=\{(\mathbf{u}(a),\mathbf{u}(b))\,|\,&(u_p(a)\cdot u_q(b)=0)\vee(u_p(a)\cdot u_q(b)=1)\Rightarrow\\&\Rightarrow((u_p(\cdot),u_q(\cdot))\equiv R_n\vee(p=q))\vee((u_q(a)\cdot u_p(b)=1)\wedge(z_{ij}<0)),\\&\mathbf{u}(a),\mathbf{u}(b)\in\mathbf{U}_f,\ p,q\in\mathscr{P}\}.\end{aligned} \quad (8.50)$$

Example 8.6

α) Determine whether, in Example 8.2, the ordered pair of phases

$$(\mathbf{u}(10),\mathbf{u}(13))=\left(\begin{bmatrix}0\\1\\0\\0\\1\\0\end{bmatrix},\begin{bmatrix}0\\0\\0\\0\\1\\1\end{bmatrix}\right)=((0,1,0,0,1,0)^T,(0,0,0,0,1,1)^T)$$

belongs to R_s relation.

The answer to this question is obtained by analyzing all cases with $u_p(10)\cdot u_q(13)=1$, $(p,q\in\mathscr{P})$. The following products satisfy this condition:

$$u_2(10)\cdot u_5(13)=1$$
$$u_2(10)\cdot u_6(13)=1$$
$$u_5(10)\cdot u_5(13)=1$$
$$u_5(10)\cdot u_6(13)=1$$

Ordered pairs of control variables: $(u_2(\cdot),u_5(\cdot))$, $(u_2(\cdot),u_6(\cdot))$, and $(u_5(\cdot),u_6(\cdot))$ belong to R_n relation (because pairs (D_2',D_5'), (D_2',D_6'), and (D_5',D_6') belong to the signal group compatibility relation, C_g). Therefore, $(\mathbf{u}(10),\mathbf{u}(13))\in R_s$.

β) For the intersection presented in Fig. 8.15 determine whether phase $\mathbf{u}(b)$ can follow immediately after phase $\mathbf{u}(a)$, as shown in Fig. 8.16.

The possibility of transition from $\mathbf{u}(a)$ to $\mathbf{u}(b)$ can be determined by analyzing all products $u_p(a)\,u_q(b)=1$. In this case, these are the following products:

a) $u_1(a)\cdot u_2(b)=1$

b) $u_1(a)\cdot u_3(b)=1$

c) $u_2(a)\cdot u_3(b)=1$

d) $u_2(a)\cdot u_2(b)=1$

e) $u_3(a)\cdot u_2(b)=1$

f) $u_3(a)\cdot u_3(b)=1$

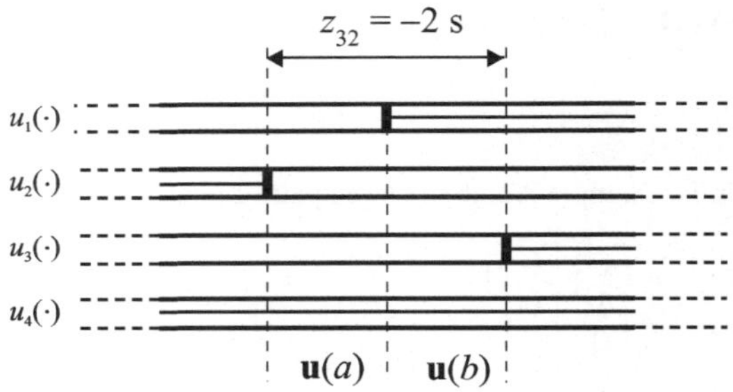

Figure 8.16

Pairs $(u_1(\cdot),u_2(\cdot))$, $(u_1(\cdot),u_3(\cdot))$, and $(u_3(\cdot),u_2(\cdot))$ (products a, b, e) belong to R_n relation. Hence, condition II.2 is satisfied. Pairs $(u_2(\cdot),u_2(\cdot))$ and $(u_3(\cdot),u_3(\cdot))$ are the pairs where $p=q$, and the condition II.1 is satisfied.

The fact that $(u_3(\cdot),u_2(\cdot))\in R_n$ and $(u_2(\cdot),u_3(\cdot))\notin R_n$ can be observed on graph $G_n=(\mathcal{U}_n,R_n)$, presented in Fig. 8.15. However, since $z_{32}<0$, condition II.3 (products c and e) is satisfied, i.e.,

$$(u_3(a)\cdot u_2(b)=1)\wedge(u_2(a)\cdot u_3(b)=1)\wedge(z_{32}<0).$$

Therefore, phase $\mathbf{u}(b)$ can follow immediately after $\mathbf{u}(a)$, i.e., $(\mathbf{u}(a),\mathbf{u}(b))\in R_s$ because all necessary conditions are satisfied.

8.6.2. The phase transition graph

The phase transition graph is defined by the set of feasible phases, $\mathbf{U}_f$, and relation, R_s, i.e.,

$$G_s = (\mathbf{U}_f, R_s). \tag{8.51}$$

By introducing the mapping, Γ_s, [9], [12]:

$$\Gamma_s : \mathbf{U}_f \to \mathcal{P}(\mathbf{U}_f),$$

such that

$$\Gamma_s \mathbf{u}(a) = \{\mathbf{u}(b) \mid (\mathbf{u}(a), \mathbf{u}(b)) \in R_s\}, \quad (\mathbf{u}(a), \mathbf{u}(b) \in \mathbf{U}_f), \tag{8.52}$$

the phase transition graph can be presented in the form:

$$G_s = (\mathbf{U}_f, \Gamma_s). \tag{8.53}$$

Nodes of this graph represent feasible phases, and an edge between two nodes exists if the pair of phases represented by these nodes belongs to R_s relation.

Graph G_s, in general case, is neither oriented nor nonoriented (Appendix I).

Relations R_n and R_s are symmetric in the case when there isn't any negative or zero-valued m.e.i.t. In this case:

$$\mathbf{u}(a) R_s \mathbf{u}(b) \Rightarrow \mathbf{u}(b) R_s \mathbf{u}(a), \quad (\mathbf{u}(a), \mathbf{u}(b) \in \mathbf{U}_f),$$

and G_s is a nonoriented graph.

In order to construct graph G_s it is necessary to determine, for each feasible phase, $\mathbf{u}(a) \in \mathbf{U}_f$, the subset of $\mathbf{U}_f$ that contains the phases that can immediately follow $\mathbf{u}(a)$, i.e.,

$$\Gamma_s \mathbf{u}(a), \quad (\mathbf{u}(a) \in \mathbf{U}_f).$$

For this construction, the definition of mapping Γ_s is sufficient.

Graph G_s can also be constructed using the procedure described in Appendix VI.

Example 8.7

α) Determine the phase transition graph, G_s, for the intersection presented in Fig. 8.7, together with the compatibility graph of signal groups. The set of signal groups is:

$$\mathcal{D}_a^1 = \{D_1^1, D_2^1, D_3^1, D_4^1, D_5^1, D_6^1\}$$
$$= \{\{\sigma_1\}, \{\sigma_2\}, \{\sigma_3\}, \{\sigma_4\}, \{\sigma_5\}, \{\sigma_6\}\}.$$

Relation R_s in this case is symmetric because there is no negative or zero m.e.i.t. Hence, graph G_s is nonoriented. The set of feasible phases (Example 8.2) is determined using the procedure presented in Section 8.4.

$$\begin{aligned}\mathbf{U}_f = \{&(0,0,0,0,0,0)^T, (1,0,0,0,0,0)^T, (0,1,0,0,0,0)^T, (0,0,1,0,0,0)^T,\\ &(0,0,0,1,0,0)^T, (0,0,0,0,1,0)^T, (0,0,0,0,0,1)^T, (1,1,0,0,0,0)^T,\\ &(1,0,1,0,0,0)^T, (1,0,0,0,1,0)^T, (0,1,0,0,1,0)^T, (0,1,0,0,0,1)^T,\\ &(0,0,0,1,1,0)^T, (0,0,0,0,1,1)^T, (1,1,0,0,1,0)^T, (0,1,0,0,1,1)^T\}.\end{aligned}$$

According to expression 8.52, mapping Γ_s of each feasible phase is determined:

$$\begin{aligned}\Gamma_s(1,1,0,0,1,0)^T = \{&(1,1,0,0,0,0)^T, (1,0,0,0,1,0)^T, (0,1,0,0,1,0)^T, (1,0,0,0,0,0)^T,\\ &(0,1,0,0,0,0)^T, (0,0,0,0,1,0)^T, (0,0,0,0,0,0)^T\}\end{aligned}$$

$$\begin{aligned}\Gamma_s(1,1,0,0,0,0)^T = \{&(1,1,0,0,1,0)^T, (1,0,0,0,1,0)^T, (0,1,0,0,1,0)^T, (1,0,0,0,0,0)^T,\\ &(0,1,0,0,0,0)^T, (0,0,0,0,1,0)^T, (0,0,0,0,0,0)^T\}\end{aligned}$$

$$\Gamma_s(0,0,0,0,0,0)^T = \mathbf{U}_f \setminus \{(0,0,0,0,0,0)^T\}$$

$$\begin{aligned}\Gamma_s(1,0,0,0,0,0)^T = \{&(1,1,0,0,1,0)^T, (1,1,0,0,0,0)^T, (1,0,0,0,1,0)^T,\\ &(0,1,0,0,1,0)^T, (0,0,0,0,1,0)^T, (0,1,0,0,0,0)^T,\\ &(1,0,1,0,0,0)^T, (0,0,1,0,0,0)^T, (0,0,0,0,0,0)^T\}\end{aligned}$$

$$\begin{aligned}\Gamma_s(0,1,0,0,0,0)^T = \{&(1,1,0,0,1,0)^T, (1,1,0,0,0,0)^T, (1,0,0,0,1,0)^T, (0,1,0,0,1,0)^T,\\ &(0,0,0,0,1,0)^T, (1,0,0,0,0,0)^T, (0,0,0,0,1,1)^T, (0,0,0,0,0,1)^T,\\ &(0,1,0,0,0,1)^T, (0,1,0,0,1,1)^T, (0,0,0,0,0,0)^T\}\end{aligned}$$

$$\Gamma_s(0,0,1,0,0,0)^T = \{(1,0,1,0,0,0)^T, (1,0,0,0,0,0)^T, (0,0,0,0,0,0)^T\}$$

$$\Gamma_s(0,0,0,1,0,0)^T = \{(0,0,0,1,1,0)^T, (0,0,0,0,1,0)^T, (0,0,0,0,0,0)^T\}$$

$$\Gamma_s(0,0,0,0,1,0)^T = \{(1,1,0,0,1,0)^T,(1,1,0,0,0,0)^T,(1,0,0,0,1,0)^T,(0,1,0,0,1,0)^T,\\ (1,0,0,0,0,0)^T,(0,1,0,0,0,0)^T,(0,0,0,1,1,0)^T,(0,0,0,1,0,0)^T,\\ (0,1,0,0,1,1)^T,(0,1,0,0,0,1)^T,(0,0,0,0,1,1)^T,(0,0,0,0,0,0)^T\}$$

$$\Gamma_s(0,0,0,0,0,1)^T = \{(0,1,0,0,1,1)^T,(0,1,0,0,0,1)^T,(0,0,0,0,1,1)^T,(0,0,0,0,1,0)^T,\\ (0,1,0,0,1,0)^T,(0,1,0,0,0,0)^T,(0,0,0,0,0,0)^T\}$$

$$\Gamma_s(1,0,1,0,0,0)^T = \{(0,0,1,0,0,0)^T,(1,0,0,0,0,0)^T,(0,0,0,0,0,0)^T\}$$

$$\Gamma_s(1,0,0,0,1,0)^T = \{(1,1,0,0,1,0)^T,(1,1,0,0,0,0)^T,(0,1,0,0,1,0)^T,(1,0,0,0,0,0)^T,\\ (0,1,0,0,0,0)^T,(0,0,0,0,1,0)^T,(0,0,0,0,0,0)^T\}$$

$$\Gamma_s(0,1,0,0,1,0)^T = \{(1,1,0,0,1,0)^T,(1,1,0,0,0,0)^T,(1,0,0,0,1,0)^T,(0,1,0,0,0,0)^T,\\ (0,1,0,0,0,0)^T,(0,0,0,0,1,0)^T,(0,1,0,0,1,1)^T,(0,1,0,0,0,1)^T,\\ (0,0,0,0,1,1)^T,(0,0,0,0,0,1)^T,(0,0,0,0,0,0)^T\}$$

$$\Gamma_s(0,1,0,0,0,1)^T = \{(0,1,0,0,1,1)^T,(0,1,0,0,1,0)^T,(0,0,0,0,1,1)^T,(0,1,0,0,0,0)^T,\\ (0,0,0,0,0,1)^T,(0,0,0,0,1,0)^T,(0,0,0,0,0,0)^T\}$$

$$\Gamma_s(0,0,0,1,1,0)^T = \{(0,0,0,1,0,0)^T,(0,0,0,0,1,0)^T,(0,0,0,0,0,0)^T\}$$

$$\Gamma_s(0,0,0,0,1,1)^T = \{(0,1,0,0,1,1)^T,(0,1,0,0,0,1)^T,(0,1,0,0,1,0)^T,(0,1,0,0,0,0)^T,\\ (0,0,0,0,1,0)^T,(0,0,0,0,0,1)^T,(0,0,0,0,0,0)^T\}$$

$$\Gamma_s(0,1,0,0,1,1)^T = \{(0,1,0,0,1,0)^T,(0,1,0,0,0,1)^T,(0,0,0,0,1,1)^T,(0,1,0,0,0,0)^T,\\ (0,0,0,0,1,0)^T,(0,0,0,0,0,1)^T,(0,0,0,0,0,0)^T\}.$$

The graph of phase transitions, $G_s = (\mathbf{U}_f, \Gamma_s)$, is given in Fig. 8.17.

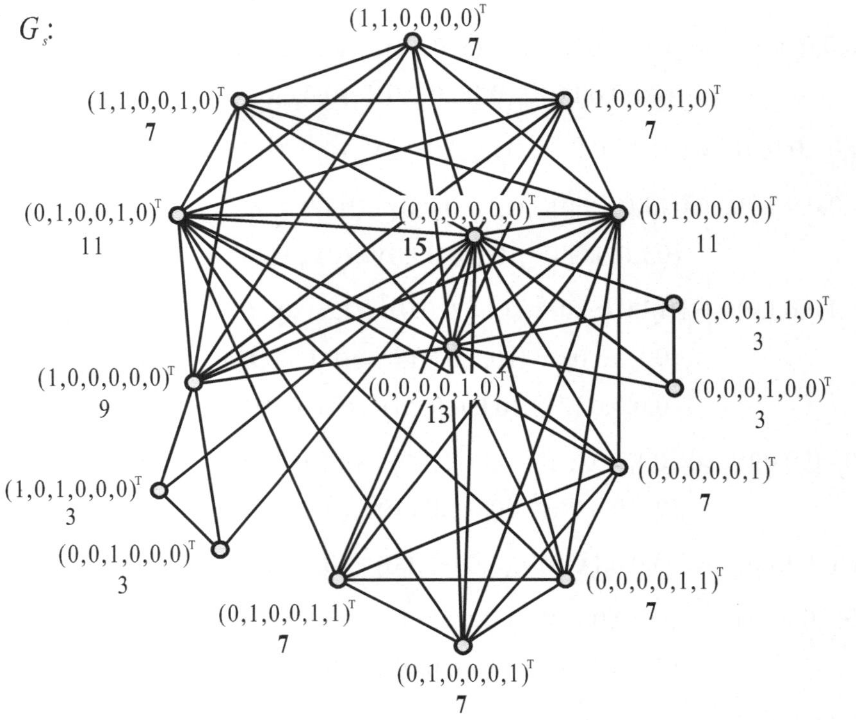

G_s:
$(1,1,0,0,0,0)^T$
7
$(1,1,0,0,1,0)^T$
7
$(1,0,0,0,1,0)^T$
7
$(0,1,0,0,1,0)^T$
11
$(0,0,0,0,0,0)^T$
15
$(0,1,0,0,0,0)^T$
11
$(0,0,0,1,1,0)^T$
3
$(1,0,0,0,0,0)^T$
9
$(0,0,0,0,1,0)^T$
13
$(0,0,0,1,0,0)^T$
3
$(0,0,0,0,0,1)^T$
7
$(1,0,1,0,0,0)^T$
3
$(0,0,1,0,0,0)^T$
3
$(0,1,0,0,1,1)^T$
7
$(0,0,0,0,1,1)^T$
7
$(0,1,0,0,0,1)^T$
7

Figure 8.17

β) Determine the phase transition graph for the same intersection (as in Example 8.7α), but for the following set of signal groups:

$$\mathcal{D}_a^2 = \{D_1^2, D_2^2, D_3^2, D_4^2, D_5^2\}$$
$$= \{\{\sigma_3\}, \{\sigma_4\}, \{\sigma_5\}, \{\sigma_6\}, \{\sigma_1, \sigma_2\}\}.$$

The graph of signal group compatibility and the phase transition graph are presented in Fig. 8.18. The phase transition graph, in this case, is a subgraph of G_s presented in Example 8.7α.

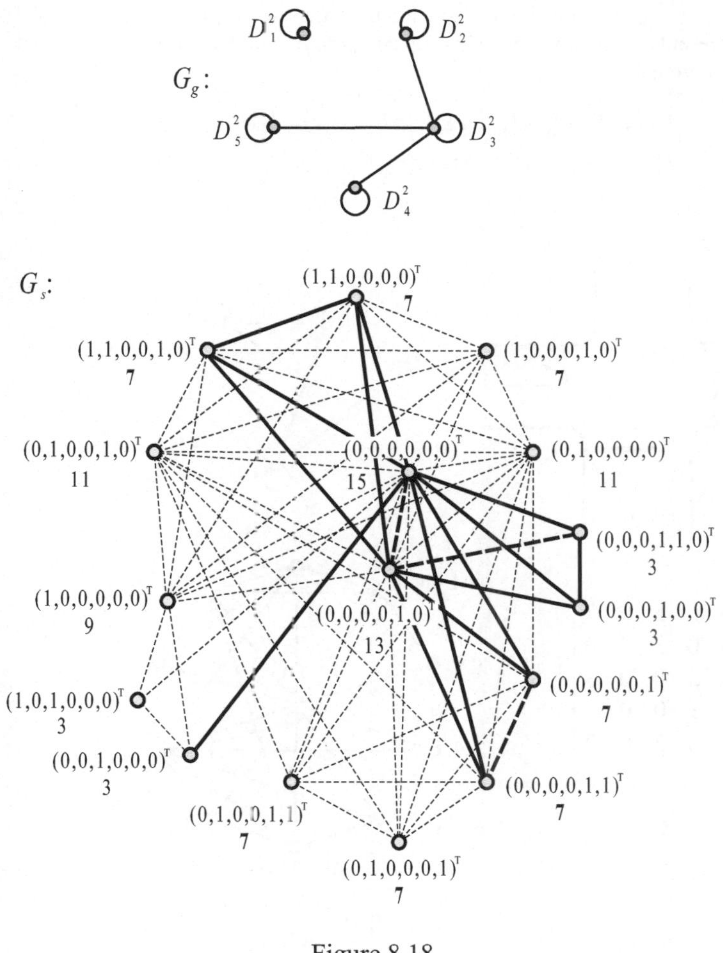

Figure 8.18

Omitted are the nodes that represent phases giving the right-of-way to D_1^1 only, or D_2^1 only, and the edges incident to these nodes. The nodes representing the phases that give the simultaneous right-of-way to D_1^1 and D_2^1 remained because pair (D_1^1, D_2^1) in this example represents one signal group, D_5^2. Thus, from the feasible set of phases, $\mathbf{U}_f$, determined in Example 8.7α, the following phases are omitted: $(0,1,0,0,1,0)^T$, $(1,0,0,0,1,0)^T$, $(0,1,0,0,0,0)^T$, $(0,1,0,0,0,1)^T$, $(0,1,0,0,1,1)^T$, $(1,0,1,0,0,0)^T$, $(1,0,0,0,0,0)^T$. This graph is also nonoriented because relation R_s is symmetric.

γ) Determine the phase transition graph for the intersection presented in Fig. 8.19. The same figure presents the graph of signal group compatibility and the m.e.i.t. matrix. The set of feasible signal groups is:

$$\mathcal{D}_a^1 = \{D_1^1, D_2^1, D_3^1, D_4^1, D_5^1\}$$
$$= \{\{\sigma_1\}, \{\sigma_2\}, \{\sigma_3\}, \{\sigma_4\}, \{\sigma_5\}\}.$$

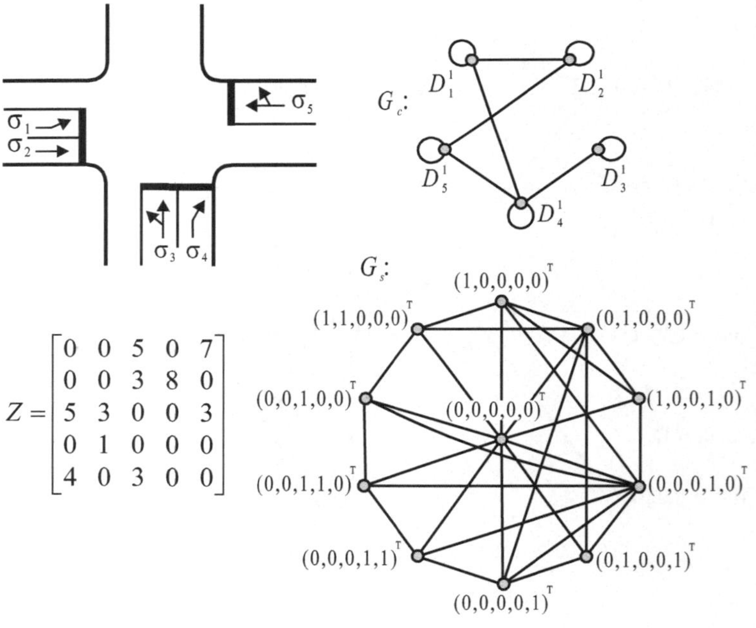

Figure 8.19

The set of feasible phases is defined by the following expression:

$$\mathbf{U}_f = \{(1,1,0,0,0)^T, (1,0,0,0,0)^T, (0,1,0,0,0)^T, (1,0,0,1,0)^T, \\ (0,0,0,1,0)^T, (0,1,0,0,1)^T, (0,0,0,0,1)^T, (0,0,0,1,1)^T, \\ (0,0,1,1,0)^T, (0,0,1,0,0)^T, (0,0,0,0,0)^T\}.$$

Graph $G_s = (\mathbf{U}_f, R_s) = (\mathbf{U}_f, \Gamma_s)$ is presented in Fig. 8.19. The phase transition graph is nonoriented in this case, also.

8.6.3. Some features of the phase transition graph

a) Graph G_s is a connected graph, i.e., any two nodes can be connected by a path (or by a chain, if G_s contains oriented edges as well).

b) Graph G_s contains an articulation node if the graph of signal group compatibility, G_g, is disconnected. The articulation node always represents phase $(0,0,...,0)^T$. If an isolated node is a connected component of graph G_g (Fig. 8.20, Fig. 8.21), then graph G_s contains the articulation node and a pending edge.

   If the compatibility graph has n connected components, phase $(0,0,...,0)^T$ ("all red") will appear at least n times ($n > 1$) in the signal plan.

c) The nodes of graph G_s (nonoriented) can be classified according to the node degree as either internal or connection nodes.

   I – Internal nodes

   "Internal nodes" are the nodes that have the node degree equal to the cardinal number of set $\mathbf{U}_{f\pi}$ to whom they belong. The node

   $$\mathbf{u}(s) \in \mathbf{U}_{f\pi}$$

   is internal if the following condition is satisfied:

   $$d(\mathbf{u}(s)) = card\ \mathbf{U}_{f\pi}. \tag{8.54}$$

   $d(\mathbf{u}(s))$ is the degree of node $\mathbf{u}(s)$.

   Appendix VI presents the procedure for determining all phases that are generated from a maximal phase, which corresponds to the maximal clique of graph G_g, i.e., the way of obtaining the set $\mathbf{U}_{f\pi} = P_m(\mathbf{u}_a^{\pi})$.

   The phase represented by an internal node can be followed only by a phase that belongs to the same set $\mathbf{U}_{f\pi}$.

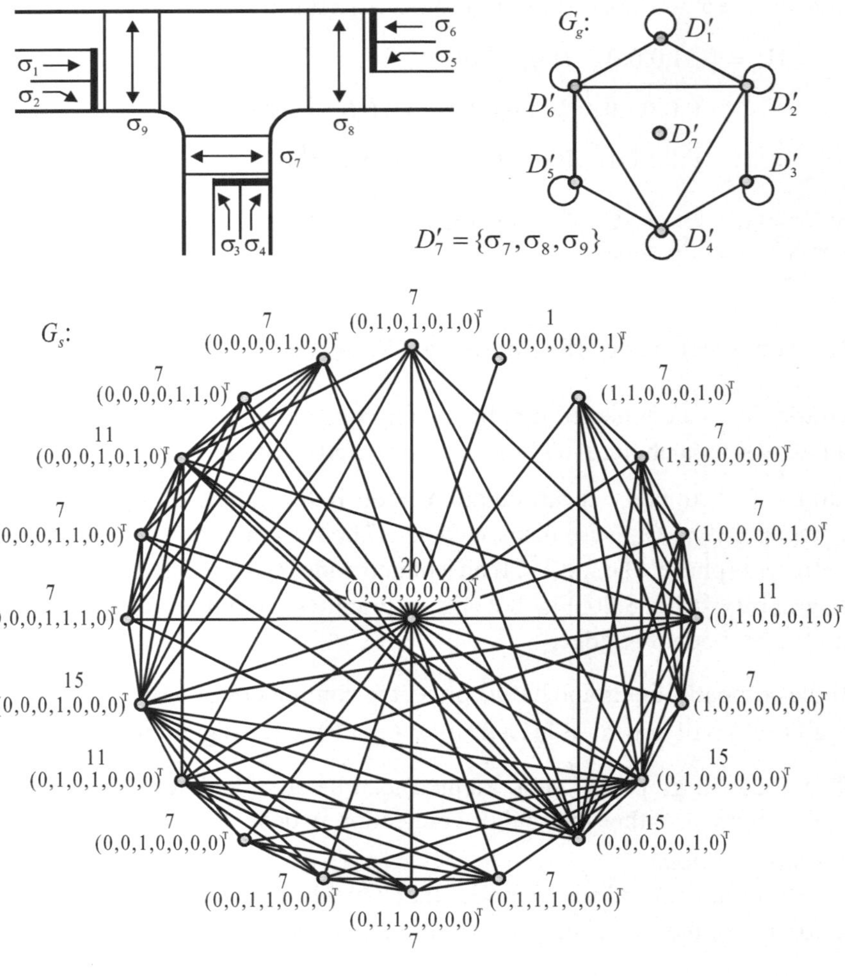

Figure 8.20

Set $\mathbf{U}_{f\pi}$ is the set of phases "generated" by a maximal clique, $\overline{\mathfrak{D}}^{\pi}_{am}$, of the signal group compatibility graph, G_g, i.e.,

$$\mathbf{U}_{f\pi} = P_{am}(\mathbf{u}^{\eth}_{a}), \quad (\pi \in \overline{\Pi}). \tag{8.55}$$

Internal nodes exist if the maximal clique $\overline{\mathfrak{D}}^{\pi}_{am}$, whose elements they are, contains at least one signal group that is not present in other maximal cliques. Namely there exist maximal cliques whose all nodes are elements of other maximal cliques.

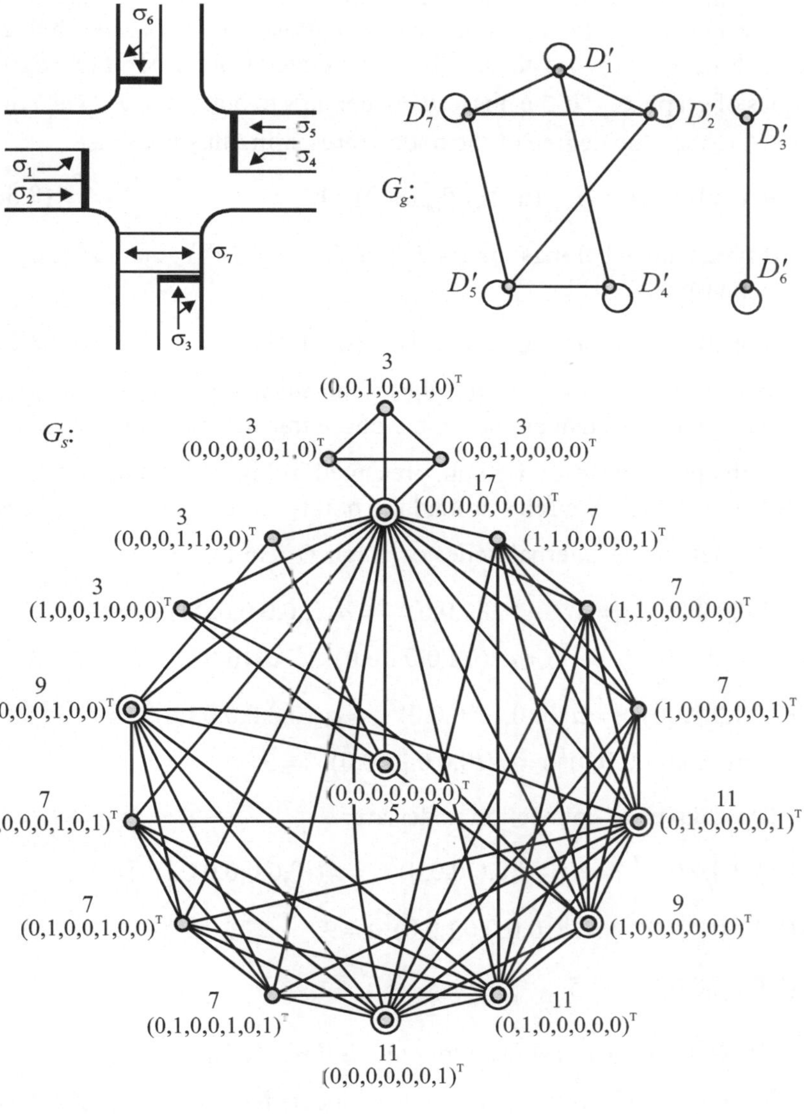
σ_6
σ_5
σ_4
σ_1
σ_2
σ_7
σ_3
D_1'
D_7'
D_2'
D_3'
G_g:
D_5'
D_4'
D_6'
G_s:
3
$(0,0,1,0,0,1,0)^T$
3
$(0,0,0,0,0,1,0)^T$
3
$(0,0,1,0,0,0,0)^T$
17
$(0,0,0,0,0,0,0)^T$
3
$(0,0,0,1,1,0,0)^T$
7
$(1,1,0,0,0,0,1)^T$
3
$(1,0,0,1,0,0,0)^T$
7
$(1,1,0,0,0,0,0)^T$
9
$(0,0,0,0,1,0,0)^T$
7
$(1,0,0,0,0,0,1)^T$
$(0,0,0,0,0,0,0)^T$
5
7
$(0,0,0,0,1,0,1)^T$
11
$(0,1,0,0,0,0,1)^T$
7
$(0,1,0,0,1,0,0)^T$
9
$(1,0,0,0,0,0,0)^T$
7
$(0,1,0,0,1,0,1)^T$
11
$(0,1,0,0,0,0,0)^T$
11
$(0,0,0,0,0,0,1)^T$

Figure 8.21

II – Connection nodes

"Connection nodes" in the phase transition graph, G_s, are the nodes that are connected to the nodes which correspond to phases that are elements of two or more phase subsets generated from different maximal cliques of graph G_g. If the phase $\mathbf{u}(v)$ belongs to both sets, $P_{am}(\mathbf{u}_a^\rho)$ and $P_{am}(\mathbf{u}_a^\lambda)$, then the degree of the node representing this phase is:

$$d(\mathbf{u}(v)) = card\ (P_{am}(\mathbf{u}_a^\rho) \cup P_{am}(\mathbf{u}_a^\lambda)) - 1. \tag{8.56}$$

If phase $\mathbf{u}(v)$ belongs to sets $P_{am}(\mathbf{u}_a^\rho), \ldots, P_{am}(\mathbf{u}_a^\varphi)$, the node degree of this phase is:

$$d(\mathbf{u}(v) = card\ (P_{am}(\mathbf{u}_a^\rho) \cup \ldots \cup P_{am}(\mathbf{u}_a^\varphi)) - 1. \tag{8.57}$$

In a signal plan structure, the phases represented by connection nodes separate the phases that are not in the phase transition relation.

In the phase transition graph, presented in Fig. 8.21, the connection nodes are marked by circles around the points representing these nodes.

The degrees of internal nodes of graph G_s, in Fig. 8.21, are:

$$d((1,1,0,0,0,0,1)^T) = d((1,1,0,0,0,0,0)^T) = d((1,0,0,0,0,0,1)^T) = \\ d((0,1,0,0,1,0,1)^T) = d((0,1,0,0,1,0,0)^T) = d((0,0,0,0,1,0,1)^T) = 7$$

$$d((0,0,1,0,0,1,0)^T) = d((0,0,1,0,0,0,0)^T) = d((0,0,0,0,0,1,0)^T) = \\ d((1,0,0,1,0,0,0)^T) = d((0,0,0,1,1,0,0)^T) = 3$$

The degrees of connection nodes are:

$$d((0,1,0,0,0,0,1)^T) = d((0,1,0,0,0,0,0)^T) = d((0,0,0,0,0,0,1)^T) = 1$$

$$d((1,0,0,0,0,0,0)^T) = d((0,0,0,0,1,0,0)^T) = 9$$

$$d((0,0,0,1,0,0,0)^T) = 5$$

III – Node $(0,0,\ldots,0)^T$ representing the "all red" phase

The degree of this node, which is connected to all other nodes if G_s is nonoriented, is:

$$d((0,0,\ldots,0)^T) = card\ \mathbf{U}_f - 1. \tag{8.58}$$

The procedure for determining the number of elements in set $\mathbf{U}_f$ is given in Appendix VI.

Node $(0,0,...,0)^T$ is a connection node, as well.

The degree of node $(0,0,...,0)^T$ in Fig. 8.21 is:

$d((0,0,...,0)^T) = card\ \mathbf{U}_f - 1 = 15$.

d) The diameter of graph G_c

The diameter of a graph is the length of the shortest path between the most distant nodes (Appendix I). The length between adjacent nodes is 1.

The node that represents phase $(0,0,...,0)^T$ – all red, in a nonoriented graph, is connected to all other nodes. The diameter of this graph is, therefore, equal to 2.

8.6.4. Structural constraints on phase transition

The signal plan structure is defined (Subsection 4.2.2) as a sequence of control vectors, i.e.,

$$\mathbf{u} = [\mathbf{u}^1, \mathbf{u}^2, \ldots, \mathbf{u}^k, \ldots \mathbf{u}^{\mathrm{K}}].$$

A signal plan structure has to satisfy certain constraints. Some of these constraints refer to any signal plan structure, while others reflect special requirements that exist only in some problems of signal plan choice.

a) Phase transition constraints that have to be satisfied by any feasible signal plan

Phase transition constraints that have to be satisfied by any two adjacent phases represent the constraints on the structure of a feasible signal plan. A feasible structure is represented on the phase transition graph, G_s, by the path

$$\mu = [(\mathbf{u}^1, \mathbf{u}^2), (\mathbf{u}^2, \mathbf{u}^3), \ldots, (\mathbf{u}^{\mathrm{K}-1}, \mathbf{u}^{\mathrm{K}})],$$

where the ordered pair $(\mathbf{u}^k, \mathbf{u}^{k+1})$ represents the edge between nodes $\mathbf{u}^k$ and $\mathbf{u}^{k+1}$. Hence, this path has the following feature:

$$\mathbf{u}^{k+1} \in \Gamma_s \mathbf{u}^k.$$

This also holds if $k = \mathrm{K}$ (K denotes the index of the last phase in the signal plan, i.e., this is the number of phases in the signal plan). Since the control is a periodic function of time, it is necessary that after phase $\mathbf{u}^{\mathrm{K}}$, phase $\mathbf{u}^1$ begins, i.e.,

$$\mathbf{u}^1 \in \Gamma_s \mathbf{u}^K . \tag{8.59}$$

The path

$$\mu' = [(\mathbf{u}^1, \mathbf{u}^2), (\mathbf{u}^2, \mathbf{u}^3), \ldots, (\mathbf{u}^{K-1}, \mathbf{u}^K), (\mathbf{u}^K, \mathbf{u}^1)],$$

therefore, represents a closed path in graph G_s [20]. This path, in a general case, is not an elementary path, i.e., one phase can appear several times in a signal plan.

The constraints that have to be satisfied by signal plan structure **u**, resulting from the phase transition relation, are given by the following expression:

$$\mathbf{u}^{k(\mathrm{mod}\,K)+1} \in \Gamma_s \mathbf{u}^k , \quad (k \in \mathscr{K}) . \tag{8.60}$$

Thus, any feasible signal plan structure can be represented by a closed path in graph G_s.

Constraints (8.60) are related only to the variables contained in the signal plan structure, not to the time variables τ^k. Therefore, these constraints are *structural constraints*.

b) Special structural constraints

In some problems of signal plan choice it is necessary to include special structural constraints. Such cases are, for example, the problems with permitted "filtering" of one traffic stream through another. In these problems, as mentioned in Subsection 2.2.4, it is better to give the right-of-way first to the opposing stream, and after its queue is discharged, to both streams, and, finally, only to the opposed stream.

If the opposing stream belongs to signal group D'_a, controlled by control variable $u_a(\cdot)$, and the opposed stream to signal group D'_b, controlled by $u_b(\cdot)$, then, for the mentioned sequence, it is necessary that phases with $u_a^k = 1$ and $u_b^k = 0$, precede phases with $u_a^{k+1} = 1$ and $u_b^{k+1} = 1$, followed by phases with $u_a^{k+2} = 0$ and $u_b^{k+2} = 1$. Therefore, on the graph of phase transitions, it is necessary to prevent transition from phases with $u_a^{k+1} = 1$ and $u_b^{k+1} = 1$, to phases with $u_a^{k+2} = 1$ and $u_b^{k+2} = 0$, and also from phases with $u_a^k = 0$ and $u_b^k = 1$, to phases with $u_a^{k+1} = 1$ and $u_b^{k+1} = 1$. This is achieved by introducing oriented edges in graph G_s, such that the mentioned "banned" transitions are impossible.

Example 8.8

An intersection with five traffic streams is presented in Fig. 8.22, together with the compatibility graph G_g. Each signal group controls a single traffic stream. Edge (σ_1,σ_4) exists if streams σ_1 and σ_4 are allowed to have the right-of-way simultaneously. The opposing stream is σ_1, and the opposed stream is σ_4. Determine the phase transition graph, G_s.

The set of feasible phases, $\mathbf{U}_f$, is:

$$\mathbf{U}_f = \{(1,0,0,0,1)^T, (1,0,0,0,0)^T, (0,0,0,0,1)^T, (0,1,0,1,1)^T, (0,1,0,1,0)^T, (0,1,0,0,1)^T, (0,0,0,1,1)^T, (0,1,0,0,0)^T, (0,0,0,1,0)^T, (0,0,1,0,0)^T, (0,0,0,0,0)^T\} \cup \{(1,0,0,1,1)^T, (1,0,0,1,0)^T\}.$$

The phase transition graph, G_s, is presented in Fig. 8.22. Certain edges in this graph are oriented to prevent undesirable phase sequences and provide for desirable phase sequences. This figure also presents, by bold lines in graph G_s, one structure that satisfies the structural constraints.

In Example 12.6 and in Fig. 12.10 another case with special structure constraints is presented.

Besides the described structural constraints, there exists another type of structural constraints—the constraints of one period of green indication in a cycle (Subsection 8.2.1).

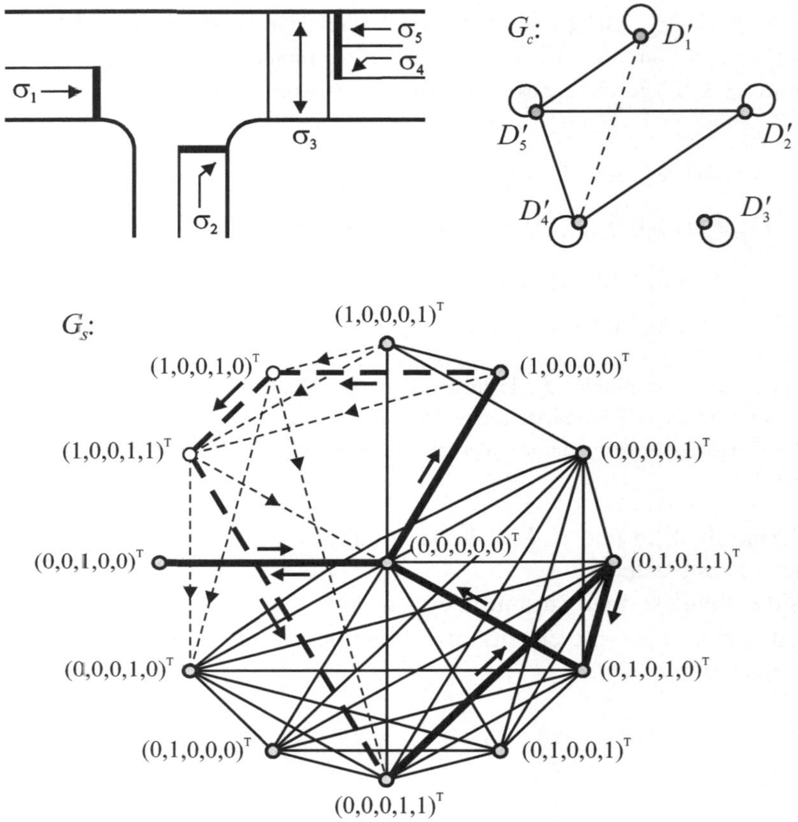

$$\mathbf{u} = \begin{bmatrix} \mathbf{u}_1 \\ \mathbf{u}_2 \\ \mathbf{u}_3 \\ \mathbf{u}_4 \\ \mathbf{u}_5 \end{bmatrix} = \overset{\begin{matrix} \mathbf{u}^1 & \mathbf{u}^2 & \mathbf{u}^3 & \mathbf{u}^4 & \mathbf{u}^5 & \mathbf{u}^6 & \mathbf{u}^7 & \mathbf{u}^8 & \mathbf{u}^9 \end{matrix}}{\begin{bmatrix} 1 & 1 & 1 & 0 & 0 & 0 & 0 & 0 & 0 \\ 0 & 0 & 0 & 0 & 1 & 1 & 0 & 0 & 0 \\ 0 & 0 & 0 & 0 & 0 & 0 & 0 & 1 & 0 \\ 0 & 1 & 1 & 1 & 1 & 1 & 0 & 0 & 0 \\ 0 & 0 & 1 & 1 & 1 & 0 & 0 & 0 & 0 \end{bmatrix}}$$

Figure 8.22

8.7. Minimal intergreen time constraints for phases

The constraints of minimal intergreen times for pairs of signal groups were analyzed in Subsection 8.4.2. The method of their determination was discussed, and the matrix of minimal effective green times was used for their representation:

$$Z = [z_{pq}]_{\mathrm{P}\times\mathrm{P}}.$$

A signal plan has to satisfy constraints that are the consequence of m.e.i.t. existence. Namely, if a part of the structure is considered, e.g., $\mathbf{u}^{k-1}, \mathbf{u}^{k}, \mathbf{u}^{k+1}$, the duration, τ^k, of phase $\mathbf{u}^k$ has to satisfy these constraints if pair $(\mathbf{u}^{k-1}, \mathbf{u}^{k+1})$ is not an element of the phase transition relation. This also holds if between the phases, which are not elements of the phase transition relation, several other phases exist. If the pair $(\mathbf{u}^{k-2}, \mathbf{u}^{k+1})$ in structure $\mathbf{u}^{k-2}, \mathbf{u}^{k-1}, \mathbf{u}^{k}, \mathbf{u}^{k+1}$, does not belong to the phase transition relation, then the sum $\tau^{k-1} + \tau^k$ has to satisfy the m.e.i.t. constraints.

The duration of phases situated in the structure between two phases that are not in the phase transition relation has to be greater than the maximal value of m.e.i.t. between incompatible signal groups contained in these two phases.

The following fact has to be taken into consideration: If u_p^{k-1} cuts the right-of-way in this phase, i.e., $u_p^{k-1} = 1$, $u_p^k = 0$, and u_q^{k+1} starts giving the right-of-way, i.e., $u_q^k = 0$, $u_q^{k+1} = 1$, then z_{pq} has to be included in determination of m.e.i.t. constraints.

The analytical expression of these constraints is:

$$\sum_{l=0}^{l=\alpha} \tau^{\mathrm{K}-(\mathrm{K}+l-k)(\mathrm{mod\,K})} \geq \max\,\{z_{pq} \mid$$
$$(u_p^{\mathrm{K}-(\mathrm{K}+\alpha-k)(\mathrm{mod\,K})}\, u_q^{k(\mathrm{mod\,K})+1} \cdot \bar{u}_p^{\mathrm{K}-(\mathrm{K}+\alpha-k-1)(\mathrm{mod\,K})} \cdot \bar{u}_q^{k}) = 1,$$
$$\mathbf{u}^{k(\mathrm{mod\,K})+1} \notin \Gamma_s(\mathbf{u}^{(\mathrm{K}+\alpha-k)(\mathrm{mod\,K})}), \quad p,q \in \mathscr{P}\} \tag{8.61}$$

$(k \in \mathscr{K})$ and $(\alpha \in \mathscr{A}'(\mathbf{u}^k, \mathbf{u}^{k(\mathrm{mod\,K})+1}))$.

The symbols not used before have the following meanings:

$\bar{u}_q^s = 1 - u_q^s$

α – the number of phases whose durations have to be taken into account when formulating the m.e.i.t. constraints

$\mathscr{A}'(\mathbf{u}^k, \mathbf{u}^{k(\mathrm{mod K})+1}) = \{1, 2, \ldots, \alpha, \ldots, \alpha_{\mathrm{M}}^k(\mathbf{u}^k, \mathbf{u}^{k(\mathrm{mod K})+1})\}$ – the set of values assigned to α

These constraints have to be formulated for each $\alpha \in \mathscr{A}'(\mathbf{u}^k, \mathbf{u}^{k(\mathrm{mod K})+1})$. The total number of m.e.i.t. constraints for one k is $\alpha_{\mathrm{M}}^k(\mathbf{u}^k, \mathbf{u}^{k(\mathrm{mod K})+1})$. The constraint formulated for $\alpha = 1$ contains one time variable (τ^k). The constraint formulated for $\alpha = 2$ contains the sum of two time variables. Finally, the constraint formulated for $\alpha = \alpha_{\mathrm{M}}^k(\mathbf{u}^k, \mathbf{u}^{k(\mathrm{mod K})+1})$ contains the maximal number of time variables.

The maximal number of phases, α_{M}^k, that have to be taken into account when formulating m.e.i.t. for one value of *k*, depends on signal groups that gain the right-of-way in phase $\mathbf{u}^{k(\mathrm{mod K})+1}$, i.e., the groups for which:

$$u_q^{k(\mathrm{mod K})+1} = 1 \text{ and } u_q^k = 0 .$$

The number of intergreen time constraints, α_{M}^k, for one k is equal to the maximal number of phases preceding phase $\mathbf{u}^{k(\mathrm{mod K})+1}$, whose durations have to be considered in formulating m.e.i.t. constraints.

The following facts have to be taken into account when determining α_{M}^k:

a) The shortest phase duration is 1 s. Therefore, the maximal number of phases, preceding phase $\mathbf{u}^{k(\mathrm{mod K})+1}$, whose durations have to be included in intergreen constraints, is equal to the maximal value of m.e.i.t. between any signal group and the group that receives the right-of-way by phase $\mathbf{u}^{k(\mathrm{mod K})+1}$. This number is, thus, equal or less than the maximal value z_{pq} in the columns of *Z* matrix corresponding to the *q* for which $u_q^{k(\mathrm{mod K})+1} = 1$, $u_q^k = 0$, i.e., this number is equal to:

$$\max\{z_{pq} \mid (u_q^{k(\mathrm{mod K})+1} = 1) \wedge (u_q^k = 0), \; p, q \in \mathscr{P}\} . \tag{8.62}$$

b) The maximal number of phases included in the m.e.i.t. constraints cannot be greater than $\mathrm{K} - 2$. If the following phase sequence is considered

$$\mathbf{u}^1, \mathbf{u}^2, \ldots, \mathbf{u}^k, \mathbf{u}^{k+1}, \mathbf{u}^{k+2}, \underbrace{\ldots, \mathbf{u}^{\mathrm{K}}, \mathbf{u}^1, \mathbf{u}^2, \ldots, \mathbf{u}^{k-1}, \mathbf{u}^k}_{\mathrm{K}-2}, \mathbf{u}^{k+1}, \ldots, \mathbf{u}^{\mathrm{K}}, \ldots,$$

it can be observed that the inclusion of phase $\mathbf{u}^{k+2}$ duration in this constraint means that intergreen time would be calculated between phase $\mathbf{u}^{k+1}$ and the same phase in the next cycle, which does not make any sense.

Therefore, α_{M}^{k} can be determined using the following expression:

$$\begin{aligned}\alpha_{\mathrm{M}}^{k} &= \alpha_{\mathrm{P}}(\mathbf{u}^{k}, \mathbf{u}^{k(\mathrm{mod\,K})+1}) \\ &= \min\{\max\{z_{pq} \mid (u_q^{k(\mathrm{mod\,K})+1} = 1) \wedge (u_q^k = 0),\ p, q \in \mathscr{P}\}, \mathrm{K} - 2\}. \qquad (8.63)\end{aligned}$$

When using expression (8.61), it is necessary to determine the values of z_{pq} that will be used in cases when they are not defined. These cases arise when $(D'_p, D'_q) \in C_g$ and when $p = q$. The value $z_{pq} = 0$ will be used for these cases, i.e.,

$$((D'_p, D'_q) \in C_g) \vee (p = q) \quad \Rightarrow (z_{pq} = 0). \qquad (8.64)$$

If $z_{pq} < 0$, then the duration, τ^k, of phase $\mathbf{u}^k$ that lies between the phase $\mathbf{u}^{k-1}$, in which signal group D'_p lost the right-of-way $(u_p^{k-1} = 1, u_p^k = 0)$, and phase $\mathbf{u}^{k+1}$, in which signal group D'_q gained the right-of-way $(u_q^k = 0, u_q^{k+1} = 1)$, has to satisfy the constraint:

$$\tau^k \geq z_{pq} < 0.$$

Since expression (8.63) relates to phase duration, τ^k, which satisfies the constraint:

$$\tau^k \geq 0\ , \quad (k \in \mathscr{K}),$$

then, in the case when $z_{pq} < 0$, the m.e.i.t. will be assumed $z_{pq} = 0$ in this expression.

The constraints for duration of phases that give the right-of-way to both signal groups, D'_p and D'_q, ($u_p^k = u_q^k = 1$), valid in the case $z_{pq} < 0$, are given by expression (8.40).

Therefore, when expression (8.61) is used, it is necessary to take into consideration that:

$$((D'_p, D'_q) \in C_g) \vee (p = q) \vee (z_{pq} < 0) \Rightarrow (z_{pq} = 0). \qquad (8.65)$$

Intergreen times for all pairs of feasible phases can be calculated in advance. They will be elements of the matrix of minimal effective intergreen times between phases, Z^f.

8.8. The constraint on the sum of phase durations

The control is, as mentioned in Part I, a periodic time function. Because of this fact, the sum of all control vector (phase) durations has to be equal to the cycle time. The cycle time, c, is given in some problems of signal plan choice. In some other problems, the cycle time results from problem solution, as the sum of duration of all phases that constitute the signal plan. Anyway, the cycle time has to be less than a maximal value, $c_{\max}$, which is usually prescribed by standards in any country. This value is 120 s in most countries.

The constraint on phase durations is defined by the following expression:

$$\sum_{k=1}^{\mathrm{K}} \tau^k = c \text{, or} \tag{8.66}$$

$$\sum_{k=1}^{\mathrm{K}} \tau^k \le c_{\max} . \tag{8.67}$$

8.9. Mathematical expressions of signal plan constraints

Mathematical expressions for constraints, discussed in this part, relate to control vectors assigned to particular signal groups. These expressions are valid for a complete set of signal groups. However, when starting with problem solution, the available data are related to traffic streams and relations between them. Therefore, it is necessary to show how the data related to traffic streams can be transformed into constraints on control variables for signal groups.

In formulating the constraints by which the set of feasible controls (signal plans) is defined, the following data are needed:

a) *The system data*

- The set of traffic streams:

  $$\mathscr{S} = \{\sigma_1, \sigma_2, \ldots, \sigma_{\mathrm{I}}\} .$$

- The vector of saturation flow volumes of vehicle traffic streams:

  $$s = (s_1, s_2, \ldots, s_i, \ldots, s_{\mathrm{I'}}) .$$

- The compatibility relation of traffic streams, C.
- The chosen, complete set of signal groups:

  $$\mathcal{D}_a = \{D'_1, D'_2, \ldots, D'_p, \ldots, D'_{\mathrm{P}}\} ,$$

  which are controlled by control variables:

  $$u_1(\cdot), u_2(\cdot), \ldots, u_p(\cdot), \ldots, u_{\mathrm{P}}(\cdot) .$$

- The set of vehicle signal groups:

  $$\mathcal{D}'_a = \{D'_1, D'_2, \ldots, D'_p, \ldots, D'_{P'}\}.$$

- The set of feasible phases, $\mathbf{U}_f$.
- The phase transition relation, R_s.
- The vector of minimal effective green times of signal groups:

  $$g_m = (g_{m1}, g_{m2}, \ldots, g_{mp}, \ldots, g_{mP}).$$

- The vector of maximal effective red times of certain vehicle signal groups:

  $$r_M = (r_{M1}, r_{M2}, \ldots, r_{Mp}, \ldots, r_{MP_R}),$$

  $$p \in \mathcal{P}^M = \{1, 2, \ldots, p, \ldots P_R\}.$$

- The matrix of minimal effective intergreen times for all pairs of signal groups:

  $$Z = [z_{pq}]_{P \times P}.$$

- The cycle duration, c, or the maximum cycle time, $c_{\max}$.
- Input–output matrix A, containing data about participation of arrival traffic stream volumes in output traffic flows.

b) *The data about volumes of vehicle traffic streams*

- The average values of vehicle traffic stream volumes, q_i, $(i \in \mathcal{J}')$.
- The maximum acceptable saturation degrees, ρ_i, for all vehicle traffic streams $(i \in \mathcal{J}')$.

Based on these data, the constraints that define the set of feasible controls (signal plans) are formulated as follows:

α. *The constraints related to control variables*

α.1 The constraint of one interval of green indication in a cycle:

$$\sum_{k=1}^{K} (u_p^k + u_p^{k(\mathrm{mod}\,K)+1})(\mathrm{mod}\,2) = 2, \quad (p \in \mathcal{P}).$$

α.2 The constraints of minimal effective green times for signal groups:

$$\sum_{k=1}^{K} u_p^k \sigma^k \geq g_{mp}, \quad (p \in \mathcal{P}).$$

α.3 The constraints of maximal effective red times for signal groups:

$$\sum_{k=1}^{K}(1-u_q^k)\tau^k \le r_{Mq}\,.$$

α.4 The flow balance (capacity) constraints

α.4.1 The flow balance constraint when filtering is not permitted:

$$\sum_{k=1}^{K} u_p^k \tau^k \ge \gamma_p\,, \quad (p \in \mathscr{P}')\text{, when } c \text{ is given,}$$

$$\sum_{k=1}^{K} u_p^k \tau^k \ge \sum_{k=1}^{K} \tau^k \max\left\{\frac{q_i}{\overline{\rho}_i\, s_i} \mid \sigma_i \in D'_p\right\} \quad , \quad (p \in \mathscr{P}')\,,$$

when c is not given.

α.4.2 The flow balance constraint when filtering is permitted:

I The flow balance constraint for the opposing stream σ_a (the streams σ_a and σ_b are at the same time signal groups; $D'_a = \{\sigma_a\}$, $D'_b = \{\sigma_b\}$):

$$\sum_{k=1}^{K} (s_a u_a^k (1-u_b^k) + q_a u_a^k u_b^k)\tau^k \ge c\, q_a\,.$$

II The condition of queue discharge of stream σ_a, before giving the simultaneous right-of-way to both signal groups, D'_a and D'_b:

$$\sum_{k=1}^{K} ((1-u_a^k u_b^k) q_a - u_a^k (1-u_b^k) s_a)\tau^k \ge 0\,.$$

III The flow balance constraint for the opposed stream, σ_b, that filters through stream σ_a:

$$\sum_{k=1}^{K} \left(u_a^k u_b^k \frac{q_a e^{-\alpha' q_a}}{1-e^{-\beta_b q_a}} + u_b^k (1-u_a^k) s_{bf} \right) \tau^k \ge q_b\, c\,.$$

β. *The constraint on control vectors composition*

Components $u_p(r)$ of any feasible phase $\mathbf{u}(r) \in \mathbf{U}_f$ have to satisfy the constraints given by the following expression:

$$u_p(r) = \begin{cases} 1, \text{ if } D'_p \in \overline{\mathfrak{D}}_a^r \\ 0, \text{ if } D'_p \notin \overline{\mathfrak{D}}_a^r \end{cases}, \quad (p \in \mathscr{P})\,,$$

where $\overline{\mathfrak{D}}_a^r$ is a clique of graph $G_g = (\mathfrak{D}_a, C_g)$.

γ. *The constraints related to phase sequences*

γ.1 Phase sequence constraints:

$$\mathbf{u}^{k(\text{mod K})+1} \in \Gamma_s \mathbf{u}^k, \quad (k \in \mathscr{K}).$$

γ.2 Minimal effective intergreen time constraints:

$$\sum_{l=1}^{l=\alpha-1} \tau^{\text{K}-(\text{K}+l-k)\,\text{mod K}} \geq z^f(\mathbf{u}^{\text{K}-(\text{K}+\alpha-k)(\text{mod K})}, \mathbf{u}^{k(\text{mod K})+1}), \quad (k \in \mathscr{K})$$

$$\alpha \in \mathscr{A}'(\mathbf{u}^{\text{K}-(\text{K}+\alpha-k)(\text{mod K})}, \mathbf{u}^{k(\text{mod K})+1}).$$

δ. *The constraint on the duration of interval in which the simultaneous right-of-way is given to signal groups having negative minimal effective intergreen times between them*

$$\sum_{k=1}^{\text{K}} (u_v^k u_m^k)\tau^k \leq |z_{vm}|, \ (z_{vm} < 0), (v, m \in \mathscr{P}).$$

ε. *The constraint on the sum of phase durations*

ε.1 $\sum_{k=1}^{\text{K}} \tau^k = c$.

ε.2 $\sum_{k=1}^{\text{K}} \tau^k \leq c_{\max}$.

The listed constraints relate to control variables, which control signal groups, and to control vectors (phases) in particular intervals during a cycle. Primary data used in formulating these constraints relate to traffic streams.

The way of obtaining constraints related to control variables, which control signal groups, was explained when the constraints were formulated. The summary of these expressions is given below.

- The minimal effective green time constraints for signal groups, α.2, are obtained from minimal effective green times of signal groups using (8.4) (Subsection 8.2.2):

  $$g_{mp} = \max\{g_i^m \mid \sigma_i \in D_p'\}, \quad (p \in \mathscr{P}),$$

  where:

  g_i^m – the minimal effective green time for traffic stream τ_i,

  g_{mp} – the minimal effective green time of signal group D_p'.

- Maximal effective red time constraints for certain signal groups, α.3, are obtained from maximal effective red times of traffic streams controlled by these signal groups using expression (8.8) (Subsection 8.2.3):

  $$r_{Mq} = \min\{r_j^M \mid \sigma_j \in D'_q\} \ , \quad (q \in \mathscr{P}^M) ,$$

  where:

  r_{Mq} – the maximal effective red time of signal group D'_q,

  r_j^M – the maximal effective red time for traffic stream τ_j,

  $\mathscr{P}^M$ – the index set of the set of signal groups for which it is necessary to introduce the maximal effective red time constraints.

- In flow balance constraints, α.4.1, the right side of the constraints, γ_p, is determined from traffic streams data using expression (8.14) (Subsection 8.2.4):

  $$\gamma_p = \max\left\{\frac{cq_i}{\bar{\rho}_i s_i} \mid (\sigma_i \in D'_p)\right\}, \quad (p \in \mathscr{P}') .$$

- The set of feasible control vectors (phases) is obtained starting from the compatibility relation, C_g, in the following way:

  a) In the case when no negative intergreen times exist (Section 8.3):

  $$\mathbf{U}'_f = \{\mathbf{u}(1), \mathbf{u}(2), \ldots, \mathbf{u}(r), \ldots, \mathbf{u}(\mathrm{R})\} \cup \{(0,0,\ldots,0)\} ,$$

  where:

  $$\mathbf{u}(r) = [u_1(r), u_2(r), \ldots, u_p(r), \ldots, u_{\mathrm{P}}(r)]^T \text{, and}$$

  $$u_p(r) = \begin{cases} 1, \text{ if } D'_p \in \bar{\mathfrak{D}}_a^r \\ 0, \text{ if } D'_p \notin \bar{\mathfrak{D}}_a^r \end{cases}, \quad (p \in \mathscr{P}, r \in \mathscr{R}) ,$$

  $$\bar{\mathfrak{D}}_a^r \in \bar{\mathfrak{D}}_a ,$$

  $$\bar{\mathfrak{D}}_a = \{\bar{\mathfrak{D}}_a^r \mid (D'_p \in \bar{\mathfrak{D}}_a^r) \wedge (D'_q \in \bar{\mathfrak{D}}_a^r) \Rightarrow D'_q \in \Gamma_g D'_p \ ; \ p,q \in \mathscr{P}\} .$$

  The signal group compatibility graph

  $$G_g = (\mathfrak{D}_a, \Gamma_g) = (\mathfrak{D}_a, C_g)$$

  is obtained using compatibility relation C in the set of traffic streams.

  Relation C_g can be determined by the following expression:

  $$C_g = \{(D'_p, D'_q) \mid (\sigma_i \in D'_p) \wedge (\sigma_j \in D'_q) \Rightarrow (\sigma_i, \sigma_j) \in C\} .$$

b) In the case when there exist negative intergreen times (Subsection 8.4.3):

- The extended set of feasible phases:

$$\mathbf{U}''_f = \mathbf{U}_f \cup \mathbf{U}'_f ,$$

where:

$$\mathbf{U}'_f = \{\mathbf{u}(s) \mid \mathbf{u}(s) = [u_1(s), u_2(s), \ldots, u_p(s), \ldots, u_q(s), \ldots, u_{\mathrm{P}}(s)]^T ,$$
$$u_p(s) = u_q(s) = 1,\ z_{pq} < 0,$$
$$s = \overline{R} + 2, \ldots, \overline{R} + \mathrm{N}, (D'_p, D'_q) \in C'_g \},$$

$$C'_g = \{(D'_p, D'_q) \mid z_{pq} < 0, \quad D'_p, D'_q \in \mathcal{D}_a \} ,$$

$$G'_g = (D'_a, C'_g) ,$$

$$G''_g = G_g \cup G'_g = (\mathcal{D}_a, C_g) \cup (\mathcal{D}_a, C'_g) =$$
$$= (\mathcal{D}_a \cup \mathcal{D}_a, C_g \cup C'_g) = (\mathcal{D}_a, C''_g) = (\mathcal{D}_a, \Gamma''_g) .$$

- When using expression (8.61) for determination of

$$z^f(\mathbf{u}^{\mathrm{K}-(\mathrm{K}+\alpha-k)(\mathrm{mod\,K})}, \mathbf{u}^{k(\mathrm{mod\,K})+1}),$$

it is assumed that $z_{pq} = 0$ if

$$((D'_p, D'_q) \in C_g) \vee (p = q) \vee (z_{pq} < 0) .$$

The values z_{pq} in expressions (8.61) and (8.63) are minimal effective intergreen times, defined for each pair of control variables, $(u_p(\cdot), u_q(\cdot))$, controlling signal groups D'_p and D'_q. These values are obtained from z'_{ij}, defined for each pair of traffic streams, (σ_i, σ_j), using the following expression:

$$z_{pq} = \begin{cases} \max\{z'_{ij} \mid \sigma_i \in D'_p, \sigma_j \in D'_q, (D'_p, D'_q) \in (\mathcal{D}_a \times \mathcal{D}_a) \setminus C_g\} \\ 0, \quad (D'_p, D'_q) \in C_g . \end{cases}$$

Formulation of m.e.i.t. constraints starts with determination of the matrix

$$Z' = [z'_{ij}]_{\mathrm{I} \times \mathrm{I}} ,$$

whose elements are the minimal effective intergreen times between traffic streams. The matrix of m.e.i.t. between signal groups,

$$Z = [z_{pq}]_{\mathrm{P} \times \mathrm{P}} ,$$

is obtained using data from Z'. Finally, the elements from Z are used for determining m.e.i.t. between phases, i.e., for determining function Z^f,

$$Z^f: \quad \mathbf{U}_f \times \mathbf{U}_f \to \mathrm{R}.$$

- The constraints of minimal effective intergreen times, γ.2, are defined by the expression:

$$\sum_{l=0}^{l=\alpha} \tau^{\mathrm{K}-(\mathrm{K}+l-k)(\mathrm{mod\,K})} \geq \max\, \{z_{pq} \mid$$

$$(u_p^{\mathrm{K}-(\mathrm{K}+\alpha-k)(\mathrm{mod\,K})}\, u_q^{k(\mathrm{mod\,K})+1} \cdot \overline{u}_p^{\mathrm{K}-(\mathrm{K}+\alpha-k-1)(\mathrm{mod\,K})} \cdot \overline{u}_q^{k}) = 1,$$

$$\mathbf{u}^{k(\mathrm{mod\,K})+1} \notin \Gamma_s(\mathbf{u}^{(\mathrm{K}+\alpha-k)(\mathrm{mod\,K})}), \quad p,q \in \mathscr{P}\}$$

$$(k \in \mathscr{K}) \text{ and } (\alpha \in \mathscr{A}'(\mathbf{u}^k, \mathbf{u}^{k(\mathrm{mod K})+1})),$$

where α is the number of phases participating in the formulation of constraints γ.2. One constraint is formulated for each $\alpha \in \mathscr{A}'(\mathbf{u}^k, \mathbf{u}^{k(\mathrm{modK})+1})$, where:

$$\mathscr{A}'(\mathbf{u}^k, \mathbf{u}^{k(\mathrm{modK})+1}) = \{1, 2, \ldots, \alpha, \ldots, \alpha_{\mathrm{M}}(\mathbf{u}^k, \mathbf{u}^{k(\mathrm{modK})+1})\}, \text{ and}$$

$$\alpha_{\mathrm{M}}(\mathbf{u}^k, \mathbf{u}^{k(\mathrm{mod\,K})+1}) = \min\{\max\{z_{qr} \mid (u_r^{k(\mathrm{mod\,K})+1} = 1) \wedge (u_r^k = 0)$$
$$q, r \in \mathscr{P}\}, \mathrm{K}-2\}.$$

In expression (8.61), $z^f(\mathbf{u}^{\mathrm{K}-(\mathrm{K}+\alpha-k)(\mathrm{modK})}, \mathbf{u}^{k(\mathrm{modK})+1}) \geq 0$.

Example 8.9

Formulate all constraints related to phase durations for the intersection presented in Fig. 8.10. The phase transition graph is given in Fig. 8.17. The signal plan structure is given, and every traffic stream makes one signal group.

The signal plan structure is:

$$\mathbf{u} = \begin{bmatrix} \mathbf{u}_1 \\ \mathbf{u}_2 \\ \mathbf{u}_3 \\ \mathbf{u}_4 \\ \mathbf{u}_5 \\ \mathbf{u}_6 \end{bmatrix} = \overset{\begin{matrix}\mathbf{u}^1 & \mathbf{u}^2 & \mathbf{u}^3 & \mathbf{u}^4 & \mathbf{u}^5 & \mathbf{u}^6 & \mathbf{u}^7 & \mathbf{u}^8 & \mathbf{u}^9 & \mathbf{u}^{10}\end{matrix}}{\begin{bmatrix} 1 & 0 & 0 & 0 & 0 & 0 & 0 & 0 & 1 & 1 \\ 0 & 0 & 0 & 0 & 0 & 0 & 1 & 1 & 1 & 0 \\ 1 & 1 & 0 & 0 & 0 & 0 & 0 & 0 & 0 & 0 \\ 0 & 0 & 0 & 1 & 1 & 0 & 0 & 0 & 0 & 0 \\ 0 & 0 & 0 & 0 & 1 & 1 & 1 & 1 & 1 & 0 \\ 0 & 0 & 0 & 0 & 0 & 0 & 1 & 0 & 0 & 0 \end{bmatrix}}.$$

This structure is marked by a bold line on graph G_s, presented in Fig. 8.23. The structure satisfies constraints $\alpha.1$ and $\alpha.2$. Feasible phases are represented by nodes of graph G_s.

The values of saturation flow volumes, average volumes of vehicle traffic streams, minimal effective green times, and maximal effective red times are given in Table 8.1.

Table 8.1

$i = p$	**1**	**2**	**3**	**4**	**5**	**6**
s_i[PCU/h]	1850	1650	1620	1650	1600	/
g_{mp}[s]	20	15	15	15	15	16
r_{Mp}[s]	/	/	/	60	75	/
q_i[PCU/h]	185	165	324	165	160	/

The matrix of minimal effective intergreen times is:

$$Z = \begin{bmatrix} 0 & 0 & 0 & 4 & 0 & 4 \\ 0 & 0 & 3 & 5 & 0 & 0 \\ 0 & 3 & 0 & 3 & 5 & 2 \\ 2 & 1 & 2 & 0 & 0 & 2 \\ 0 & 0 & 1 & 0 & 0 & 0 \\ 8 & 0 & 4 & 8 & 0 & 0 \end{bmatrix}.$$

The cycle time is $c = 90$ s.

The acceptable saturation degree is the same for all vehicle signal groups, $\overline{\rho}_i = 0.9$ $(i = 1,\ldots,5)$.

The time constraints for this structure are:

I – The constraints of minimal effective green times, α.2:

1) $\tau^1 + \tau^9 + \tau^{10} \geq 20$

2) $\tau^7 + \tau^8 + \tau^9 \geq 15$

3) $\tau^1 + \tau^2 \geq 15$

4) $\tau^4 + \tau^5 \geq 15$

5) $\tau^5 + \tau^6 + \tau^7 + \tau^8 + \tau^9 \geq 15$

6) $\tau^7 \geq 16$

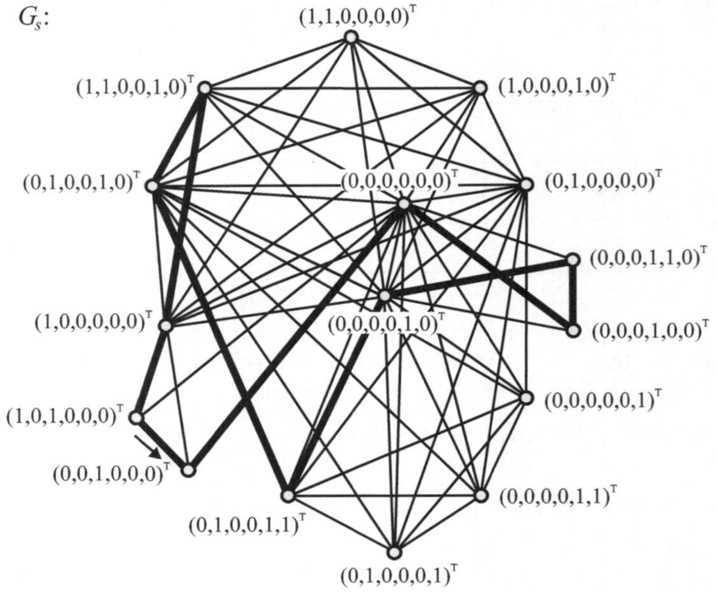

Figure 8.23

II – The constraints of maximal effective red times, α.3:

$$7)\quad \tau^1+\tau^2+\tau^3+\tau^6+\tau^7+\tau^8+\tau^9+\tau^{10}\le 60$$

$$8)\quad \tau^1+\tau^2+\tau^3+\tau^4+\tau^{10}\le 75$$

III – The flow balance constraints, α.4:

$$9)\quad \tau^1+\tau^9+\tau^{10}\ge \frac{c\cdot q_1}{\bar{\rho}_1\cdot s_1}=10$$

$$10)\quad \tau^7+\tau^8+\tau^9\ge 10$$

$$11)\quad \tau^1+\tau^2\ge 20$$

$$12)\quad \tau^4+\tau^5\ge 10$$

$$13)\quad \tau^5+\tau^6+\tau^7+\tau^8+\tau^9\ge 10$$

IV – The constraints of minimal effective intergreen times, γ.2:

These constraints exist only if the conditions given in expression (8.61) are fulfilled.

Using this expression, the constraints are formulated for each value of k $(k = 1,2,\ldots,10)$ as follows:

$k = 1$

$u_3^2 = 1$, but $u_3^1 = 1$, meaning that in phase $\mathbf{u}^2$ no signal group starts with its right-of-way.

$k = 2$

$u_1^3 = u_2^3 = \ldots = u_6^3 = 0$, meaning that in phase $\mathbf{u}^3$ no signal group starts with its right-of-way.

$k = 3$

$u_4^4 = 1;\ u_4^3 = 0$

$\alpha_M = 8$

$\alpha = 1 \quad \mathbf{u}^4 \notin \Gamma_s \mathbf{u}^2;\ u_3^2 = 1,\ u_3^3 = 0$

$\tau^3 \geq 3$

$\alpha = 2 \quad \mathbf{u}^4 \notin \Gamma_s \mathbf{u}^1;\ u_1^1 = 1,\ u_1^2 = 0$

$\tau^2 + \tau^3 \geq 4$

$\alpha = 3 \quad \mathbf{u}^4 \notin \Gamma_s \mathbf{u}^{10}$, but $u_1^{10} = 1,\ u_1^1 = 1$

$\alpha = 4 \quad \mathbf{u}^4 \notin \Gamma_s \mathbf{u}^9;\ u_2^9 = u_5^9 = 1,\ u_2^{10} = u_5^{10} = 0$

$\tau^1 + \tau^2 + \tau^3 + \tau^{10} \geq 5$

$\alpha = 5 \quad \mathbf{u}^4 \notin \Gamma_s \mathbf{u}^8;\ u_1^8 = u_5^8 = 1$, but $u_1^9 = u_5^9 = 1$

$\alpha = 6 \quad \mathbf{u}^4 \notin \Gamma_s \mathbf{u}^7;\ u_6^7 = 1,\ u_6^8 = 0$

$\tau^1 + \tau^2 + \tau^3 + \tau^8 + \tau^9 + \tau^{10} \geq 8$

$\alpha = 7 \quad \mathbf{u}^4 \in \Gamma_s \mathbf{u}^6$

$\alpha = 8 \quad \mathbf{u}^4 \in \Gamma_s \mathbf{u}^5$

$k = 4$

$u_5^5 = 1;\ u_5^4 = 0$

$\alpha_M = 5$

$\alpha = 1 \quad \mathbf{u}^5 \in \Gamma_s \mathbf{u}^3$

$\alpha = 2 \quad \mathbf{u}^5 \notin \Gamma_s \mathbf{u}^2;\ u_3^2 = 1,\ u_3^3 = 0$

$\tau^3 + \tau^4 \geq 5$

$\alpha = 3 \quad \mathbf{u}^5 \notin \Gamma_s \mathbf{u}^1;\ u_1^1 = 1,\ u_1^2 = 0$

$\tau^2 + \tau^3 + \tau^4 \geq 0$

$\alpha = 4 \quad \mathbf{u}^5 \notin \Gamma_s \mathbf{u}^{10};\ u_1^{10} = 1,\ u_1^1 = 1$

$\alpha = 5 \quad \mathbf{u}^5 \notin \Gamma_s \mathbf{u}^9;\ u_2^9 = u_5^9 = 1,\ u_2^{10} = u_5^{10} = 0$

$\tau^1 + \tau^2 + \tau^3 + \tau^4 + \tau^{10} \geq 0$

$k = 5$

$u_5^6 = 1$, but $u_5^5 = 1$

$k = 6$

$u_2^7 = u_6^7 = 1;\ u_2^6 = u_6^6 = 0$

$\alpha_M = 4$

$\alpha = 1 \quad \mathbf{u}^7 \notin \Gamma_s \mathbf{u}^5;\ u_4^5 = 1,\ u_4^6 = 0$

$\tau^6 \geq 2$

$\alpha = 2 \quad \mathbf{u}^7 \notin \Gamma_s \mathbf{u}^4;\ u_4^4 = 1$, but $u_4^5 = 1$

$\alpha = 3 \quad \mathbf{u}^7 \in \Gamma_s \mathbf{u}^3$

$\alpha = 4 \quad \mathbf{u}^7 \notin \Gamma_s \mathbf{u}^2;\ u_3^2 = 1,\ u_3^3 = 0$

$\tau^3 + \tau^4 + \tau^5 + \tau^6 \geq 3$

$k = 7$

$u_1^8 = u_5^8 = 1$ and $u_1^7 = u_5^7 = 1$

$k = 8$

$u_1^9 = 1;\ u_1^8 = 0$

$\alpha_M = 8$

$\alpha = 1 \quad \mathbf{u}^9 \notin \Gamma_s \mathbf{u}^7;\ u_6^7 = 1,\ u_6^8 = 0$

$\tau^8 \geq 8$

$\alpha = 2 \quad \mathbf{u}^9 \in \Gamma_s \mathbf{u}^6$

$\alpha = 3 \quad \mathbf{u}^9 \notin \Gamma_s \mathbf{u}^5;\ u_4^5 = 1,\ u_4^6 = 0$

$\tau^6 + \tau^7 + \tau^8 \geq 2$

$\alpha = 4 \quad \mathbf{u}^9 \notin \Gamma_s \mathbf{u}^4;\ u_4^4 = 1,\ u_4^5 = 1$

$\alpha = 5 \quad \mathbf{u}^9 \in \Gamma_s \mathbf{u}^3$

$\alpha = 6 \quad \mathbf{u}^9 \notin \Gamma_s \mathbf{u}^2 ;\ u_3^2 = 1,\ u_3^3 = 0$

$\tau^3 + \tau^4 + \tau^5 + \tau^6 + \tau^7 + \tau^8 \geq 0$

$\alpha = 7 \quad \mathbf{u}^9 \notin \Gamma_s \mathbf{u}^1 ;\ u_1^1 = 1.\ u_1^2 = 0$

$\alpha = 8 \quad \mathbf{u}^9 \in \Gamma_s \mathbf{u}^{10}$

$k = 9$

$u_1^{10} = 1$, but $u_1^9 \neq 0$ (the phase $\mathbf{u}^{10}$ doesn't begin to give the right-of-way to any signal group).

$k = 10$

$u_3^1 = 1 ;\ u_3^{10} = 0$

$\alpha_M = 4$

$\alpha = 1 \quad \mathbf{u}^1 \notin \Gamma_s \mathbf{u}^9 ;\ u_2^9 = u_5^9 = 1,\ u_2^{10} = u_5^{10} = 0$

$\tau^{10} \geq 3$

$\alpha = 2 \quad \mathbf{u}^1 \notin \Gamma_s \mathbf{u}^8 ;\ u_2^8 = u_5^8 = 1,$ but $u_2^9 = u_5^9 = 1$

meaning that phase $\mathbf{u}^8$ doesn't stop giving the right-of-way to any signal group

$\alpha = 3 \quad \mathbf{u}^1 \notin \Gamma_s \mathbf{u}^7 ;\ u_6^7 = 1,\ u_6^8 = 0$

$\tau^8 + \tau^9 + \tau^{10} \geq 4$

$\alpha = 4 \quad \mathbf{u}^1 \notin \Gamma_s \mathbf{u}^6 ;\ u_5^6 = 1.\ u_5^7 = 1$

Therefore, all constraints of minimal effective intergreen times, obtained using (8.61), are:

14) $\tau^3 \geq 3$

15) $\tau^2 + \tau^3 \geq 4$

16) $\tau^1 + \tau^2 + \tau^3 + \tau^{10} \geq 5$

17) $\tau^1 + \tau^2 + \tau^3 + \tau^8 + \tau^9 + \tau^{10} \geq 8$

18) $\tau^3 + \tau^4 \geq 5$

19) $\tau^2 + \tau^3 + \tau^4 \geq 0$

20) $\tau^1 + \tau^2 + \tau^3 + \tau^4 + \tau^{10} \geq 0$

21) $\tau^6 \geq 2$

22) $\tau^3 + \tau^4 + \tau^5 + \tau^6 \geq 5$

23) $\tau^8 \geq 8$

24) $\tau^6 + \tau^7 + \tau^8 \geq 2$

25) $\tau^3 + \tau^4 + \tau^5 + \tau^6 + \tau^7 + \tau^8 \geq 0$

26) $\tau^{10} \geq 3$

27) $\tau^8 + \tau^9 + \tau^{10} \geq 4$

V – The constraint on the sum of phase durations:

28) $\sum_{k=1}^{10} \tau^k = 90$

By observing constraints I, II, III, IV, and V, the following facts can be found:

- The constraints 19, 20, and 25 are needless because $\tau^k > 0, (k \in \mathscr{K})$;
- The constraints 2, 5, and 24 are needless because the constraint 6 exists;
- The constraints 16 and 17 are needless because the constraint 3 exists;
- The constraint 27 is needless because the constraint 23 exists;
- The constraint 22 is needless because the constraint 4 exists;
- The constraint 9 is needless because the constraint 1 exists;
- The constraint 10 is needless because the constraint 2 exists;
- The constraint 3 is needless because the constraint 11 exists;
- The constraint 12 is needless because the constraint 4 exists;
- The constraint 13 is needless because the constraint 5 exists.

Accordingly, phase durations have to satisfy the following constraints:

$$
\begin{array}{l}
\tau^1 + \tau^2 + \tau^3 + \tau^4 + \tau^5 + \tau^6 + \tau^7 + \tau^8 + \tau^9 + \tau^{10} = 90 \\
\tau^1 + \tau^2 + \tau^3 + \tau^4 + \tau^{10} \leq 75 \\
\tau^1 + \tau^2 + \tau^3 + \tau^6 + \tau^7 + \tau^8 + \tau^9 + \tau^{10} \leq 60 \\
\tau^1 + \tau^9 + \tau^{10} \geq 20 \\
\tau^1 + \tau^2 \geq 20 \\
\tau^2 + \tau^3 \geq 4 \\
\tau^3 \geq 3 \\
\tau^3 + \tau^4 \geq 5 \\
\tau^4 + \tau^5 \geq 15 \\
\tau^6 \geq 2 \\
\tau^7 \geq 16 \\
\tau^8 \geq 8 \\
\tau^{10} \geq 3
\end{array}
$$

$\tau^k \geq 0, \quad (k \in \mathscr{K})$

9. OPTIMIZATION CRITERIA

The goal of traffic control on a signalized intersection is to enable safe passage of vehicles and pedestrians through the intersection. To achieve this goal, it is necessary to prevent the situation that two vehicles (or a vehicle and a pedestrian) belonging to mutually conflicting traffic streams meet at the same conflict point.

This goal can be realized in various ways, i.e., by more than one control (signal plan). The signal plans enabling achievement of this goal are all feasible signal plans, i.e., elements of set $\mathscr{U}_f$.

9.1. The general form and features of optimization criteria

For an exact formulation of the signal plan choice problem, it is necessary to define the pair $(\mathscr{U}_f, J_c)$, i.e., besides precise determination of the set of feasible controls, $\mathscr{U}_f$, it is necessary to choose the performance index, J_c, i.e., the optimality criterion.

Optimization criteria that can be used for comparing quality of various signal plans have to satisfy several requirements. The most important requirements are that:

- By choosing the control that optimizes the selected criterion, traffic conditions on the intersection improve, e.g., the delay and number of stops reduce, or the capacity increases, etc.
- The criteria are explicit functions of control and state of the traffic process.
- The criteria are measurable.
- The criteria are related to the intersection as a whole, in a given time interval.

There are many factors significant for assessment of the traffic process quality on a signalized intersection. The most important factors are: the total delay, or the waiting time of vehicles, the number of vehicle stops, the total exploitation costs, the fuel consumption, the environmental influence, e.g., the level of pollutant emissions and noise. The number of vehicles that can pass through the intersection per time unit, i.e., the capacity for one approach and the whole intersection, becomes a very important performance index, especially nowadays, with constant increase in number of cars in cities. The maximal or average queue on certain approaches, or on all of them, can also be the performance index of traffic control process on a signalized intersection.

Obviously, there are many optimization criteria suitable for mutual comparison of signal plans, and some of these criteria are very significant.

On the basis of the mentioned criteria features and expression (7.2):

$$J_c'' : \mathcal{Q}_{[t_0,t]} \times \mathcal{U}_{[t_0,t]} \times \mathcal{W}_{[t_0,t]} \to R,$$

it can be concluded that for determining optimization criterion values it is necessary to know the vector of arrival flow volumes, the signal plan, and queues during a cycle. Since in one interval the arrival flow volumes are stationary processes, the queues on particular approaches, i.e., the components of vector $w_{[t_0,t]}$, can be determined according to the expression:

$$\varphi^3_{[t_0,t]} : \mathcal{W}_{t_0} \times (\mathcal{Q}_{[t_0,t]} \times \mathcal{U}_{[t_0,t]}) \to \mathcal{W}_{[t_0,t]},$$

on the basis of the initial state, $w(t_0)$, the vector of arrival flow volumes, $q_{[t_0,t]}$, and the signal plan, $u(\cdot)$. Therefore, the value of the optimization criterion depends only on the choice of the control, i.e.,

$$J_c : \mathcal{U}_f \to \mathrm{R},$$

as defined by expression (7.4).

Different information on arrival flows, contained in vector $q_{[t_0,t]}$, is necessary for determination of values of different optimization criteria. For example, for some criteria only the arrival flow volumes are necessary, while for others it is necessary to have information on components of vector $q_{[t_0,t]}$, such as the dispersion, the dispersion index, etc.

9.2. Types of optimization criteria

Optimization criteria can be classified in several ways. The Canadian Capacity Guide for Signalized Intersections [82], for example, lists 17 basic criteria classified in three groups:

- Criteria related to capacity
- Criteria related to queuing (average overall delay, the number of stops, queue lengths, etc.)
- Operational and environmental criteria (fuel consumption, emission of pollutants, such as CO, CO_2, NO_x, RCH, etc.)

By optimizing some of these criteria, negative consequences of the traffic process will be decreased (delay, number of stops, fuel consumption, etc.), or some characteristics will increase (e.g., capacity). It would be desirable to choose the control that optimizes all of the mentioned criteria, but this is not possible. Because of that, various signal plan choice methods are used, including more than one criterion. The following methods are most frequently used:

a) The statement of the signal plan choice problem as a multicriteria optimization problem.
b) Application of different criteria for signal plan choice for different traffic process conditions [66]. Thus, in the case of low volumes, the total number of stops is used as the optimization criterion, for medium and large volumes—the total delay, and for very large volumes—the capacity of an approach or the whole intersection. In the case of congestion, besides delay, the duration of congestion or queue lengths is used as the optimization criteria [55], [68].
c) Various *synthetic* criteria, composed of several criteria.
   c.1) The most frequently used criterion of this type is the weighted sum of the total delay and the number of stops at an intersection:

$$J_c = k_0 V + k_1 H = k_0 \left(V + \frac{k_1}{k_0} H \right) = k_0 (V + k'H) = k_0 J', \tag{9.1}$$

where V is the total delay, H is the total number of stops in a given interval (usually equal to the cycle time), and k_0 and k_1 are the weighting coefficients. By reducing both elements of the sum in (9.1), some other undesirable effects are decreased, such as the total emission of pollutants, noise, fuel consumption, the total costs related to the traffic process, etc.

The pollutant emission is closely related to the delay. The amount of pollutant when vehicles wait is greater than during motion (e.g., the amount of carbon monoxide (CO) is three to four times greater).

The amount of pollutants increases with the increase in number of stops. Each vehicle stop is followed by start of the movement and acceleration, which lead to increased fuel consumption, and thus an increased emission of pollutants. The noise level also depends on the number of stops, i.e., the deceleration/acceleration connected with each stop.

The fuel consumption is related to both elements of criterion J_c (9.1). Vehicles waiting on intersection approaches, during red signal indication, cause an in-vain fuel consumption. Each stop leads to an "additional" fuel consumption due to vehicle acceleration.

The total costs of the traffic process at a signalized intersection comprise the costs related to vehicles and the costs related to "lost" time of passengers. Vehicle operational costs include the fuel and lubricants, tires and brakes wear and tear, and amortization costs. Passenger costs depend on the lost time value. For estimation of the lost time value it is necessary to take into consideration vehicle occupancy rates, the percentage of passengers traveling during working and nonworking time, national gross product per capita in the city, etc. [78].

The minimization of criterion J_c is achieved by minimizing criterion J' because k_0 is a constant. The value of k' (stop penalty) depends on the optimization problem. Thus, in the problem of fuel consumption minimization, the typical values of k' are in the range 30–60. If the total cost is the optimization criterion, the typical values of stop penalty lie in the range 10–30 [2].

c.2) Another optimization criterion [59] consists of a weighted sum of the total delay and queue lengths, i.e.,

$$J_c = k_0 V + \sum_{i=1}^{I''} k_i w_i \,, \tag{9.2}$$

where:

- k_i – the weighting coefficients, $k_i \geq 0 \ \ (i = 1,\ldots,I'')$,
- I'' – the number of selected or all vehicle stream approaches,
- V – the total delay of vehicles on the intersection,
- w_i – the length of average or maximal vehicle queues.

The control strategy can be changed by choosing appropriate weighting coefficient values. Thus, only delay can be minimized by choosing $k_0 = 1$ and $k_i = 0$ ($i = 1,\ldots,I''$), or only the sum of vehicle queues if $k_0 = 0$ and $k_i = 1$ ($i = 1,\ldots,I''$).

In some cases, it is suitable to make the queue weighting coefficients conversely proportional to the area on which the queues can form. The consideration of queues is important when flow volumes are high. The reduction of queues is very important in congested conditions.

c.3) Optimization criteria composed of vehicles delay, V_v, and pedestrian time losses, V_p, can be expressed as follows:

$$J_c = |\, k_1 V_p - k_2 V_v \,|. \tag{9.3}$$

This criterion leads to a "fair" use of the intersection, by both vehicles and pedestrians.

Although the minimization of *synthetic* criteria leads to an improvement of traffic control quality in respect to several criteria, this approach is not frequently used in practice. Determination of signal plans is still, most often, set and solved as a single-criterion optimization problem. However, even in this case, the optimization of certain criteria has multiple effects. One of such criteria is the total delay at an intersection, i.e., the sum of queue integrals on all vehicle stream approaches. Minimization of this criterion certainly reduces the delay but also reduces the amount of pollutant emissions, total fuel consumption, the vehicle costs, etc. [70].

In some cases it is necessary to determine the limit values of some signal plan elements, such as the cycle time, green time for a certain signal group, the number of signal plan phases, etc. These problems can be set as optimization problems in which the optimization criteria are the mentioned signal plan elements.

The most frequently used criteria of this type are:

a) Cycle time [64], [65], [85], [45]

The cycle time is often one of the constraints in the problems of optimal signal plan choice. If the cycle time is adopted as the optimization criterion, then the problems of its minimization and maximization can be solved. The obtained minimal and maximal cycle time values can be used as constraints in other problems.

When solving problems of network optimal control, it is necessary that cycle times on all intersections in the network have the same value. The first step in determining this value is determination of minimal cycle times for all intersections. After that, the common cycle time is selected as the maximal value of minimal cycle times.

A maximal cycle time value exists in cases when the optimization problem includes maximal red time constraints. The common cycle time for all intersections in the network cannot be greater than the minimal value of maximal cycle times for individual intersections.

Therefore, the optimization problems in which the cycle time is used as the optimization criterion have sense as minimization as well as maximization problems.

b) The sum of green times of all signal groups [22], [18], [75]

Optimization of this criterion indirectly influences intersection performance. The maximization of available green time, by all means, improves traffic conditions on the intersection, but in solving this problem no difference is made between traffic streams with low and high volumes. The minimization of this criterion is merely of a historical significance because it is one of the first traffic control problems set and solved as a problem of mathematical programming [22].

c) The green time of a signal group

The green time of a certain signal group can be used as the optimization criterion instead of the sum of green times for all signal groups. This criterion makes sense when one intersection approach is critical, i.e., having remarkably higher arrival flow volume (and longer queues) than other approaches.

d) The number of control vectors (phases) in a signal plan

Traffic control specialists, using "phase-oriented" controllers, had the intention to minimize the number of phases in a signal plan, believing that the lost time would reduce if the number of phases decreased [2]. This is true only in the case when the right-of-way to particular signal groups is given only by one phase, as is the case with phase-oriented controllers. Generally, this is not valid in traffic control systems with modern "signal–group-oriented controllers."

Optimization criteria can be also classified according to the type of mathematical programming problem that can represent the signal plan choice problem. Some optimization criteria, for a given signal plan structure, become linear functions of variables τ^k, ($k = 1,\ldots,\mathrm{K}$), while other criteria become nonlinear functions of these variables. The corresponding optimization problems, therefore, result in linear or nonlinear mathematical programming problems.

The number of optimization criteria is considerable, but only a few of them are used in practice. Most frequently used are the criteria related to capacity, total vehicle delay, and cycle time [45]. The exact expressions for these criteria are presented in further exposition.

9.3. Optimization criteria related to capacity

The condition that all vehicles of a traffic stream, arriving at an intersection approach, can leave the intersection during the same cycle is satisfied if:

$$q_i \le \frac{s_i g_p}{c}, \text{i.e., } q_i c \le s_i g_p, \quad (i \in \mathscr{I}', \sigma_i \in D'_p, p \in \mathscr{P}'), \tag{9.4}$$

where:

q_i – the average volume of traffic stream σ_i,

s_i – the saturation flow volume of traffic stream σ_i,

g_p – the effective green time of signal group D'_p to which σ_i belongs,
c – the cycle time.

9.3.1. Theoretical capacity

The maximal volume of a traffic stream that can pass through the intersection is given by the following expression:

$$\omega_i^s = \frac{s_i g_p}{c}, \quad (i \in \mathcal{J}', \sigma_i \in D'_p, p \in \mathcal{P}').$$

Volume ω_i^s is the capacity of traffic stream σ_i.
According to expression (4.19),

$$g_p = \sum_{k=1}^{K} u_p^k \tau^k, \quad (\sigma_i \in D'_p, p \in \mathcal{P}'),$$

so that

$$J_c = \omega_i^s = \frac{s_i}{c} \sum_{k=1}^{K} u_p^k \tau^k, \quad (i \in \mathcal{J}', \sigma_i \in D'_i, p \in \mathcal{P}'). \tag{9.5}$$

Since all traffic streams belonging to one signal group are controlled by the same signal indications, the *signal group capacity* can be defined by the following expression:

$$\omega_p^g = \sum_{e=1}^{E(p)} \frac{s_{pe} g_p}{c} = \frac{g_p}{c} \sum_{e=1}^{E(p)} s_{pe} = \frac{g_p s_p^g}{c}, \quad (p \in \mathcal{P}'),$$

where:

$$s_p^g = \sum_{e=1}^{E(p)} s_{pe}, \quad (p \in \mathcal{P}').$$

Using expression (4.19) the expression for signal group capacity becomes:

$$J_c = \omega_p^g = \frac{s_p^g}{c} \sum_{k=1}^{K} u_p^k \tau^k, \quad (p \in \mathcal{P}'). \tag{9.6}$$

The *intersection capacity* can be defined as the sum of the signal group capacities:

$$\Omega^g = \sum_{p=1}^{P'} \omega_p^g = \frac{1}{c} \sum_{p=1}^{P'} g_p \sum_{e=1}^{E(p)} s_{pe} = \frac{1}{c} \sum_{p=1}^{P'} g_p s_p^g,$$

or, using expression (4.19),

$$J_c = \Omega^g = \frac{1}{c}\sum_{p=1}^{P'}(s_p^g \sum_{k=1}^{K} u_p^k \tau^k). \tag{9.7}$$

In the case when each vehicle traffic stream makes a signal group, the intersection capacity will be:

$$\Omega^s = \sum_{i=1}^{I'} \omega_i^s = \sum_{i=1}^{I'} \frac{s_i g_i}{c} = \frac{1}{c}\sum_{i=1}^{I'} s_i g_i ,$$

because in this case $\mathscr{I}' = \mathscr{P}'$.

The following expression also holds:

$$J_c = \Omega^s = \frac{1}{c}\sum_{i=1}^{I'}(s_i \sum_{k=1}^{K} u_i^k \tau^k). \tag{9.8}$$

9.3.2. Practical capacity

The expressions for ω_i^s, ω_p^g, Ω^s, Ω^g are theoretical expressions, whilst somewhat lower values are used in practice. Namely, if arrival volumes assumed the theoretical capacity values, very long queues would form, and because of that, the *practical capacity* is defined [88] as follows:

- For traffic stream σ_i:

$$J_c = \tilde{\omega}_i^s = \frac{\overline{\rho}_i s_i g_p}{c}, \quad (i \in \mathscr{I}', \sigma_i \in D'_p, p \in \mathscr{P}'), \tag{9.9}$$

where $\overline{\rho}_i < 1$, $(i \in \mathscr{I}')$

- For signal group D'_p:

$$J_c = \tilde{\omega}_p^g = \sum_{e=1}^{E(p)} \frac{\overline{\rho}_{pe} s_{pe} g_p}{c} = \frac{g_p}{c}\sum_{e=1}^{E(p)} \overline{\rho}_{pe} s_{pe} = \frac{g_p \tilde{s}_p^g}{c}, \quad (p \in \mathscr{P}'), \tag{9.10}$$

where

$$\tilde{s}_p^g = \sum_{e=1}^{E(p)} \overline{\rho}_{pe} s_{pe}, \quad (p \in \mathscr{P}')$$

- For intersection:

$$J_c = \tilde{\Omega}^g = \sum_{p=1}^{P'} \tilde{\omega}_p^g = \frac{1}{c}\sum_{p=1}^{P'} g_p \sum_{e=1}^{E(p)} \overline{\rho}_{pe} s_{pe} = \frac{1}{c}\sum_{p=1}^{P'} g_p \tilde{s}_p^g \tag{9.11}$$

If $\mathscr{T}' = \mathscr{P}'$, the practical intersection capacity, Ω^s, can be defined as follows:

$$J_c = \widetilde{\Omega}^s = \sum_{i=1}^{I'} \widetilde{\omega}_i^s = \frac{1}{c} \sum_{i=1}^{I'} \overline{\rho}_i s_i g_i \,. \tag{9.12}$$

In these expressions, $\overline{\rho}_i$ is the maximally acceptable saturation degree of traffic stream σ_i. The value of $\overline{\rho}_i$ is usually estimated by traffic engineers, and can have different values for different traffic streams. Value 0.9 is used most often (as suggested by Webster and Cobbe, [88]). If longer queues can be tolerated for some approach, then the value of $\overline{\rho}_i$ can be greater than 0.9, whereas if the queues should be avoided, the value of $\overline{\rho}_i$ should be less than 0.9.

9.3.3. Capacity per cycle

The capacity per cycle [2] is often used in traffic engineering practice. It is the maximal number of vehicles of one traffic stream or one signal group that can pass through an intersection during its effective green time, g_p, in one cycle. For an intersection, the capacity per cycle is the maximal number of vehicles that can pass through the intersection during all effective green times, g_p, ($p \in \mathscr{P}'$), in one cycle.

Capacity per cycle is defined by the following expressions:

- Traffic stream capacity per cycle:

$$J_c = \overline{\omega}_i^s = c\,\omega_i^s = s_i\, g_p \,, \quad (i \in \mathscr{J}', \sigma_i \in D'_p, p \in \mathscr{P}') \tag{9.13}$$

- Signal group capacity per cycle:

$$J_c = \overline{\omega}_p^g = c\,\omega_p^g = s_p^g\, g_p \,, \quad (p \in \mathscr{P}') \tag{9.14}$$

- Intersection capacity per cycle:

$$J_c = \overline{\Omega}^s = c\,\Omega^s = \sum_{i=1}^{I'} s_i\, g_i \tag{9.15}$$

$$J'_c = \overline{\Omega}^g = c\,\Omega^g = \sum_{p=1}^{P'} s_p^g\, g_p \tag{9.16}$$

Practical capacity per cycle can be obtained using expressions (9.13), (9.14), (9.15), and (9.16) in which the saturation flow of each stream has first to be multiplied by the value of maximally acceptable saturation degree.

The expressions for capacities as functions of variables $\mathbf{u}$ and τ are obtained by substituting $g_p = \sum_{k=1}^{K} u_p^k \tau^k$ in expressions (9.13)–(9.16).

9.3.4. Saturation degree of a traffic stream

The saturation degree, ρ_i, of a vehicle traffic stream, σ_i, is the ratio of the average arrival flow volume, q_i, to the capacity, i.e.,

$$\rho_i = \frac{q_i}{\omega_i^s} = \frac{q_i}{\frac{s_i\, g_p}{c}} = \frac{q_i\, c}{s_i\, g_p} \quad , \quad (i \in \mathcal{I}', \sigma_i \in D_p', p \in \mathcal{P}') . \qquad (9.17)$$

When $\rho_i = 1$, i.e., when the arrival flow volume, q_i, is equal to the capacity, ω_i^s, very long queues can be formed, as already mentioned. Therefore, instead of the requirement

$$\rho_i \le 1, \quad (i \in \mathcal{I}'),$$

as the flow balance constraint, the following constraint should be introduced:

$$\rho_i \le \overline{\rho}_i , \quad (i \in \mathcal{I}'), \qquad (9.18)$$

where $\overline{\rho}_i$ is the maximally acceptable saturation degree.

The tendency to equalize saturation degrees of vehicle traffic streams is present from the first attempts of formulating exact statements of traffic control problems on a signalized intersection [89], [30]. This can be achieved by optimizing several optimization criteria. Some of them are given here.

a) The mean square of differences between saturation degrees of traffic streams [35]

$$J_c = \sum_{i<j} (\rho_i - \rho_j)^2 = \sum_{i<j} \left(\frac{q_i\, c}{s_i g_p} - \frac{q_j\, c}{s_j g_q} \right)^2 ,$$

$$(\sigma_i \in D_p', \sigma_j \in D_q', i, j \in \mathcal{I}', p, q \in \mathcal{P}')$$

By substituting

$$g_p = \sum_{k=1}^{K} u_p^k \tau^k \quad \text{and} \quad g_q = \sum_{k=1}^{K} u_q^k \tau^k$$

in the expression for J_c, it becomes:

$$J_c = c^2 \sum_{i<j} \left(\frac{q_i}{s_i \sum_{k=1}^{K} u_p^k \tau^k} - \frac{q_j}{s_j \sum_{k=1}^{K} u_q^k \tau^k} \right)^2 . \tag{9.19}$$

A similar effect to the saturation degrees equalization can be obtained by minimizing the following optimization criterion:

$$J_c = \sum_{i<j} \left(\frac{1}{\rho_i} - \frac{1}{\rho_j} \right)^2 , \quad (i, j \in \mathcal{T}') . \tag{9.20}$$

b) The maximal saturation degree value

Minimization of the maximal saturation degree value leads to saturation degrees equalization, in a certain sense. The optimization criterion in this case can be defined as:

$$J_c = \max\{\rho_i \mid i \in \mathcal{T}'\} = \max\left\{ \frac{q_i c}{s_i g_p} \middle| \sigma_i \in D'_p, i \in \mathcal{T}', p \in \mathcal{P}' \right\} . \tag{9.21}$$

9.3.5. Capacity factor

The capacity factor is one of the criteria related to intersection capacity. Flow balance constraints ensure that each traffic stream (signal group) receives enough green time, i.e., that the practical capacity is greater than the average flow volumes for each vehicle traffic stream (signal group). However, a signal plan has to be chosen so that this condition is satisfied even in the event of flow volume changes. The changes can be various—some traffic stream volumes can be reduced, some can be increased, while some of them may retain their values.

For practical purposes, it is assumed that the change of volumes is relatively simple, i.e., that all average volumes either increase or reduce in the same proportion [4], [5]. In this case, the criterion for signal plan choice can be a number, μ, such that the flow balance constraints, (α.4), remain satisfied when all average traffic stream volumes are multiplied by this number. Thus, the optimization criterion is:

$$J_c = \mu . \tag{9.22}$$

If the maximal capacity factor, μ^*, is greater than 1, there exists a *reserve of intersection capacity*. The quantity of this reserve can be assessed by the difference $\mu^* - 1$ or $(\mu^* - 1) \cdot 100$ (expressed as a percentage).

In the case $\mu^* < 1$, the intersection capacity is not sufficient, i.e., no signal plan exists such that the capacity constraints, (α.4), are satisfied. In this case, the measure of intersection saturation can be the difference $1-\mu^*$ or $(1-\mu^*)\cdot 100$ (expressed as a percentage).

The capacity factor is a generally accepted criterion for optimal traffic control. Using this criterion for optimal signal choice gives especially good results when saturation degrees are high and volume changes are significant. [45].

9.4. The delay at an intersection

The total vehicle delay at all intersection approaches is the most often used criterion for signal plan choice [3], [89], [13], [44]. By its minimization multiple positive effects can be achieved—economic, ecological, etc. The financial equivalent of the delay unit can be determined in a relatively simple way [78]. The delay at an intersection is defined as the sum of delays for vehicle traffic streams on all intersection approaches.

The analytical relation between the delay and control (signal plan) has been formulated in various methods used for signal plan choice. For a given function of traffic stream volume, $q_i(t)$, on approach T_i, the queue that forms from the beginning of effective red time ($t=0$) until an instant *t*, has the value:

$$w_i(t) = \int_0^t q_i(\tau)d\tau, \quad (i \in \mathcal{I}').$$

Thus, the delay accumulated in interval [0,T], on approach T_i, can be expressed as:

$$V_i = \int_0^{\mathrm{T}} w_i(t)dt, \quad (i \in \mathcal{I}').$$

The total delay for the whole intersection is defined by the following expression:

$$V = \int_0^{\mathrm{T}} w^T e\, dt = \sum_{i=1}^{\mathrm{I}'} V_i, \tag{9.23}$$

where:

$$w = (w_1, w_2, \ldots, w_{\mathrm{I}'})^T,$$

$$e = (e_1, e_2, \ldots, e_{\mathrm{I}'})^T = (1,1,\ldots,1)^T.$$

This criterion is the most frequently used criterion in signal plan optimization problems.

Some authors, instead of criterion (9.23), suggest the "quadratic" criterion:

$$V' = \int_0^T \|w\|_A^2 \, dt = \int_0^T w^T A w \, dt . \tag{9.24}$$

where $A = [a_{ij}]_{I' \times I'}$, $(a_{ij} \geq 0,\ i,j \in \mathcal{I}')$ is a properly sized matrix of weighting factors. This criterion is slightly more "democratic" than the previous one, since the solution, obtained by its optimization, leads to an approximate equalization of queue lengths on intersection approaches [68].

The total delay of pedestrians and vehicle passengers can be also adopted as the optimization criterion. In this case, the calculated delays are multiplied by different occupancy coefficients, depending on the type of the traffic stream (e.g., 1.3 for passenger vehicles, 30 for buses, 1 for pedestrians, etc.) [2], and the sum of these products is calculated for each approach and for the whole intersection.

The delay on a signalized intersection, as mentioned in Subsection 5.2.2, is caused by the following factors:

α) The alternation of red and green traffic light indications causes delay even in the case when the arriving traffic is regular (deterministic, uniform arrivals), with flow volume less than the capacity.

β) The stochastic nature of arrival flows causes increase in delay because of two reasons:
 β.1) The stochastic nature of queuing,
 β.2) Queues, in some cycles, are not fully discharged until the end of green indication, thus making the initial queues in subsequent cycles greater than zero.

When the stochastic nature of traffic process is taken into consideration, the expressions determining the delay consist of two terms. One term, the *uniform* delay, exists even in the case when arrival flow volumes are constant and equal to average volumes of real flows. The second term, the *random* delay, represents the consequence of the stochastic nature of traffic flows.

The increase in delay, due to the stochastic nature of traffic flows, even in the case when the queue fully discharges before the end of green time, can be explained using Fig. 9.1 [46]. This figure illustrates the evolution of queues in three cycles. The volume of stream σ_i is q_i in the first cycle, $q_i + \Delta q_i$ in the second, and $q_i - \Delta q_i$ in the third.

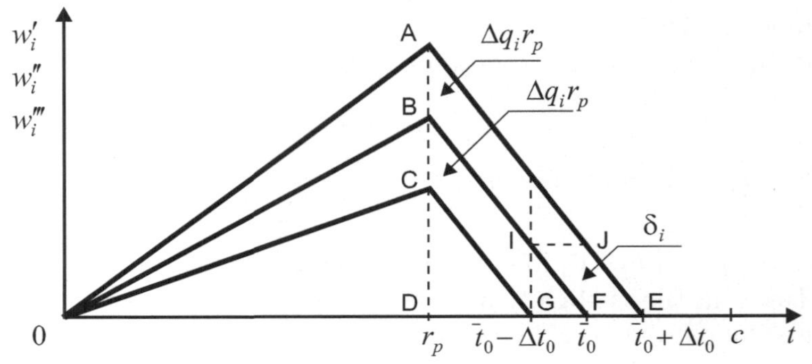

Figure 9.1

The delay is given by the following expressions:

$$V_i' = \int_0^{t_0} w_i'(t)dt\,, \quad \text{for } q_i\,,$$

$$V_i'' = \int_0^{t_0+\Delta t_0} w''(t)dt\,, \quad \text{for } q_i + \Delta q_i\,,$$

$$V_i''' = \int_0^{t_0-\Delta t_0} w_i'''(t)dt\,, \quad \text{for } q_i - \Delta q_i\,.$$

In Fig. 9.1, V_i' is represented by the area of triangle OBF, V_i'' is equal to the area of triangle OAE, and V_i''' to the area of triangle OCG. Observing these three triangles, it can be noted that the sum of their areas is greater than the threefold area of triangle OBF, which corresponds to the delay when the arrival flow volume is equal to q_i. The surplus, δ_i, is equal to the area of parallelogram FEJI. This means that the fluctuation of volume around the average, for the same amount, Δq_i, does not result in the increase and decrease of delay for the same extent—the increase is greater than the decrease. Therefore, the consequence of the stochastic nature of traffic flows is the increase of delay compared to the case when flow volumes are uniform.

The expressions for average queue lengths, for periods when traffic process is stationary, are given in Subsection 5.2.2. Mathematical expectation of the delay accumulated during one cycle is equal to the product of average queue length and cycle time. Thus, mathematical expectation of the total delay on approach T_i during one cycle is:

$$\mathrm{M}V_i = \mathrm{M}\{c\,\mathrm{M}w_i\} = c\,\mathrm{M}w_i\,, \quad (i \in \mathcal{T}')\,.$$

Mathematical expectation of the total delay on the intersection is given by the following expression:

$$\mathrm{M}V = \sum_{i=1}^{\mathrm{I}'} \mathrm{M}V_i = \sum_{i=1}^{\mathrm{I}'} c\,\mathrm{M}w_i = c\sum_{i=1}^{\mathrm{I}'} \mathrm{M}w_i \,. \tag{9.25}$$

Mathematical expectation of the total delay on approach T_i can be calculated using Newell's expression (5.11), as:

$$\mathrm{M}V_i = c(\mathrm{M}w_{\beta i} + \mathrm{M}w_{\eta i}) = c(\mathrm{M}w_{\beta i} + \mathrm{M}w'_{\eta i} + \mathrm{M}w''_{\eta i})\,. \tag{9.26}$$

In this expression, term $\mathrm{M}w_{\beta i}$ is the queue component resulting from influence of factors listed under α (alternation of red and green indications), term $\mathrm{M}w'_{\eta i}$ is the consequence of factors β.1, i.e., the stochastic nature of arrival flows. Term $\mathrm{M}w''_{\eta i}$ represents the consequence of factors listed under β.2, i.e., the fact that in some cycles (even when $\rho_i < 1$) the queue does not discharge and the initial queue is not equal to zero at the beginning of red indication. In this case,

$$\mathrm{M}w''_{\eta i} = \mathrm{M}w_{\eta i}(0)\,,$$

where $\mathrm{M}w_{\eta i}(0)$ is the queue length at the beginning of red indication, which is, also, the beginning of the cycle.

Therefore, mathematical expectation of the total delay on approach T_i during one cycle can be determined using the following expression:

$$\mathrm{M}V_i = c\left(\frac{c\,q_i(1-\lambda_E)^2}{2(1-\theta_i)} + \frac{q_i I_i(1-\lambda_p)}{2s_i(1-\theta_i)^2} + \frac{q_i I_i H(\mu_i)}{2s_i(1-\theta_i)}\right).$$

The meaning of symbols used in the expression is given in Subsection 5.2.2.

In Webster's formula (5.12), the first term, representing the uniform delay, is the same as the first term in Newell's expression. The second term, as Webster states [89], "… makes some allowance for the random nature of the arrivals. It is an expression for the delay experienced by vehicles arriving randomly in time at a 'bottleneck,' queuing up, and leaving at constant intervals." This term includes the consequences of both factors listed under β.1 and β.2, while the third term represents an empirical correction. Webster's formula can be represented by expression (5.13), having only two terms, as explained in Subsection 5.2.2. Using expression (5.13), mathematical expectation of the total delay on approach T_i can be determined as follows:

$$
\begin{aligned}
\mathrm{M}V_i &= c\,\mathrm{M}w_i = \\
&= c \cdot 0.9\left(\frac{c q_i (1-\lambda_p)^2}{2(1-\theta_i)} + \frac{\rho_i^2}{2(1-\rho_i)}\right) = \\
&= 0.9\left(\frac{c^2 q_i (1-\lambda_p)^2}{2(1-\theta_i)} + \frac{c\,\rho_i^2}{2(1-\rho_i)}\right) = \\
&= 0.9\left(\frac{q_i\, r_p^2}{2(1-\theta_i)} + \frac{c\,\rho_i^2}{2(1-\rho_i)}\right), \\
&(i \in \mathscr{T}', \sigma_i \in D'_p, p \in \mathscr{P}').
\end{aligned} \tag{9.27}
$$

The results obtained using this expression and by Newell's formula (5.11) are very similar, and close to real values of delay for almost all values of volume.

By introduction of signal plan variables, $u(t) = (\mathbf{u}, \tau)^T$, i.e., by substituting r_p and g_p by:

$$
g_p = \sum_{k=1}^{\mathrm{K}} u_p^k \tau^k, \quad (p \in \mathscr{P}'),
$$

$$
r_p = \sum_{k=1}^{\mathrm{L}} (1-u_p^k)\tau^k, \quad (p \in \mathscr{P}'),
$$

in Webster's formula, the following expressions for mathematical expectation of delay are obtained.

9.4.1. Mathematical expectation of delay when cycle time is known

a) Mathematical expectation of total delay of one signal group

Mathematical expectation of the total delay of vehicles belonging to traffic streams that are elements of signal group $D'_p \in \mathfrak{D}_a = \{D'_1, D'_2, \ldots, D'_p, \ldots, D'_{\mathrm{P}'}\}$, where $D'_p = \{\sigma_{p1}, \sigma_{p2}, \ldots, \sigma_{pe}, \ldots, \sigma_{p\mathrm{E}(p)}\}$, can be obtained by the following expression:

$$
\mathrm{M}V_p = 0.9 \sum_{e=1}^{e=\mathrm{E}(p)} \left(a_{pe}^1 \left(\sum_{k=1}^{\mathrm{K}} (1-u_p^k)\tau^k \right)^2 + \frac{a_{pe}^2}{\left(\sum_{k=1}^{\mathrm{K}} u_p^k \tau^k \right)^2 - a_{pe}^3 \sum_{k=1}^{\mathrm{K}} u_p^k \tau^k} \right),
$$

$$
(p \in \mathscr{P}'), \tag{9.28}
$$

where:

$$a_{pe}^{1} = \frac{q_{pe}\, s_{pe}}{2(s_{pe} - q_{pe})}, \quad a_{pe}^{2} = \frac{c^{3} q_{pe}}{2 s_{pe}}, \quad a_{pe}^{3} = \frac{c\, q_{pe}}{s_{pe}}.$$

b) Mathematical expectation of the total delay on an intersection

Mathematical expectation of the total delay on an intersection can be determined as the sum of average delays for all signal groups. Since one signal group, $D'_p = \{\sigma_{p1}, \sigma_{p2}, \ldots, \sigma_{pe}, \ldots, \sigma_{pE(p)}\}$, is controlled by one control variable, $u_p(\cdot)$, $p \in \mathscr{P}' = \{1, 2, \ldots, p, \ldots, \mathrm{P}'\}$, mathematical expectation of the total delay on an intersection is given by the following expression:

$$J_c = \mathrm{M}V = 0.9 \sum_{p=1}^{\mathrm{P}'} \sum_{e=1}^{e=\mathrm{E}(p)} \left(a_{pe}^{1} \left(\sum_{k=1}^{\mathrm{K}} (1 - u_p^k) \tau^k \right)^2 + \frac{a_{pe}^{2}}{\left(\sum_{k=1}^{\mathrm{K}} u_p^k \tau^k \right)^2 - a_{pe}^{3} \sum_{k=1}^{\mathrm{K}} u_p^k \tau^k} \right),$$

$(p \in \mathscr{P}')$. (9.29)

9.4.2. Mathematical expectation of delay when cycle time is not given

a) Mathematical expectation of total delay of one signal group

The mathematical expectation of total delay of vehicles belonging to traffic streams that are elements of signal group $D'_p = \{\sigma_{p1}, \sigma_{p2}, \ldots, \sigma_{pe}, \ldots, \sigma_{pE(p)}\}$ is determined by the following expression:

$$\mathrm{M}V_p = 0.9 \sum_{e=1}^{e=\mathrm{E}(p)} \left(a_{pe}^{1} \left(\sum_{k=1}^{\mathrm{K}} (1 - u_p^k) \tau^k \right)^2 + \frac{\bar{a}_{pe}^{2} \left(\sum_{k=1}^{\mathrm{K}} \tau^k \right)^3}{\left(\sum_{k=1}^{\mathrm{K}} u_p^k \tau^k \right)^2 - \bar{a}_{pe}^{3} \left(\sum_{k=1}^{\mathrm{K}} u_p^k \tau^k \right) \left(\sum_{k=1}^{\mathrm{K}} \tau^k \right)} \right)$$

$(p \in \mathscr{P}')$, (9.30)

where:

$$a_{pe}^{1} = \frac{q_{pe}\, s_{pe}}{2(s_{pe} - q_{pe})}, \quad \bar{a}_{pe}^{2} = \frac{q_{pe}^{2}}{2 s_{pe}^{2}}, \quad \bar{a}_{pe}^{3} = \frac{q_{pe}}{s_{pe}}.$$

b) Mathematical expectation of the total delay on an intersection

The mathematical expectation of the total delay on an intersection can be determined as the sum of average delays for all signal groups, i.e.,

$$J_c = \mathrm{M}V = 0.9 \sum_{p=1}^{\mathrm{P}'} \sum_{e=1}^{e=\mathrm{E}(p)} \left(a_{pe}^1 \left(\sum_{k=1}^{\mathrm{K}} (1-u_p^k)\tau^k \right)^2 + \frac{\overline{a}_{pe}^2 \left(\sum_{k=1}^{\mathrm{K}} \tau^k \right)^3}{\left(\sum_{k=1}^{\mathrm{K}} u_p^k \tau^k \right)^2 - \overline{a}_{pe}^3 \left(\sum_{k=1}^{\mathrm{K}} u_p^k \tau^k \right) \left(\sum_{k=1}^{\mathrm{K}} \tau^k \right)} \right)$$

$(p \in \mathscr{P}')$. (9.31)

9.5. The number of vehicle stops

The total number of stops of vehicles belonging to traffic stream σ_{pe} during one cycle is equal to the number of vehicles in queue at the beginning of the green interval, increased by the number of vehicles arriving while the queue exists during the green interval.

The mathematical expectation, $\overline{w}_{pe}^g$, of the queue at the beginning of green interval is given by the following expression [89], [2]:

$$\overline{w}_{pe}^g = \mathrm{M}w_{pe}^g = \max\left\{ q_{pe}\left(\frac{r_p}{2} + v_{pe} \right), q_{pe} r_{pe} \right\}, \quad (p \in \mathscr{P}'),$$

where v_{pe} is the average delay per vehicle of traffic stream σ_{pe}, i.e., $v_{pe} = \mathrm{M}v_{pe}$.

The mathematical expectation of number of stops is determined by the following expressions:

α) If the average number of vehicles in the queue at the beginning of green indication, $\overline{w}_{pe}^g$, can discharge during the green time, i.e., if

$$\frac{\overline{w}_{pe}^g}{s_{pe} - q_{pe}} < g_p,$$

then the mathematical expectation of number of stops/starts is obtained as:

$$\overline{v}_{pe} = \mathrm{M}v_{pe} = \overline{w}_{pe}^g + \frac{q_{pe}\overline{w}_{pe}^g}{s_{pe} - q_{pe}} = \frac{\overline{w}_{pe}^g s_{pe}}{s_{pe} - q_{pe}}, \tag{9.32}$$

$(p \in \mathscr{P}', e \in \mathscr{E}(p))$.

β) If the queue at the beginning of green indication, $\overline{w}_{pe}^{g}$, cannot discharge during the green time, i.e., if

$$\frac{\overline{w}_{pe}^{g}}{s_{pe}-q_{pe}} > g_p ,$$

then the mathematical expectation of number of stops/starts is obtained as:

$$\overline{v}_{pe} = \mathrm{M}v_{pe} = \overline{w}_{pe}^{g} + q_{pe}g_p , \quad (p \in \mathscr{P}', e \in \mathscr{E}(p)). \tag{9.33}$$

Besides the absolute number of stops/starts, another interesting performance index is the ratio of the number of stopped vehicles to the total number of vehicles that arrive to an approach during a cycle. If the queue is discharged during green time, then all vehicles arriving in interval $r_p + \alpha_{pe}$ leave the intersection in interval α_{pe}, where α_{pe} is the queue discharge time, i.e.,

$$q_{pe}(r_p + \alpha_{pe}) = s_{pe}\alpha_{pe} ,$$

$$\alpha_{pe} = \frac{q_{pe}r_p}{s_{pe}-q_{pe}} .$$

The ratio of number of stopped vehicles and the number of vehicles that arrive during a cycle is:

$$\begin{aligned} b_{pe} &= \frac{\overline{v}_{pe}}{c\,q_{pe}} = \frac{q_{pe}(r_p+\alpha_{pe})}{c\,q_{pe}} = \frac{1}{c}\left(r_p + \frac{q_{pe}r_p}{s_{pe}-q_{pe}}\right) \\ &= \frac{1}{c}\frac{s_{pe}r_p}{(s_{pe}-q_{pe})} = \frac{1-\lambda_p}{1-\theta_{pe}}, \quad (\sigma_{pe} \in D'_p, p \in \mathscr{P}', e \in \mathscr{E}(p)). \end{aligned} \tag{9.34}$$

If the queue is not discharged during green time, this ratio is:

$$b'_{pe} = \frac{\overline{v}'_{pe}}{c\,q_{pe}} = \frac{\overline{w}_{pe}^{g}}{c\,q_{pe}} + \lambda_p . \tag{9.35}$$

The expressions for the number of stops, (9.32) and (9.33), and the expressions for the ratio of number of stopped vehicles and the number of vehicles that arrive during a cycle, (9.34) and (9.35), usually give values that are slightly higher than the real ones. The reason lies in the fact that not all arriving vehicles stop—some of them only slow down. This phenomenon was analyzed by Australian researchers [2], and it was concluded that fairly good values, close to real ones, could be obtained if the values calculated using expressions (9.32)–(9.35) are multiplied by 0.9.

9.6. Signal plan parameters

9.6.1. The cycle time

The cycle time, c, is one of constraints in many problems of signal plan choice. However, due to the presence of other constraints, such as the constraints of minimal green times and minimal intergreen times, there exist boundary values of cycle time—its minimal and maximal value.

The cycle time cannot be shorter than a minimal cycle time, whose value is determined by solving the optimization problem with the cycle time as the optimization criterion, i.e.,

$$J_c = c = \sum_{k=1}^{K} \tau^k . \tag{9.36}$$

A maximal value of cycle time exists if the constraints of maximal red are present in the problem. The maximal cycle time value can be determined by maximization of the same criterion, (9.36).

9.6.2. Green time of a signal group

The effective green time of a signal group or the sum of effective green times of several signal groups can be also adopted as optimization criteria.

The effective green time of signal group D'_p, as the optimization criterion, is given by the expression:

$$J_c = \sum_{k=1}^{K} u_p^k \tau^k . \tag{9.37}$$

Sometimes it is necessary to maximize the total green time of several signal groups, e.g., in cases when priority is given to certain traffic streams. Thus, if the green time for signal groups D'_p and D'_q has to be maximized, the optimization criterion will have the following form:

$$J_c = \sum_{k=1}^{K} u_p^k \tau^k + \sum_{k=1}^{K} u_q^k \tau^k . \tag{9.38}$$

In the case when the total green time of all signal groups (vehicle, pedestrian, etc.) should be maximized, the optimization criterion would take the following form:

$$J_c = \sum_{p=1}^{\mathrm{P}} \sum_{k=1}^{\mathrm{K}} u_p^k \tau^k \, . \tag{9.39}$$

9.6.3. The total number of control vectors (phases)

The total number of control vectors in a signal plan structure is an interesting criterion if older, phase-oriented traffic controllers are used for traffic control. The number of control vectors in a signal plan structure is K, so that the optimization criterion, which should be minimized, becomes:

$$J_c = \mathrm{K} \, . \tag{9.40}$$

PART III

THE METHOD OF OPTIMAL TRAFFIC CONTROL DETERMINATION

The problem of optimal signal plan determination can now be formulated exactly, using the expressions for constraints and performance indices given in Part II. Various optimization problems can be formulated. The main difference between them is related to the type of optimization, which can be single or multiple criteria optimization. In this text only single criterion problems will be considered.

The determination of the optimal signal plan is formulated, in Part II, as the problem of finding the optimal closed path on graph G_s.

10. THE STATEMENT OF THE PROBLEM OF FINDING THE OPTIMAL CLOSED PATH ON GRAPH G_s

A signal plan $u(\cdot)$ is defined by its structure **u**, and cycle time split τ, i.e., $u(\cdot) = (\mathbf{u}, \tau)^T$. The optimal signal plan structure $\mathbf{u}^*$ and the optimal cycle time split τ^* correspond to the optimal signal plan $u^*(\cdot) = (\mathbf{u}^*, \tau^*)^T$.

In Part II it is noted that a feasible signal plan structure can be represented by a closed path on the control vector transition graph G_s. The determination of such a path implies determination of all components u_p^k of control vector $\mathbf{u}^k$ $(k \in \mathcal{K})$. Then, the constraints including variables τ^k and also the optimality criterion become functions of τ^k $(k \in \mathcal{K})$ variables only. The problem of determining these variables is a mathematical programming problem. Many of these problems can be stated as linear programming problems. However, the optimization criterion in delay minimization problems is a nonlinear function, the flow balance constraints in the problems with permitted conditional turnings are nonlinear as well, so that there exist control problems that are stated as problems of nonlinear mathematical programming.

The problem of optimal signal plan determination is, hence, the problem of finding the closed path on graph G_s (structure) and the values of variables τ^k (cycle time split) assigned to each node on the path so that the chosen optimization criterion achieves the optimal value.

The set of feasible controls $\mathcal{U}_f$ is defined by the constraints listed in Part II. If the chosen optimization criterion is given by the expression

$$J_c : \mathcal{U}_f \to \mathrm{R},$$

where R is the set of real numbers, and the optimum value J_c^* of this criterion is defined as:

$$J_c^* = \operatorname*{opt}_{u(\cdot)\in\mathcal{U}_f} J_c(u(\cdot)), \tag{10.1}$$

then the problem of the optimal signal plan determination can be stated as follows: Determine the set of optimal controls

$$\mathcal{U}^* = \{u^*(\cdot) \mid u^*(\cdot) \in \mathcal{U}_f, J_c(u^*(\cdot)) = J_c^*\}, \tag{10.2}$$

where

$$u^*(\cdot) = (\mathbf{u}^*_; \ \tau^*)^T .$$

The optimal structure $\mathbf{u}^*$ can be represented by the closed path on graph G_s, and τ^* is the optimal cycle time split vector.

11. THE METHOD OF FINDING THE OPTIMAL CLOSED PATH ON GRAPH G_s

The algorithms for solving problem (10.2) have been developed, based on the general method of combinatorial optimization—branch-and-bound method. In developing these algorithms the axiomatic system was used, defining the basic, general elements [62], [8], [71]. The method of signal plan determination using this approach is presented in papers [37], [38], [39].

According to the axiomatic approach, when using the branch-and-bound method the following elements are defined:

a) *The relaxation*
containing the steps:
a.1 Introduction of the superset of the set of feasible solutions
a.2 Introduction of functions that perform mapping of the superset elements into the set of real numbers

b) *The elimination criterion*

c) *The branching rule*

d) *The bounding rule*
d.1 The lower bound rule
d.2 The upper bound rule

e) *The branch-and-bound recursive operation*

The branch-and-bound method actually represents a recursive operation of branching and bounding. This procedure leads, step by step, to the optimal solution.

Elements of the superset are mapped, by a suitable function, to the set of real numbers. This mapping enables calculation of lower and upper bounds for subsets of the superset. These bounds are used in the branch-and-bound recursive operation.

11.1. Elements of the method

The optimal solution $u^*(\cdot)$ of problem (10.2) is in some problems the minimal and in others the maximal value of the optimization criterion. Namely, the optimal criterion (10.1) value

$$J_c^* = \underset{u(\cdot)\in\mathcal{U}_f}{\text{opt}} \; J_c(u(\cdot))$$

is defined in some problems as:

$$J_c^* = \sup_{u(\cdot)\in\mathcal{U}_f} J_c(u(\cdot)),$$

and in some other problems as:

$$J_c^* = \inf_{u(\cdot)\in\mathcal{U}_f} J_c(u(\cdot)).$$

Certain elements of branch-and-bound method are common to all methods of optimal signal plan determination. These common elements will be first defined here to the level of details suitable for application to any problem of determining the optimal signal plan. Afterwards, this method will be applied for solving the specific problem of maximizing the intersection capacity, and all elements of the method will be precisely defined. In Part IV the method will be applied for solving some other, typical traffic control problems on a signalized intersection.

11.1.1. The relaxation and extension of $J_c(u(\cdot))$ function

The relaxation is composed of:

a) The introduction of superset $\mathcal{U}^s$, whose subset is the set of feasible signal plans $\mathcal{U}_f$, i.e.,

$$\mathcal{U}_f \subset \mathcal{U}^s. \tag{11.1}$$

In the method there are also used:

- The partitive set ξ of the set $\mathcal{U}^s$

$$\xi = \mathcal{P}(\mathcal{U}^s). \tag{11.2}$$

- Elements $\mathcal{U}_\alpha$ of the set ξ that are subsets of the set $\mathcal{U}^s$, i.e.,

$$\mathcal{U}_\alpha \in \xi,\ \mathcal{U}_\alpha \subset \mathcal{U}^s. \tag{11.3}$$

- The partitive set $\overline{\mathscr{U}}$ of set ξ

  $$\overline{\mathscr{U}} = \mathcal{P}(\xi). \tag{11.4}$$

- Elements $\overline{u}$ of set $\overline{\mathscr{U}}$, which are subsets of set ξ, i.e.,

  $$\overline{u} \in \overline{\mathscr{U}}, \ \overline{u} \subset \xi \ . \tag{11.5}$$

- The union $\underline{\mathscr{U}}(\overline{u})$ of all subsets of collection $\overline{u}$, i.e.,

  $$\underline{\mathscr{U}}(\overline{u}) = \bigcup_{\mathscr{U}_\alpha \in \overline{u}} \mathscr{U}_\alpha \, . \tag{11.6}$$

- The collection of optimal solutions.
  The set of optimal solutions, $\mathscr{U}^*$, is in fact one collection $\overline{u}^*$. Any element of $\overline{u}^*$ is a singleton set, i.e.,

  $$\overline{u}^* = \{\{u^*(\cdot)\} \mid u^*(\cdot) \in \mathscr{U}^*\}. \tag{11.7}$$

It means that the aim of the branch-and-bound procedure is to find collection $\overline{u}^*$, which is the same as determining the set $\mathscr{U}^*$ of optimal solutions.

Superset $\mathscr{U}^s$, besides feasible solutions $u(\cdot) \in \mathscr{U}_f$, contains solutions whose length, κ, is shorter than K—the length of a feasible signal plan structure. The *length of the structure* is the number of control vectors in the signal plan.

Since the structure of a feasible signal plan satisfies constraints α.1, it means that the number of changes of each control variable value during a cycle is 2. Thus, the total number of changes of values of control variables during a cycle is 2P (because the number of variables is P). Therefore, the structure with the number of changes of control variables less than 2P is the structure of an infeasible signal plan with the length $\kappa < \mathrm{K}$.

The constraints that have to be satisfied by elements of set $\mathscr{U}^s$ are given by the following expressions:

α'.1 $\quad \sum_{k=1}^{\kappa} (u_p^k + u_p^{k(\mathrm{mod}\,\mathrm{K})+1})(\mathrm{mod}\,2) = \theta_p, \quad (p \in \mathscr{P}),$

and

$$(\kappa < \mathrm{K}) \Rightarrow (\exists p \in \mathscr{P}), \theta_p < 2),$$

$$(\kappa = \mathrm{K}) \Rightarrow (\theta_p = 2, \, p \in \mathscr{P}).$$

α'.2 $$\sum_{k=1}^{\mathrm{K}} u_p^k \tau^k \geq \begin{cases} 0, \text{ if } \sum_{k=1}^{\mathrm{K}} (u_p^k + u_p^{k+1}) (\mathrm{mod}\, 2) < 2 \quad \text{or} \\ \qquad \left(\sum_{k=1}^{\mathrm{K}} (u_p^k + u_p^{k(\mathrm{mod}\, \mathrm{K})+1}) (\mathrm{mod}\, 2) = 2 \right) \wedge (\kappa \neq \mathrm{K}) \\ g_{mp}, \text{ if } \left(\sum_{k=1}^{\mathrm{K}} (u_p^k + u_p^{k(\mathrm{mod}\, \mathrm{K})+1}) (\mathrm{mod}\, 2) = 2 \right) \wedge (\kappa = \mathrm{K}) \end{cases}$$

$(p \in \mathscr{P})$.

α'.3 $$\sum_{k=1}^{\mathrm{K}} (1 - u_q^k) \tau^k \leq r_q^M, \quad (q \in \mathscr{P}^M \subset \mathscr{P}').$$

α'.4.1 $$\sum_{k=1}^{\mathrm{K}} u_p^k \tau^k \geq \begin{cases} 0, \text{ if } \sum_{k=1}^{\mathrm{K}} (u_p^k + u_p^{k+1}) (\mathrm{mod}\, 2) < 2 \text{ or} \\ \qquad \left(\sum_{k=1}^{\mathrm{K}} (u_p^k + u_p^{k(\mathrm{mod}\, \mathrm{K})+1}) (\mathrm{mod}\, 2) = 2 \right) \wedge (\kappa \neq \mathrm{K}) \\ \gamma_p, \text{ if } \left(\sum_{k=1}^{\mathrm{K}} (u_p^k + u_p^{k(\mathrm{mod}\, \mathrm{K})+1}) (\mathrm{mod}\, 2) = 2 \right) \wedge (\kappa = \mathrm{K}) \end{cases}$$

$(p \in \mathscr{P}')$.

α'.4.2 I $$\sum_{k=1}^{\mathrm{K}} (s_a u_a^k (1 - u_b^k) + q_a u_a^k u_b^k)\, \tau^k \geq$$

$$\geq \begin{cases} 0, \text{ if } \left(\sum_{k=1}^{\mathrm{K}} (u_a^k + u_a^{k+1}) (\mathrm{mod}\, 2) \leq 2 \right) \wedge (\kappa \neq \mathrm{K}) \\ c q_a, \text{ if } \left(\sum_{k=1}^{\mathrm{K}} (u_a^k + u_a^{k(\mathrm{mod}\, \mathrm{K})+1}) (\mathrm{mod}\, 2) = 2 \right) \wedge (\kappa = \mathrm{K}) \end{cases}$$

$(a, b \in \mathscr{P}')$.

II $$\sum_{k=1}^{\mathrm{K}} (u_a^k (1 - u_b^k) s_a - (1 - u_a^k u_b^k) q_a)\, \tau^k \geq 0.$$

$$\text{III}\qquad \sum_{k=1}^{\kappa}\left(u_a^k u_b^k \frac{q_a e^{-\alpha' \varsigma_a}}{1-e^{-\beta_b q_a}} + u_b^k(1-u_a^k)s_b\right) \geq$$

$$\geq \begin{cases} 0, \text{ if } \left(\sum_{k=1}^{\kappa}(u_p^k + u_p^{k(\mathrm{mod}\,\mathrm{K})+1})(\mathrm{mod}\,2) = 2\right) \wedge (\kappa \neq \mathrm{K}) \\ cq_b, \text{ if } \left(\sum_{k=1}^{\kappa}(u_b^k + u_b^{k(\mathrm{mod}\,\mathrm{K})+1})(\mathrm{mod}\,2) = 2\right) \wedge (\kappa = \mathrm{K}), \end{cases}$$

$$(a,b \in \mathscr{P}').$$

$$\gamma'.1 \qquad \mathbf{u}^{k(\mathrm{mod}\,\mathrm{K})+1} \in \Gamma_s \mathbf{u}^k, \quad (k \in \mathscr{K}').$$

$$\gamma'.2 \qquad \sum_{l=0}^{\omega-1} \tau^{\kappa-(\kappa+l-k)(\mathrm{mod}\,\mathrm{K})} \geq \max\{z_{pq} \mid (u_p^{\kappa-(\kappa+\omega-k)(\mathrm{mod}\,\mathrm{K})} \cdot u_q^{k(\mathrm{mod}\,\mathrm{K})+1} = 1)$$

$$\wedge (u_p^{\kappa-(\kappa+\omega-k-1)(\mathrm{mod}\,\mathrm{K})} = u_q^k = 0);\ p,q \in \mathscr{P}\}$$

$$(k \in \mathscr{K}',\ p \in \Omega'(\mathbf{u}^k, \mathbf{u}^{k(\mathrm{mod}\,\mathrm{K})+1})).$$

$$\delta'. \qquad \sum_{k=1}^{\kappa}(u_r^k u_m^k)\tau^k \leq |z_{rm}|, \quad (z_{rm} < 0;\ r,m \in \mathscr{P}'').$$

$$\varepsilon'.1 \qquad \sum_{k=1}^{\kappa}\tau^k \leq c \ \text{ or } \sum_{k=1}^{\kappa}\tau^k = c.$$

$$\varepsilon'.2 \qquad \sum_{k=1}^{\kappa}\tau^k \leq c_{\max}.$$

where:

$$\mathscr{K}' = \{1,2,\ldots,k,\ldots,\kappa\}, \quad (\kappa \in \mathscr{K})$$

$$\Omega'(\mathbf{u}^k, \mathbf{u}^{k(\mathrm{mod}\,\mathrm{K})+1}) = \{1,2,\ldots,\omega,\ldots,\omega'_{\mathrm{M}}(\mathbf{u}^k, \mathbf{u}^{k(\mathrm{mod}\,\mathrm{K})+1})\}$$

– the set containing ω.

$$\omega'_{\mathrm{M}}(\mathbf{u}^k, \mathbf{u}^{k(\mathrm{mod}\,\mathrm{K})+1}) =$$

$$= \min\{\max\{z_{pq} \mid (u_q^{k(\mathrm{mod}\,\mathrm{K})+1} = 1) \wedge (u_q^k = 0); p,q \in \mathscr{P}\},$$

$$(\mathrm{K} + (k - \mathrm{K} - 1)\,sign(\mathrm{K} - \kappa)), (\mathrm{K} - 2)\}$$

– the number of intergreen constraints for a given value of k.

If c is not given, constraints ε'.1 do not exist, and c in constraints α'.4 is substituted by $\sum_{k=1}^{\kappa} \tau^k$.

When using formula γ'.2 to determine the intergreen time constraints, the constraints for τ^k ($k \le \kappa \le \mathrm{K}$) cannot include duration τ^1 of control vector $\mathbf{u}^1$ nor duration of vectors $\mathbf{u}^{\mathrm{K}}$, $\mathbf{u}^{\mathrm{K}-1}$, etc. that precede $\mathbf{u}^1$. This is because $\mathbf{u}^{\mathrm{K}}$, $\mathbf{u}^{\mathrm{K}-1}$, etc. are not known until the feasible solution is obtained. For example, let the formed part of the structure be $\mathbf{u}' = [\mathbf{u}^1, \mathbf{u}^2, \ldots, \mathbf{u}^5]$. It means that $\kappa = 5$ ($\kappa < \mathrm{K}$). The constraints for $k = 4$ obtained by formula γ'.2 are related to the following intervals: τ^4 – between control vectors $\mathbf{u}^3$ and $\mathbf{u}^5$ ($\omega = 1$), $\tau^3 + \tau^4$ – between control vectors $\mathbf{u}^2$ and $\mathbf{u}^5$ ($\omega = 2$), and $\tau^2 + \tau^3 + \tau^4$ – between control vectors $\mathbf{u}^1$ and $\mathbf{u}^5$ ($\omega = \mathrm{K} + (4 - \mathrm{K} - 1)\, sign(\mathrm{K} - 5)$). The constraints for intervals $\tau^1 + \tau^2 + \tau^3 + \tau^4$, $\tau^{\mathrm{K}} + \tau^1 + \tau^2 + \tau^3 + \tau^4$, etc. will not be formed using γ'.2. These constraints, as well as all other constraints that have to be satisfied by an optimal signal plan, will be formed when κ becomes K.

For $\kappa = \mathrm{K}$ constraints α' to ε', defining superset $\mathscr{U}^s$, become constraints α to ε, by which the set of feasible solutions $\mathscr{U}_f$ is defined. Accordingly, the constraints α' to ε' hold for all elements of set $\mathscr{U}^s$.

One solution, $u(\cdot)$, with the feature

$$(u(\cdot) \in \mathscr{U}^s) \wedge (u(\cdot) \notin \mathscr{U}_f)$$

is presented in Fig. 11.1. Graph G_s used in this example is given in Fig. 8.17.

Example 11.1

Structure $\mathbf{u}'$ of one solution (which is not feasible) is presented on graph G_s from Example 8.9. The structure of solution $\mathbf{u}'$, presented in Fig. 11.1, is determined by the following expression:

$$\mathbf{u}' = [\mathbf{u}^1, \mathbf{u}^2, \ldots, \mathbf{u}^7] = \begin{bmatrix} \mathbf{u}_1 \\ \mathbf{u}_2 \\ \mathbf{u}_3 \\ \mathbf{u}_4 \\ \mathbf{u}_5 \\ \mathbf{u}_6 \end{bmatrix} = \overset{\begin{matrix} \mathbf{u}^1 & \mathbf{u}^2 & \mathbf{u}^3 & \mathbf{u}^4 & \mathbf{u}^5 & \mathbf{u}^6 & \mathbf{u}^7 \end{matrix}}{\begin{bmatrix} 1 & 1 & 1 & 1 & 0 & 0 & 0 \\ 1 & 0 & 0 & 0 & 0 & 0 & 0 \\ 0 & 0 & 0 & 1 & 1 & 0 & 0 \\ 0 & 0 & 0 & 0 & 0 & 0 & 1 \\ 1 & 1 & 0 & 0 & 0 & 0 & 0 \\ 0 & 0 & 0 & 0 & 0 & 0 & 0 \end{bmatrix}}. \tag{11.8}$$

The initial control vector is $\mathbf{u}^1 = (1,1,0,0,1,0)^T$, and $\kappa = 7$.

Determine constraints α'.2, γ'.2, and ε'.1.

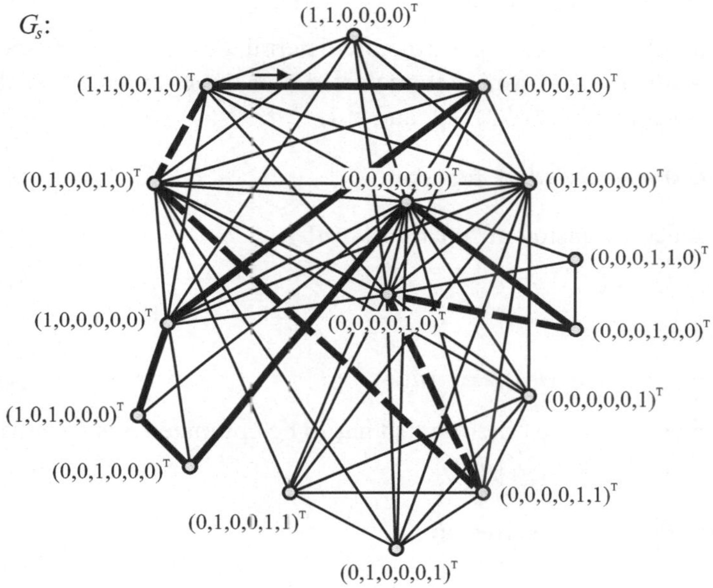

Figure 11.1

It is obvious (Fig. 11.1) that the structural constraints α'.1 and γ'.1 are satisfied. Constraints α'.2, γ'.2, and ε'.1 are:

$$
\begin{array}{lccccccccl}
\alpha'.2 & & & & \tau^4 & + \ \tau^5 & & & \geq & 15 \\
\gamma'.2 & & \tau^2 & + \ \tau^3 & & & & & \geq & 3 \\
 & & & \tau^3 & & & & & \geq & 1 \\
 & & & & & \tau^5 & + \ \tau^6 & & \geq & 4 \\
 & & & & & & \tau^6 & & \geq & 3 \\
\varepsilon'.2 & \tau^1 & + \ \tau^2 & + \ \tau^3 & + \ \tau^4 & + \ \tau^5 & + \ \tau^6 & + \ \tau^7 & = & 90
\end{array}
\qquad (11.9)
$$

Constraints δ' do not exist because none of minimal intergreen times is negative in this example.

The solutions $u(\cdot) = (\mathbf{u}, \tau)^T$, with the structure $\mathbf{u}'$ presented in Fig. 11.1, satisfy the structural constraints α'.1 and γ'.1. However, in order to have $u(\cdot) \in \mathcal{U}^s$, it is necessary that the components of τ, i.e., $\tau^1, \tau^2, \ldots, \tau^7$, satisfy constraints α'.2, γ'.2, and ε'.1. For example, if

$$\tau = (10, 10, 10, 10, 10, 10, 30), \tag{11.10}$$

constraints $\alpha'.2$, $\gamma'.2$, and $\varepsilon'.1$ are satisfied.

Thus, the solution $u(\cdot)$ with the structure given by expression (11.8) and the cycle time split given by (11.10), belongs to set $\mathscr{U}^s$.

This example shows that one structure determines one subset of set $\mathscr{U}^s$ containing elements with different vectors τ, but the components of these vectors satisfy constraints $\alpha'.2$, $\gamma'.2$, and $\varepsilon'.1$.

b) The extension of $J_c(u(\cdot))$ function

The bounded extension of function $J_c(u(\cdot))$

$$J_c' : \mathscr{U}^s \to \mathrm{R}, \tag{11.11}$$

has the feature

$$u(\cdot) \in \mathscr{U}_f \Rightarrow J_c'(u(\cdot)) = J_c(u(\cdot)). \tag{11.12}$$

For each performance criterion that has to be optimized it is necessary to define a suitable function $J_c'(u(\cdot))$.

11.1.2. The elimination criterion

The elimination criterion is used to identify unfeasible solutions and nonoptimal feasible solutions.

If $u(\cdot)$ is unfeasible, then:

$$(u(\cdot) \in \mathscr{U}^s) \wedge (u(\cdot) \notin \mathscr{U}_f), \text{ i.e.,}$$

$$u(\cdot) \in \mathscr{U}^s \setminus \mathscr{U}_f.$$

If collection $\bar{\mathscr{u}}_0 \in \overline{\mathscr{U}}$ is introduced, such that

$$\underline{\mathscr{U}}(\bar{\mathscr{u}}_0) \subset \mathscr{U}^s \setminus \mathscr{U}^*, \tag{11.13}$$

where:

$$\underline{\mathscr{U}}(\bar{\mathscr{u}}_0) = \bigcup_{\mathscr{U}_\alpha \in \bar{\mathscr{u}}_0} \mathscr{U}_\alpha, \tag{11.14}$$

then the unfeasible $u(\cdot)$ has the feature

$$u(\,) \in \mathscr{U}^s \setminus \mathscr{U}_f \;\Rightarrow \{\,\{u(\cdot)\}\,\} \in \bar{\mathscr{u}}_0. \tag{11.15}$$

The subsets containing feasible but nonoptimal solutions belong, also, to the collection $\bar{\mathscr{u}}_0$ [71].

11.1.3. Branching rules

The branching rule is the function

$$\beta : \overline{\mathcal{U}} \rightarrow \overline{\mathcal{U}} \tag{11.16}$$

with the following properties:

(B.1) $\underline{\mathcal{U}}(\beta(\bar{u})) = \underline{\mathcal{U}}(\bar{u})$ (11.17)

(B.2) If $\mathcal{U}_\alpha \in \beta(\bar{u}) \setminus \bar{u}$, then there exist $\mathcal{U}_{\alpha'} \in \bar{u} \setminus \beta(\bar{u})$ with

$$\mathcal{U}_\alpha \subset \mathcal{U}_{\alpha'}$$

(B.3) If there exist $\mathcal{U}_\alpha \in \bar{u}$ with $|\mathcal{U}_\alpha| > 1$, then

$$\beta(\bar{u}) \setminus \bar{u} \neq \varnothing$$

The branching rule is applied to collection $\bar{u}^n \in \overline{\mathcal{U}}$. Collection $\bar{u}^n$ is obtained by applying the recursive branch-and-bound operation $n-1$ times, starting from collection $\bar{u}^1 = \{\mathcal{U}^s\}$.

In order to define mapping β, it is necessary to:

- Determine the way to choose the subset $\mathcal{U}_\gamma \in \bar{u}^n$ that will be further divided into subsets
- Determine the way of dividing the chosen subset $\mathcal{U}_\gamma$ into subsets
- Recognize, if possible, the subsets of $\mathcal{U}_\gamma$ containing unfeasible solutions; such subsets shall be eliminated from further procedure and the upper bounds will not be calculated for them
- Specify the way of determining $\beta(\bar{u}^n)$ for a given collection $\bar{u}^n$

a) Choosing the subset $\mathcal{U}_\gamma \in \bar{u}^n$ that will be divided in subsets

The upper bound $B(\mathcal{U}_\alpha)$ is determined for each element $\mathcal{U}_\alpha$ of collection $\bar{u}^n$ (as described in Subsection 11.1.4). In general description of the method the maximization will be assumed as the optimization problem.

The subset $\mathcal{U}_\gamma$ having the maximal upper bound is chosen for further division, i.e., the subset $\mathcal{U}_\gamma$ with the property

$$B(\mathcal{U}_\gamma) \geq B(\mathcal{U}_\alpha) \quad (\mathcal{U}_\alpha \in \bar{u}^n). \tag{11.18}$$

For $n = 1$

$$B(\mathscr{U}_\gamma) \geq B(\mathscr{U}_\alpha) \qquad (\mathscr{U}_\alpha \in \bar{u}^1 = \{\mathscr{U}^s\}),$$

i.e.,

$$B(\mathscr{U}_\gamma) = B(\mathscr{U}^s). \tag{11.19}$$

This means that $\mathscr{U}_\gamma = \mathscr{U}^s$ for $n = 1$.

b) Dividing the chosen subset $\mathscr{U}_\gamma$ in subsets

After choosing the subset $\mathscr{U}_\gamma$ that will be branched, i.e., divided in subsets, it is necessary to determine the method of dividing $\mathscr{U}_\gamma$.

Subset $\mathscr{U}_\gamma$ will be divided in such a way that all elements belonging to one subset have the same initial part of the structure. Thus, after the first division of $\mathscr{U}_\gamma$, the subsets are obtained such that all elements of a subset have the same sequence of first two control vectors: $\mathbf{u}^1$, $\mathbf{u}^2$. By further divisions new subsets are obtained having the same sequence of first three control vectors, and so on. The initial sequence, same for all elements of a subset, is the initial part of the structure of all elements belonging to that subset. This sequence can be represented by one path on graph G_s (Fig. 11.1).

Let $\mathscr{U}_\gamma$ be the subset chosen for branching according to rule (11.18). All elements of this subset have the same initial part of the structure. If $u(\cdot) \in \mathscr{U}_\gamma$ and $u(\cdot) = (\mathbf{u}, \tau)^T$, the structure $\mathbf{u}$ can be represented as:

$$\mathbf{u} = [\underline{\mathbf{u}'}, \underline{\mathbf{u}''}], \tag{11.20}$$

where $\underline{\mathbf{u}'}$ is the initial part of the structure of each element that belongs to $\mathscr{U}_\gamma$:

$$\underline{\mathbf{u}'} = [\mathbf{u}^1, \mathbf{u}^2, \ldots, \mathbf{u}^\kappa]. \tag{11.21}$$

The fact that $\underline{\mathbf{u}'}$ and $\underline{\mathbf{u}''}$ are parts of the structure of $\mathscr{U}_\gamma$ elements can be expressed as follows:

$$\mathbf{u}(\mathscr{U}_\gamma) = [\underline{\mathbf{u}'}(\mathscr{U}_\gamma), \underline{\mathbf{u}''}(\mathscr{U}_\gamma)].$$

The part $\underline{\mathbf{u}''}$ of the structure is different for different elements of $\mathscr{U}_\gamma$. This part of the structure can be expressed in the following way:

$$\underline{\mathbf{u}''} = [\mathbf{u}^{\kappa(\mathrm{mod}K)+1}, \underline{\mathbf{u}'''}], \tag{11.22}$$

where $\mathbf{u}^{\kappa(\mathrm{mod}K)+1}$ is the control vector following $\mathbf{u}^\kappa$, i.e., $\mathbf{u}^{\kappa(\mathrm{mod}K)+1} \in \Gamma_s \mathbf{u}^\kappa$, and $\underline{\mathbf{u}'''}$ is the remaining part of the structure.

One of the subsets of $\mathscr{U}_\gamma$ can be the subset $\mathscr{U}'_\gamma$ whose elements have the structure with the property $\mathbf{u}^1 \in \Gamma_s \mathbf{u}^\kappa$ and the constraint $\alpha.1$ satisfied.

Two cases can arise in further procedure, depending on the initial part of the structure $\underline{\mathbf{u}}' = [\mathbf{u}^1, \mathbf{u}^2, \ldots, \mathbf{u}^\kappa]$:

b.1) The first case: $\mathbf{u}^1 \in \Gamma_s \mathbf{u}^\kappa$ and

$$\sum_{k=1}^{\kappa} (u_p^{k(\mathrm{mod}\,\kappa)+1} + u_p^1)(\mathrm{mod}\,2) = 2\,, \quad (p \in \mathscr{P})\,. \tag{11.23}$$

In this case two possibilities can be distinguished:

α) There exists only one value of $p \in \mathscr{P}$ with the property:

$$\sum_{k=1}^{\kappa} (u_p^{k-1} + u_p^k)(\mathrm{mod}\,2) = 1\,, \tag{11.24}$$

and for all other values of $p \in \mathscr{P}$ the sum in expression (11.24) equals 2. This means that in the transition from $\mathbf{u}^\kappa$ to $\mathbf{u}^1$ only one component of vector $\mathbf{u}^\kappa$ changes its value. In this case $\underline{\mathbf{u}}'$ satisfies the constraints $\alpha.1$ and $\gamma.1$. If other constraints are also satisfied, then the structure $\underline{\mathbf{u}}'$ is the structure of a feasible solution, i.e.,

$$\underline{\mathbf{u}}' = \mathbf{u}\,, \kappa = \mathrm{K}\,, \quad (u(\cdot) = (\mathbf{u}, \tau)^T\,, u(\cdot) \in \mathscr{U}_f)\,.$$

All elements of subset $\mathscr{U}'_\gamma$ have the same structure $\mathbf{u} = \underline{\mathbf{u}}'$ in this case, and they differ by control vectors durations. Further branching of this subset will give a singleton set, whose element is the feasible solution.

β) $\sum_{k=1}^{\kappa} (u_p^{k-1} + u_p^k)(\mathrm{mod}\,2) < 2$

for more values of p, rather than only one.

In this case, by branching subset $\mathscr{U}'_\gamma$, several subsets will be obtained.

When subset $\mathscr{U}'_\gamma$ is chosen for further branching, that means that the lower bound $B(\mathscr{U}'_\gamma)$ is better than the lower bound of any other element of the collection to which $\mathscr{U}'_\gamma$ belongs. Since the structures of the elements of $\mathscr{U}'_\gamma$ satisfy constraints $\alpha.1$ and $\gamma.1$, which have to be satisfied by any feasible solution, and there exists the bound $B(\mathscr{U}'_\gamma)$, this means that a feasible solution $u^b(\cdot)$ exists, with the property

$$B(\mathscr{U}'_\gamma) = \max\{J'_c(u(\cdot)) \,|\, u(\cdot) \in \mathscr{U}'_\gamma\} = J'_c(u^b(\cdot))\,,$$

where $u^b(\cdot) = (\mathbf{u}^b, \tau^b)^T$. In this case:

$$\beta'(\mathscr{U}'_\gamma) = \{\{u(\cdot)^b\}\} \cup \{\widetilde{\mathscr{U}}'_\gamma\}. \tag{11.25}$$

Subset $\widetilde{\mathscr{U}}'_\gamma$ contains nonoptimal solutions (with cycle time splits τ^b that are not optimal).

If there does not exist a feasible cycle time split τ^b for structure $\mathbf{u}^b$, then $\{u^b(\cdot)\} = \varnothing$.

In the case $\mathscr{U}_\gamma = \{u^b(\cdot)\}$, mapping $\beta'(\mathscr{U}_\gamma)$ is defined as follows:

$$\beta'(\{u^b(\cdot)\}) = \{\{u^b(\cdot)\} \mid (u^b(\cdot) \in \{u^b(\cdot)\}) \Rightarrow (u^b(\cdot) = (\mathbf{u}^b, \tau^b)^T) \wedge$$
$$\wedge (\mathbf{u}^1 \in \Gamma_s \mathbf{u}^\kappa) \wedge (\sum_{k=1}^{\kappa} (u_p^k + u_p^{k(\mathrm{mod}\,\mathrm{K})+1})(\mathrm{mod}\,2) = 2),$$
$$(p \in \mathscr{P})\} = \{\{u^b(\cdot)\}\}. \tag{11.26}$$

b.2) The second case: $\kappa < \mathrm{K}$

Since the structure length is $\kappa < \mathrm{K}$, $\underline{\mathbf{u}}'$ cannot be the structure of a feasible solution. All elements with structure $\underline{\mathbf{u}}'$ are infeasible.

A function β will define the rules of branching subset $\mathscr{U}_\gamma$ whose elements have a structure that can be represented in the form:

$$\mathbf{u} = [\underline{\mathbf{u}}', \underline{\mathbf{u}}''] = [\mathbf{u}^1, \mathbf{u}^2, \ldots, \mathbf{u}^\kappa, \underline{\mathbf{u}}''] = [\underline{\mathbf{u}}', \mathbf{u}^{\kappa+1}, \underline{\mathbf{u}}'''],$$

where $\mathbf{u}^{\kappa+1} \in \Gamma_s \mathbf{u}^\kappa$.

Branching of subset $\mathscr{U}_\gamma$, which contains elements with the structure length $\delta(\mathbf{u})$ greater than κ, and of subset $\mathscr{U}'_\gamma$, with elements structure of length κ, is performed in different ways.

Function β', defining the branching of a chosen subset into subsets, is given by the mapping

$$\beta' : \xi \rightarrow \overline{\mathscr{U}} \tag{11.27}$$

and defined by the following expression:

$$\beta'(\mathscr{U}_\gamma)=\{\mathscr{U}_{\gamma r}\mid(u(\cdot)\in\mathscr{U}_{\gamma r})\Rightarrow(u(\cdot)=([\underline{\mathbf{u}}'(\mathscr{U}_\gamma),\mathbf{u}^{\kappa+1},\underline{\mathbf{u}}'''(\mathscr{U}_\gamma)],\tau)^T)\wedge$$

$$\wedge(\mathbf{u}^{\kappa+1}\in\Gamma_s\mathbf{u}^\kappa)\wedge(\sum_{k=1}^{\kappa}(u_p^k+u_p^{k+1})(\mathrm{mod}\,2)\le 2)$$

$$\wedge((\exists q\in\mathscr{P})\Rightarrow\sum_{k=1}^{\kappa}(u_q^k+u_q^{k+1})(\mathrm{mod}\,2)<2),$$

$$(p\in\mathscr{P})\}\bigcup\{\mathscr{U}'_\gamma\}. \qquad (11.28)$$

Thus, $\beta'(\mathscr{U}_\gamma)$ is the collection with elements whose initial part of the structure is $[\underline{\mathbf{u}}',\mathbf{u}^{\kappa+1}]$. The part $\underline{\mathbf{u}}'$ is the same for all elements of the subsets $\mathscr{U}_{\gamma r}$.

The number of subsets $\mathscr{U}_{\gamma r}$ is $card(\Gamma_s\mathbf{u}^\kappa)+1$.

c) Identification of subsets containing infeasible solutions

Among the subsets that are elements of collection $\beta'(\mathscr{U}_\gamma)$ there are some subsets that can be identified, even before determining their lower bounds, as the subsets having no feasible solutions. Such subsets can be eliminated from further procedure.

Collection $\beta''(\mathscr{U}_\gamma)\subset\beta'(\mathscr{U}_\gamma)$, which can be excluded from further procedure, is formed by the following mapping:

$$\beta'':\overline{\mathscr{U}}\to\overline{\mathscr{U}}. \qquad (11.29)$$

Collection $\beta''(\overline{\omega})$ is a subset of collection $\overline{\omega}_0$, whose elements are all subsets containing infeasible or nonoptimal solutions. It means that the subsets which are elements of β'' belong also to collection $\overline{\omega}_0$.

The subsets belonging to collection β'' can be recognized in the next cases:

α) Some subsets $\mathscr{U}_{\gamma r}$ satisfy constraints α'.1 and γ'.1, but the structure having the initial part $\underline{\mathbf{u}}'$, which is the same for all elements of subset $\mathscr{U}_\gamma$, does not satisfy constraints α.1 and γ.1. Such subsets can be eliminated from further procedure.

Constraint α.1, which has to be satisfied by any feasible solution, states that the sum of changes of each control variable (from 0 to 1 and vice versa) has to be 2. If the values of a control variable $u_p(\cdot)$, until the control vector $\mathbf{u}^{\kappa'+1}$, change as shown in Fig. 11.2, it can be concluded that the value of $u_p(\cdot)$ in all subsequent control vectors, from $\mathbf{u}^{\kappa'+1}$ until the end of the cycle, has to be 1. Otherwise, if $u_p(\cdot)$ changed once more, from 1 to 0, the sum of changes of $u_p(\cdot)$ values would be 3 and constraint $\alpha.1$ would not be satisfied.

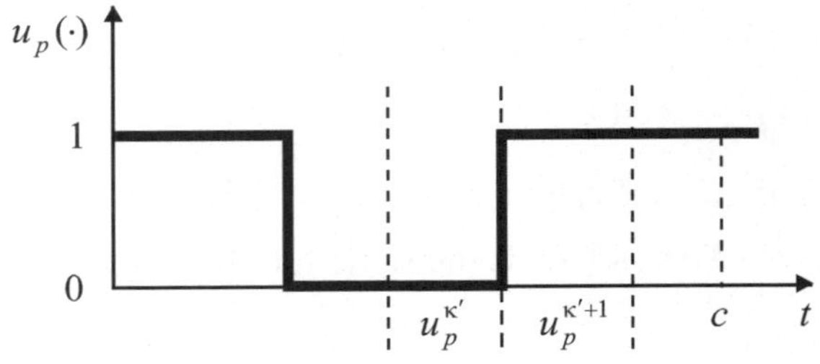

Figure 11.2

If a signal group D_l' is incompatible with D_p', and if the value of the variable $u_l(\cdot)$ from $t = 0$ until the beginning of control vector $\mathbf{u}^{\kappa'+1}$ is 0 (the group did not get the right-of-way), then this group cannot get the right-of-way at all because $u_p(\cdot)$ must have value 1 until the end of the cycle, and $u_l(\cdot)$ cannot have value 1 simultaneously with $u_p(\cdot)$.
In this case, collection β" is defined as follows:

$$\begin{aligned}
\beta_1''(\mathscr{U}) &= \beta''(\beta'(\mathscr{U}_\gamma)) \\
&= \{\mathscr{U}_{\gamma q} \in \beta'(\mathscr{U}_\gamma) \mid ([\mathbf{u}^1, \mathbf{u}^2, \ldots, \mathbf{u}^{\kappa'}, \mathbf{u}^{\kappa'+1}, \underline{\mathbf{u}'''}], \tau)^T \in \mathscr{U}_{\gamma q} \Rightarrow \\
&\quad \Rightarrow ((\exists p \in \mathscr{P}) \wedge (\exists l \in \mathscr{P}) \wedge (\, D_p', D_l') \notin C_g) \wedge \\
&\quad \wedge (\sum_{k=1}^{\kappa'} (u_p^k + u_p^{k+1})(\mathrm{mod}\, 2) = 2) \wedge (u_p^{\kappa'} = 0) \wedge \\
&\quad \wedge (u_p^{\kappa'+1} = 1) \wedge (\sum_{k=1}^{\kappa'} u_l^k = 0))\}. \qquad (11.30)
\end{aligned}$$

β) A subset $\mathscr{U}_\gamma'$, whose elements have the structure $\underline{\mathbf{u}'} = \mathbf{u}$ that cannot be extended because constraint $\alpha.1$ is already satisfied, can be eliminated from further procedure if constraint $\gamma.1$ is not satisfied. In this case the upper bound won't be calculated for the subset. However, if constraint $\gamma.1$ is satisfied, the subset remains in the procedure, and the upper bound will be calculated for the subset.

Subset $\mathscr{U}_\gamma'$ that can be eliminated from further procedure is defined by the following expression:

$$\mathcal{U}'_\gamma = \{([\mathbf{u}^1, \mathbf{u}^2, \ldots, \mathbf{u}^\kappa], \tau)^T \mid (\mathbf{u}^1 \notin \Gamma_s \mathbf{u}^\kappa) \wedge (\sum_{k=1}^{\kappa} (u_p^k + u_p^{k+1})(\mathrm{mod}\,2) = 2) \wedge (p \in \mathcal{P})\} = \mathcal{U}'_{\gamma h} = \beta''_2(\bar{u}). \tag{11.31}$$

Subset $\mathcal{U}'_\gamma$ remains in the procedure if constraints α.1 and γ.1 are satisfied. In this case this subset can be defined as:

$$\mathcal{U}'_\gamma = \{([\mathbf{u}^1, \mathbf{u}^2, \ldots, \mathbf{u}^\kappa], \tau)^T \mid (\mathbf{u}^1 \in \Gamma_s \mathbf{u}^\kappa) \wedge \wedge ((\sum_{k=1}^{\kappa} (u_p^k + u_p^1)(\mathrm{mod}\,2) \le 2), (p \in \mathcal{P}))\} = \mathcal{U}'_{\gamma f}. \tag{11.32}$$

The elements of $\mathcal{U}'_\gamma$ in this case have the same structure but different cycle time split τ.

If subset $\mathcal{U}'_{\gamma h}$ is not an empty set, then

$$\beta''_2 = \mathcal{U}'_{\gamma h}.$$

γ) Subset $\widetilde{\mathcal{U}}'_\varepsilon$ is defined by the expression:

$$\widetilde{\mathcal{U}}'_\varepsilon = \mathcal{U}'_\varepsilon \setminus \{\mathbf{u}^b\} = \beta''_3(\bar{u}), \tag{11.33}$$

where $\mathbf{u}^b$ is the optimal solution.

The elements of this subset are the solutions that are either nonoptimal or infeasible.

Collection β" containing the subsets that are elements of collection $\bar{u}_0$ can be defined by the expression:

$$\beta''(\bar{u}) = \beta''_1(\bar{u}) \cup \beta''_2(\bar{u}) \cup \beta''_3(\bar{u}). \tag{11.34}$$

Obviously, $\beta''(\bar{u}) \subset \bar{u}_0$.

d) Obtaining $\beta(\bar{u}^n)$ collection

Starting with collection $\bar{u}^n$, a new collection $\beta(\bar{u}^n)$ is obtained by applying the branching function (11.16).

The new collection, $\beta(\bar{u}^n)$, is formed by excluding from collection $\bar{u}^n$ the subset $\mathcal{U}_\gamma$ that was divided to subsets using the branching rule. After that, the union of collections $(\bar{u}^n \setminus \{\mathcal{U}_\gamma\})$ and $\beta_1(\bar{u}^n) = \beta'(\mathcal{U}_{\mathrm{a}}) \setminus \beta''(\bar{u}^n)$ is formed. It means that to collection $\bar{u}^n$, from which subset $\mathcal{U}_\gamma$ is excluded, another collection $\beta_1(\bar{u}^n)$ is added. Collection $\beta_1(\bar{u}^n)$ is obtained when collection $\beta''(\overline{\mathcal{U}}^n)$, containing the subsets whose elements cannot be optimal solutions, is excluded from collection $\beta'(\mathcal{U}_\gamma)$ obtained by branching subset $\mathcal{U}_\gamma$.

Therefore,

$$\beta(\overline{\mathcal{U}}^n) = (\overline{\mathcal{U}}^n \setminus \{\mathcal{U}_\gamma\}) \cup \beta_1(\overline{\mathcal{U}}^n), \tag{11.35}$$

where

$$\beta_1(\overline{\mathcal{U}}^n) = \beta'(\mathcal{U}_{\bar{a}}) \setminus \beta''(\overline{\mathcal{U}}^n).$$

Mapping β is illustrated in Fig. 11.3.

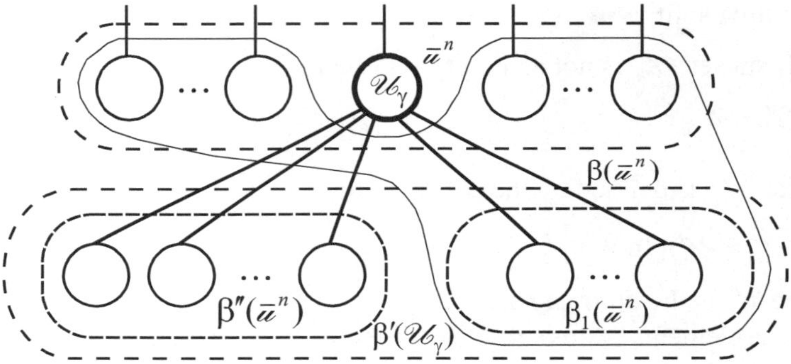

Figure 11.3

Indices of new sets $\mathcal{U}_{\gamma r}$, obtained by branching of subsets $\mathcal{U}_\gamma$, will be changed after constructing collection $\overline{\mathcal{U}}^{n+1}$. The new subsets will be also marked by indices, whereas the values of these indices will be the numbers following the number of the greatest index in $\overline{\mathcal{U}}^n$.

Example 11.2

For the intersection whose graph G_s is presented in Fig. 8.17, determine collection $\beta(\overline{\mathcal{U}}^n)$ if collection $\overline{\mathcal{U}}^n$ is known, and its subset $\mathcal{U}_\gamma$ is chosen for branching. Subset $\mathcal{U}_\gamma$ contains elements of the form

$$(\mathbf{u}, \tau)^T = ([\underline{\mathbf{u}}', \underline{\mathbf{u}}''], \tau)^T \in \mathcal{U}_\gamma,$$

where

$$\mathbf{u}'(\mathscr{U}_\gamma)=[\mathbf{u}^1,\mathbf{u}^2,\ldots,\mathbf{u}^\kappa]=[\mathbf{u}^1,\mathbf{u}^2,\ldots,\mathbf{u}^8]=\begin{array}{c} \begin{array}{cccccccc} \mathbf{u}^1 & \mathbf{u}^2 & \mathbf{u}^3 & \mathbf{u}^4 & \mathbf{u}^5 & \mathbf{u}^6 & \mathbf{u}^7 & \mathbf{u}^8 \end{array} \\ \begin{bmatrix} 1 & 1 & 1 & 1 & 0 & 0 & 0 & 0 \\ 1 & 0 & 0 & 0 & 0 & 0 & 0 & 0 \\ 0 & 0 & 0 & 1 & 1 & 0 & 0 & 0 \\ 0 & 0 & 0 & 0 & 0 & 0 & 1 & 0 \\ 1 & 1 & 0 & 0 & 0 & 0 & 0 & 1 \\ 0 & 0 & 0 & 0 & 0 & 0 & 0 & 0 \end{bmatrix} \end{array}.$$

Part $\underline{\mathbf{u}}'(\mathscr{U}_\gamma)$ of the structure is present in all elements of subset $\mathscr{U}_\gamma$. This part of the structure is marked by the bold line on graph G_s in Fig. 11.4.

When determining collection $\beta(\bar{\mathscr{u}}^n)$, according to expression (11.35), it is necessary to determine $\beta_1(\bar{\mathscr{u}}^n)$, and to determine that collection, first $\beta'(\mathscr{U}_\gamma)$ and $\beta''(\bar{\mathscr{u}}^n)$, have to be determined.

α) *Collection* $\beta'(\mathscr{U}_\gamma)$ *determination*

According to expression (11.23):

$$\beta'(\mathscr{U}_\gamma)=\{\mathscr{U}_{\gamma r}\,|\,(u(\cdot)\in\mathscr{U}_\gamma)\Rightarrow(([\underline{\mathbf{u}}',\mathbf{u}^9,\underline{\mathbf{u}'''}],\tau)^T\wedge(\mathbf{u}^9\in\Gamma_s\mathbf{u}^8)\wedge$$

$$\wedge((\sum_{k=1}^{8}(u_p^k+u_p^{k+1})(\mathrm{mod}\,2)\le 2)$$

$$\wedge((\exists q\in\mathscr{P})\Rightarrow\sum_{k=1}^{\kappa}(u_q^k+u_q^{k+1})(\mathrm{mod}\,2)<2,$$

$$(p\in\mathscr{P}=\{1,2,\ldots,6\}))\}\cup\{\mathscr{U}'_\gamma\}.$$

It is necessary to determine the set of control vectors, represented by nodes of graph G_s, such that they are adjacent to the node that represents control vector $\mathbf{u}^8=(0,0,0,0,1,0)^T$ and structure $[\underline{\mathbf{u}}'(\mathscr{U}_\gamma),\mathbf{u}^9]$ satisfies constraint α'.1.

The set of nodes adjacent to the node representing $\mathbf{u}^8$ is (Fig. 11.4):

$$\Gamma_s(\mathbf{u}^8)=\Gamma_s(0,0,0,0,1,0)^T=\{(0,0,0,0,0,0)^T,(1,0,0,0,1,0)^T,(0,0,0,1,0,0)^T,$$
$$(0,0,0,0,0,1)^T,(0,0,0,0,1,1)^T,(0,1,0,0,0,0)^T,(0,0,0,1,1,0)^T,$$
$$(0,1,0,0,0,1)^T,(0,1,0,0,1,1)^T,(1,0,0,0,0,0)^T,(0,1,0,0,1,0)^T,$$
$$(1,1,0,0,1,0)^T,(1,1,0,0,0,0)^T\}.$$

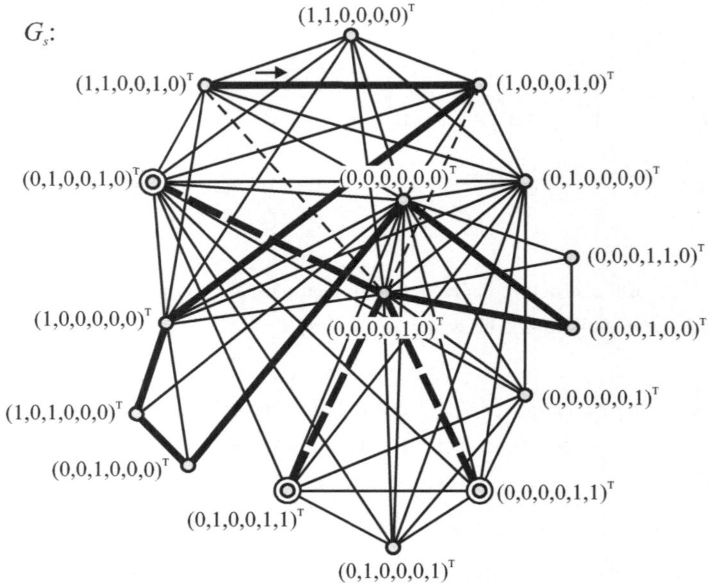

Figure 11.4

The constraint

$$\sum_{k=1}^{8}(u_p^k + u_p^{k+1})(\bmod 2) \le 2, \quad (p \in \mathscr{P})$$

is satisfied only if

$$\mathbf{u}^9 \in \{(0,0,0,0,1,1)^T, (0,1,0,0,1,1)^T, (0,1,0,0,1,0)^T, \\ (1,0,0,0,1,0)^T, (1,1,0,0,1,0)^T\} = \mathbf{U}_b^9 .$$

Hence,

$$\beta'(\mathscr{U}_\gamma) = \{\mathscr{U}_{\gamma 1}, \mathscr{U}_{\gamma 2}, \ldots, \mathscr{U}_{\gamma 5}, \mathscr{U}'_\gamma\} .$$

The structures of the elements belonging to these subsets (except to subset $\mathscr{U}'_\gamma$) are:

$$\mathbf{u}(\mathscr{U}_{\gamma r}) = [\underline{\mathbf{u}}'(\mathscr{U}_{\gamma r}), \underline{\mathbf{u}}''(\mathscr{U}_{\gamma r})] \\ = [\underline{\mathbf{u}}'(\mathscr{U}_{\gamma r}), \mathbf{u}^9, \underline{\mathbf{u}}'''(\mathscr{U}_{\gamma r})]$$

where $\underline{\mathbf{u}}'(\mathscr{U}_\gamma) = [\mathbf{u}^1, \mathbf{u}^2, \ldots, \mathbf{u}^8]$, and $\mathbf{u}^9 \in \mathbf{U}_b^9$, with $r \in \{1,2,\ldots,5\}$.

The structure of the elements of subset $\mathscr{U}'_\gamma$ is

$$\mathbf{u}(\mathscr{U}'_\gamma) = [\mathbf{u}^1, \mathbf{u}^2, \ldots, \mathbf{u}^8],$$

i.e., this structure does not have the "extension" $\underline{\mathbf{u}}''(\mathscr{U}'_\gamma)$.

For this structure it is satisfied that $\mathbf{u}^1 \in \Gamma_s \mathbf{u}^8$; however, the constraint

$$\sum_{k=1}^{8} (u_p^k + u_p^{k(\mathrm{mod}8)+1})(\mathrm{mod}\,2) = 2, \quad (p \in \{1,2,\ldots,6\})$$

is not satisfied. Therefore, $\mathscr{U}'_\gamma$ contains infeasible solutions.

In Fig. 11.4 the initial parts $[\underline{\mathbf{u}}'(\mathscr{U}_\gamma), \mathbf{u}^9]$ of the structures of elements belonging to subsets $\mathscr{U}_{\gamma r}$, obtained by branching are marked. Bold continuous line represents the structure $\underline{\mathbf{u}}'(\mathscr{U}_\gamma)$. Node $\mathbf{u}^8$ is connected to the "extensions" $\mathbf{u}^9 \in \mathbf{U}_b^9$ by dotted lines.

β) *Collection* $\beta''(\overline{\mathscr{u}}^n) = \beta_1'' \cup \beta_2'' \cup \beta_3''$ *determination*

This collection contains elements of collection $\beta'(\mathscr{U}_\gamma)$ having the initial part of their structures $[\underline{\mathbf{u}}'(\mathscr{U}_\gamma), \mathbf{u}^9]$ such that their "extension" further in the procedure cannot lead to subsets that contain feasible solutions.

Collection $\beta''(\overline{\mathscr{u}}^n)$ is determined using expression (11.34). When applying this formula, it can be noted that the initial part of the structure of collection $\beta'(\mathscr{U}_\gamma)$ elements, i.e., the structure $[\underline{\mathbf{u}}'(\mathscr{U}_\gamma), \mathbf{u}^9]$ has the following property:

$$\sum_{k=1}^{8} u_6^k = 0.$$

It means that control variable $u_6(\cdot)$, which controls signal group D_6', will obtain value 1 only in subsequent control vectors.

There can also be noted that

$$\sum_{k=1}^{8} (u_p^k + u_p^{k+1})(\mathrm{mod}\,2) = 2$$

in the following cases:

– for $p = 1$ if $\mathbf{u}^9 \in \{(1,0,0,0,1,0)^T, (1,1,0,0,1,0)^T\}$,

– for $p = 2$ if $\mathbf{u}^9 \in \{(0,1,0,0,1,1)^T, (0,1,0,0,1,0)^T, (1,1,0,0,1,0)^T\}$,

– for $p = 3$, $p = 4$, $p = 5$ if $\mathbf{u}^9 \equiv \mathbf{U}_b^9$.

If $\mathbf{u}^9$ were the element of set $\{(1,0,0,0,1,0)^T, (1,1,0,0,1,0)^T\}$, then u_1^k should be 1 until the end of the cycle in order to satisfy constraint α.1. In this case signal group D_6' could not gain the right-of-way because $(D_1', D_6') \notin C_g$. In all elements of set $\mathbf{U}_b^9$ the value of u_3^9 and u_4^9 is zero, and has to remain zero until the end of the cycle so that constraint α.1 is satisfied.

The initial part of the structure of $\mathscr{U}_\gamma$ elements, i.e., $\underline{\mathbf{u}}' = [\mathbf{u}^1, \mathbf{u}^2, \ldots, \mathbf{u}^8]$, is not the structure of a feasible element because $\mathbf{u}^1$ cannot be the next control vector in the sequence since signal group D_6' has not gained the right-of-way.

Therefore,

$$\beta'' = \{\mathscr{U}_{\gamma 4}, \mathscr{U}_{\gamma 5}, \mathscr{U}_\gamma'\}.$$

Here we also have

$$\underline{\mathbf{u}}(\mathscr{U}_{\gamma 4}) = [\underline{\mathbf{u}}'(\mathscr{U}_\gamma), (1,0,0,0,1,0)^T, \underline{\mathbf{u}}''(\mathscr{U}_{\gamma 4})],$$

$$\underline{\mathbf{u}}(\mathscr{U}_{\gamma 5}) = [\underline{\mathbf{u}}'(\mathscr{U}_\gamma), (1,1,0,0,1,0)^T, \underline{\mathbf{u}}''(\mathscr{U}_{\gamma 5})],$$

$$\underline{\mathbf{u}}'(\mathscr{U}_\gamma') = \underline{\mathbf{u}}'(\mathscr{U}_\gamma),$$

and $\underline{\mathbf{u}}''(\mathscr{U}_\gamma)$ does not exist as an "extension" of structure $\underline{\mathbf{u}}'(\mathscr{U}_\gamma')$.

γ) *Collection* $\beta_1(\bar{u}^n)$ *determination*

The elements excluded from further procedure are the elements of collection β", so that the elements taking part in further procedure are elements of collection

$$\begin{aligned}\beta_1(\bar{u}^n) &= \beta'(\mathscr{U}_\gamma) \setminus \beta''(\bar{u}^n) \\ &= \{\mathscr{U}_{\gamma 1}, \mathscr{U}_{\gamma 2}, \ldots, \mathscr{U}_{\gamma 5}, \mathscr{U}_\gamma'\} \setminus \{\mathscr{U}_{\gamma 4}, \mathscr{U}_{\gamma 5}, \mathscr{U}_\gamma'\} \\ &= \{\mathscr{U}_{\gamma 1}, \mathscr{U}_{\gamma 2}, \mathscr{U}_{\gamma 3}\}.\end{aligned}$$

The initial parts of the structure of elements that belong to subsets $\mathscr{U}_{\gamma 1}$, $\mathscr{U}_{\gamma 2}$, and $\mathscr{U}_{\gamma 3}$, can be represented in the form

$$\underline{\mathbf{u}}'(\mathscr{U}_{\gamma r}) = [\underline{\mathbf{u}}'(\mathscr{U}_\gamma), \mathbf{u}^9]$$

where $r \in \{1,2,3\}$, and

$$\mathbf{u}^9 \in \{(0,0,0,0,1,1)^T, (0,1,0,0,1,0)^T, (0,1,0,0,1,1)^T\}.$$

Control vectors $\mathbf{u}^9$ are marked in Fig. 11.4 by circles around the nodes representing them in graph G_s, and these nodes are connected to the node representing $\mathbf{u}^8$ by bold dashed lines.

δ) *Collection* $\beta(\bar{\mathscr{u}}^n)$ *determination*

The new collection $\beta(\bar{\mathscr{u}}^n)$ is cbtained by excluding from collection $\bar{\mathscr{u}}^n$ the subset $\mathscr{U}_\gamma$ for which the function $\beta_1(\bar{\mathscr{u}}^n)$ is determined, and including collection β_1 in the remaining part of $\bar{\mathscr{u}}^n$. This mapping can be expressed as follows:

$$\beta(\bar{\mathscr{u}}^n) = (\bar{\mathscr{u}}^n \setminus \{\mathscr{U}_\gamma\}) \cup \beta_1(\bar{\mathscr{u}}^n)$$
$$= (\bar{\mathscr{u}}^n \setminus \{\mathscr{U}_\gamma\}) \cup \{\mathscr{U}_{\gamma 1}, \mathscr{U}_{\gamma 2}, \mathscr{U}_{\gamma 3}\}.$$

11.1.4. Bounding rules

Branch-and-bounding procedures employ two types of bounds—a lower bound on J_c, and upper bounds on value of J'_c over subsets of $\mathscr{U}^s$.

a) The upper bound

The upper bounding rule is defined by the function

$$B : \xi \to \mathrm{R}, \tag{11.36}$$

where $\xi = \mathscr{P}(\mathscr{U}^s)$.

The properties of this function are:

$$J'_c(u(\cdot)) \le B(\mathscr{U}_\alpha), \quad (u(\cdot) \in \mathscr{U}_\alpha,\ \mathscr{U}_\alpha \in \beta'(\mathscr{U}_\gamma)), \tag{11.37}$$

$$B(\{u(\cdot)\}) = J'_c(u(\cdot)). \tag{11.38}$$

Function J'_c is an extension of the criterion function, with the property

$$u(\cdot) \in \mathscr{U}_f \Rightarrow J'_c(u(\cdot)) = J_c(u(\cdot)).$$

Thus, in each particular case, for a chosen criterion function J_c its extension J'_c has to be determined.

b) The lower bound

The lower bounding rule is the function

$$b : \overline{\mathscr{U}} \to \mathrm{R}, \tag{11.39}$$

with the following properties for any collection $\bar{\mathscr{u}}$:

$$\alpha)\quad b(\bar{\mathscr{u}}) \le J_c^*, \tag{11.40}$$

$$\beta)\quad ((u(\cdot) \in \mathscr{U}_f) \wedge (\{u(\cdot)\} \in \bar{\mathscr{u}})) \Rightarrow b(\bar{\mathscr{u}}) \ge J_c(u(\cdot)). \tag{11.41}$$

If during the procedure a collection $\overline{\mathcal{U}}$ is obtained, which contains several singleton feasible subsets, the lower bound will be equal to the best criterion function value, i.e.,

$$\begin{aligned} b(\overline{\mathcal{U}}) &= \max\{J_c'(u(\cdot)) \mid (\{u(\cdot)\} \in \overline{\mathcal{U}}) \wedge (u(\cdot) \in \mathcal{U}_f)\} \\ &= \max\{J_c(u(\cdot)) \mid (\{u(\cdot)\} \in \overline{\mathcal{U}})\}. \end{aligned} \tag{11.42}$$

If $b(\overline{\mathcal{U}})$ is the lower bound for collection $\overline{\mathcal{U}}$, then it can be claimed that the subset $\mathcal{U}_\delta \in \overline{\mathcal{U}}$ does not contain the optimal solution if

$$B(\mathcal{U}_\delta) < b(\overline{\mathcal{U}}). \tag{11.43}$$

From the lower bound definition, it can be noted that the lower bound cannot be determined for every collection $\overline{\mathcal{U}}$. Namely, the lower bound can be determined only for collections containing a singleton subset $\mathcal{U}_\eta = \{u(\cdot)\}$, with the property $u(\cdot) \in \mathcal{U}_f$.

The lower bound can be used to eliminate from further procedure the subsets whose upper bound is less than the lower bound of the collection to which the subset belongs.

11.1.5. Branch-and-bound recursive operation

Branch-and-bound recursive operation $\overline{B}$ uses the results of the previous steps to obtain a new collection $\overline{B}(\overline{\mathcal{U}})$ from collection $\overline{\mathcal{U}}$.

After applying branching and bounding rules and determining lower and upper bounds for the subsets obtained by branching, it can be noted that there exist subsets that contain infeasible or nonoptimal solutions. These subsets form the collection

$$\overline{\mathcal{U}}_- \subset \overline{\mathcal{U}} \tag{11.44}$$

that should be excluded from further procedure.

Branch-and-bound recursive operation is a function:

$$\overline{B} : \overline{\mathcal{U}} \to \overline{\mathcal{U}}, \tag{11.45}$$

where

$$\overline{B}(\overline{\mathcal{U}}) = \beta(\overline{\mathcal{U}}) \setminus \beta(\overline{\mathcal{U}})_-, \tag{11.46}$$

and $\beta(\overline{\mathcal{U}})_-$ is defined by (11.50).

a) Determination of collection $\overline{\mathcal{U}}_-$

Collection $\overline{\mathcal{U}}_-$ is a subset of collection $\overline{\mathcal{U}}_0$ containing all subsets with infeasible and nonoptimal solutions, i.e.,

$$\bar{\mathscr{u}}_- \subset \bar{\mathscr{u}}_0, \text{ and} \tag{11.47}$$

$$\underline{\mathscr{U}}(\bar{\mathscr{u}}_0) \subset \mathscr{U}^s \setminus \mathscr{U}^*. \tag{11.48}$$

Therefore, for each collection $\bar{\mathscr{u}}$, collection $\bar{\mathscr{u}}_-$ can be defined as follows:

$$\bar{\mathscr{u}}_- = (\bar{\mathscr{u}} \cap \bar{\mathscr{u}}_0) \cup \{\mathscr{U}_\alpha \mid (\mathscr{U}_\alpha \in \bar{\mathscr{u}}) \wedge (B(\mathscr{U}_\alpha) < b(\bar{\mathscr{u}}))\}. \tag{11.49}$$

By branch-and-bound recursive operation, collection $\bar{\mathscr{u}}_-$ is excluded from further procedure.

According to expression (11.49), we have:

$$\begin{aligned} \beta(\bar{\mathscr{u}})_- &= (\beta(\bar{\mathscr{u}}) \cap \bar{\mathscr{u}}_0) \cup \{\mathscr{U}_\alpha \mid (\mathscr{U}_\alpha \in \beta(\bar{\mathscr{u}}) \wedge (B(\mathscr{U}_\alpha) < b(\beta(\bar{\mathscr{u}}))\} \\ &= \beta''(\bar{\mathscr{u}}) \cup \{\mathscr{U}_\alpha \mid (\mathscr{U}_\alpha \in \beta(\bar{\mathscr{u}}) \wedge (B(\mathscr{U}_\alpha) < b(\beta(\bar{\mathscr{u}}))\}. \end{aligned} \tag{11.50}$$

It was possible to identify some elements of collection $\bar{\mathscr{u}}_-$ during branching operation, i.e., even before determining bounds for each subset obtained by branching. However, when determining bounds, it could be noted that for some subsets bounds cannot be determined. Namely, bounds are obtained by solving optimization problems, so that if no feasible solution exists, it means that some initial parts of structures cannot satisfy all constraints, which become functions of time variables after the initial part of the structure is determined.

Subsets $\mathscr{U}_\alpha$, with property $B(\mathscr{U}_\alpha) < b(\bar{\mathscr{u}})$, containing nonoptimal solutions and thus belonging to collection $\bar{\mathscr{u}}_-$, can be determined after the calculation of bounds.

Therefore, a collection is excluded from further procedure in the following cases:

a.1. The collection contains subsets with infeasible solutions.

a.2. The collection contains subsets with no optimal solution.

a.3. The collection contains subsets whose elements are members of other subsets also.

a1) Identification of collection $\bar{\mathscr{u}}_-^a$ containing infeasible solutions

During the procedure, when an initial part of signal plan structure is formed, it is possible to conclude that the given cycle time won't be long enough for the signal plan. Therefore, the subsets of solutions having that initial part of the structure can be eliminated from further procedure.

For example, one of possible control vector sequences in Example 11.1, which can be the initial part $\underline{\mathbf{u}}'$ of one subset of solutions satisfying constraints α'.1 and γ'.1, can be:

$$\underline{\mathbf{u}}' = [\mathbf{u}^1, \mathbf{u}^2, \ldots, \mathbf{u}^8] = \begin{array}{c} \begin{array}{cccccccc} \mathbf{u}^1 & \mathbf{u}^2 & \mathbf{u}^3 & \mathbf{u}^4 & \mathbf{u}^5 & \mathbf{u}^6 & \mathbf{u}^7 & \mathbf{u}^8 \end{array} \\ \begin{bmatrix} 1 & 0 & 0 & 0 & 0 & 0 & 0 & 0 \\ 0 & 0 & 0 & 0 & 0 & 0 & 1 & 0 \\ 0 & 0 & 0 & 0 & 0 & 0 & 0 & 0 \\ 0 & 0 & 0 & 0 & 1 & 0 & 0 & 0 \\ 0 & 0 & 0 & 0 & 0 & 0 & 0 & 0 \\ 0 & 0 & 1 & 0 & 0 & 0 & 0 & 0 \end{bmatrix} \end{array}.$$

The minimal cycle time necessary for any signal plan with $\underline{\mathbf{u}}'$ as the initial part of its structure must be greater than

$$c' = \tau^2_{\min} + \tau^3_{\min} + \ldots + \tau^7_{\min} = 4 + 16 + 8 + 15 + 1 + 15 = 59,$$

where terms in the sum are the minimal durations of control vectors $\mathbf{u}^2, \mathbf{u}^3, \ldots, \mathbf{u}^7$. Durations of $\mathbf{u}^1$ and $\mathbf{u}^8$ cannot be determined because for that it is necessary to know duration of control vectors that precede $\mathbf{u}^1$ and follow $\mathbf{u}^8$, and these vectors are not known yet.

Since $\underline{\mathbf{u}}'$ is not a complete structure ($8 < \mathrm{K}$), the cycle time c of any signal plan containing $\underline{\mathbf{u}}'$ in its structure cannot be shorter than c', i.e., feasible signal plans with $\underline{\mathbf{u}}'$ as the initial part of their structure have the property:

$$c > c' = 59 \text{ s}.$$

Therefore, if the given cycle time is less than 59 s, the signal plans having $\underline{\mathbf{u}}'$ as the initial part of their structure will not be feasible solutions. This fact is established when no solution can be found when determining the upper bound.

If the value of cycle time is not given, the minimal cycle time can be determined and no feasible signal plan can have its cycle time less than this value. It is obvious that a minimal cycle time has to exist because a feasible signal plan has to satisfy the constraints of minimal effective green time for each signal group and the minimal effective intergreen time constraints.

The minimum cycle time can be determined by solving the optimization problem whose statement and solution are presented in Section 14.2.

Collection $\bar{\underline{u}}^a$ that contains subsets with infeasible solutions, i.e., the solutions that need longer cycle time than the available c, can be defined as follows:

$$\overline{\mathscr{U}}_{-}^{a} = \{\mathscr{U}_{\delta} \mid ((u(\cdot) \in \mathscr{U}_{\delta}) \wedge$$
$$\wedge (u(\cdot) = ([\underline{\mathbf{u}}', \underline{\mathbf{u}}''], \tau)^T = ([(\mathbf{u}^1, \mathbf{u}^2, \ldots, \mathbf{u}^{\kappa}), \underline{\mathbf{u}}''], \tau)^T) \Rightarrow$$
$$\Rightarrow \min_{\tau^2, \ldots, \tau^{\kappa-1}} \sum_{k=1}^{\kappa-1} \tau^k > c\}. \quad (11.51)$$

a2) Determination of collection $\overline{\mathscr{U}}_{-}^{b}$ containing nonoptimal solutions

In Subsection 11.1.4. it is stated that the subset $\mathscr{U}_{\alpha} \subset \overline{\mathscr{U}}$, with the property

$$B(\mathscr{U}_{\alpha}) < b(\overline{\mathscr{U}})$$

does not contain optimal solutions. Thus, collection $\overline{\mathscr{U}}_{-}^{b}$, containing such type of subsets, can be defined as:

$$\overline{\mathscr{U}}_{-}^{b} = \{\mathscr{U}_{\alpha} \mid (\mathscr{U}_{\alpha} \in \overline{\mathscr{U}}) \wedge (B(\mathscr{U}_{\alpha}) < b(\overline{\mathscr{U}}))\}. \quad (11.52)$$

From expression (11.41), by which the lower bound is defined, it is clear that this bound cannot be determined at the beginning of the procedure. It can be determined only when singleton feasible subsets are obtained by branching operations. Thus, at the beginning of the procedure, until a feasible solution $u(\cdot) \in \mathscr{U}_f$ is obtained, we will have

$$\overline{\mathscr{U}}_{-}^{b} = \varnothing.$$

It means that this property cannot be used for elimination of some subsets from further procedure in the beginning of the procedure.

a3) Determination of collection $\overline{\mathscr{U}}_{-}^{c}$ containing subsets whose elements are members of other subsets also

Sometimes, when solving the optimization problem of determining the upper bound $B(\mathscr{U}_{\alpha})$ for a subset $\mathscr{U}_{\alpha} \subset \overline{\mathscr{U}}$, the solution may contain a control vector $\mathbf{u}^k$, the duration of which is zero. If it can be estimated that the duration of this control vector will remain zero in solutions of upper bound problems for subsets created by branching of $\mathscr{U}_{\alpha}$, and further branching of these subsets until the optimal solution is obtained (if it exists), then the subset $\mathscr{U}_{\alpha}$ can be eliminated from further procedure.

If elements $u(\cdot)$ of subset $\mathscr{U}_{\alpha}$ can be presented in the form:

$$u(\cdot) = ([\underline{\mathbf{u}}', \underline{\mathbf{u}}''], \tau)^T \in \mathscr{U}_{\alpha},$$

where $\underline{\mathbf{u}'} = [\mathbf{u}^1, \mathbf{u}^2, \ldots, \mathbf{u}^{k-1}, \mathbf{u}^k, \mathbf{u}^{k+1}, \ldots, \mathbf{u}^\kappa]$, then in the case $\tau^k = 0$, $\underline{\mathbf{u}'}$ becomes

$$\underline{\mathbf{u}'} = [\mathbf{u}^1, \mathbf{u}^2, \ldots, \mathbf{u}^{k-1}, \mathbf{u}^{k+1}, \ldots, \mathbf{u}^\kappa].$$

The fact that the duration of control vector $\mathbf{u}^k$ can be zero leads to the conclusion that the sequence $(\mathbf{u}^{k-1}, \mathbf{u}^{k+1})$ has the property:

$$\mathbf{u}^{k+1} \in \Gamma_s \mathbf{u}^{k-1}, \tag{11.53}$$

i.e., constraint $\gamma'.1$ is satisfied for this sequence. It means that control vectors $\mathbf{u}^k$ and $\mathbf{u}^{k+1}$ belong to set $\Gamma_s \mathbf{u}^{k-1}$. When it becomes certain that $\tau^k = 0$, which is possible only when determining $B(\mathscr{U}_\alpha)$ for values $\kappa > k$, then branching of such a subset can be terminated.

When determining bounds $B(\mathscr{U}_\alpha)$, the value 0 can be obtained for τ^k, but later, when determining bounds B for subsets created by branching of $\mathscr{U}_\alpha$, it may happen that the value of τ^k, as an element of the optimal solution, becomes different than zero. Thus, if we consider $\mathscr{U}_\alpha$ as a candidate for elimination from further procedure, there has to be certified that τ^k value will remain 0 until the optimal solution is obtained, if it exists.

On the basis of stated properties of collection $\overline{\mathscr{u}}_-^c$, this collection can be defined by the following expression:

$$\begin{aligned}\overline{\mathscr{u}}_-^c = \{\mathscr{U}_\alpha \mid (u(\cdot) \in \mathscr{U}_\alpha) \wedge \\ \wedge (u(\cdot) = ([\underline{\mathbf{u}'}, \underline{\mathbf{u}''}], (\tau', \tau''))^T = \\ = ([[\mathbf{u}^1, \mathbf{u}^2, \ldots, \mathbf{u}^\kappa], \underline{\mathbf{u}''}], ((\tau^1, \tau^2, \ldots, \tau^\kappa), \tau''))) \\ \wedge ((\exists (u(\cdot) = (\mathbf{u}, \tau)^T)) \Rightarrow (\tau^k = 0) \wedge (k < \kappa))\}.\end{aligned} \tag{11.54}$$

Upper bounds are determined as solutions of optimization problems, the type of which depends on the adopted optimality criterion and constraints. The criterion and constraints can be linear, or criterion or constraints, or both, can be nonlinear. Hence, the problems of upper bound determination can be stated as problems of linear or nonlinear mathematical programming.

a4) Obtaining collection $\overline{\mathscr{u}}_-$

As already mentioned, collection $\overline{\mathscr{u}}_-$ contains subsets that should be eliminated from further branch-and-bound procedure. This collection consists of collection $\overline{\mathscr{u}}_-^a$, which contains infeasible solutions, collection

$\bar{u}_-^b$, containing nonoptimal solutions, and collection $\bar{u}_-^c$, containing subsets whose elements are members of other subsets also. Therefore,

$$\bar{u}_- = \bar{u}_-^a \cup \bar{u}_-^b \cup \bar{u}_-^c . \qquad (11.55)$$

Since $\bar{u}_-^a \subset \bar{u}_0$ and $\bar{u}_-^b \subset \bar{u}_0$, expression (11.55) can be written as follows:

$$\bar{u}_- = (\bar{u} \cap \bar{u}_0) \cup \bar{u}_-^c .$$

b) Obtaining the sequence of collections by branch-and-bound recursive operation and determination of conditions for its termination

b1) Obtaining the sequence of collections

By applying the operation $\overline{B}(\bar{u})$ (11.45) to an initial collection $\bar{u}^1$, collection $\bar{u}^2$ is obtained. Further applications of $\overline{B}(\bar{u})$ yield the sequence $[\bar{u}^1, \bar{u}^2, \ldots, \bar{u}^n, \ldots, \bar{u}^\nu]$. The elements of the sequence are related according to the following expression:

$$\bar{u}^{n+1} = \overline{B}(\bar{u}^n) = \beta(\bar{u}^n) \setminus \beta(\bar{u}^n)_- , \qquad (11.56)$$

where $\bar{u}^1 = \{\mathcal{U}^s\} \in \overline{\mathcal{U}}$.

b2) Conditions for termination of branch-and-bound procedure

Branch-and-bound procedure terminates when one of the following conditions is met:

α. Collection $\bar{u}^\nu$ is the empty set:

$$\bar{u}^\nu = \overline{B}(\bar{u}^{\nu-1}) = \beta(\bar{u}^{\nu-1}) \setminus \beta(\bar{u}^{\nu-1})_- = \varnothing . \qquad (11.57)$$

It follows, further:

$$\beta(\bar{u}^{\nu-1}) = \beta(\bar{u}^{\nu-1})_- = \bar{u}_-^\nu . \qquad (11.58)$$

Obviously, if the branching operation yields a collection $\beta(\bar{u}^{\nu-1})$, which contains only infeasible solutions, the procedure should terminate and the problem has no solution.

β. Collection $\bar{u}^\nu$ contains only singleton subsets

If collection $\bar{u}^\nu$ contains only singleton subsets, then this collection contains subsets whose elements are optimal solutions:

$$\bar{u}^* = \{\{u^*(\cdot)\} \mid u^*(\cdot) \in \mathcal{U}^*\} . \qquad (11.59)$$

This statement is true because of the following reasons:

I Collection $\bar{\mathcal{U}}^{\nu}$ contains subsets with feasible solutions because the elements of this collection are singleton subsets, whereas infeasible elements are members of subsets included in collection $\bar{\mathcal{U}}_{-}^{\nu}$ that was excluded from the procedure.

II The upper bound $B(\{u(\cdot)\} \in \bar{\mathcal{U}}_{-}^{\nu})$ is equal to the lower bound $b(\bar{\mathcal{U}}^{\nu})$, and for all elements of collection $\bar{\mathcal{U}}^{\nu}$ these bounds are the same. If some bound $B(\{u(\cdot)\})$ were less than $b(\bar{\mathcal{U}}^{\nu})$, such a subset would be a member of collection $\bar{\mathcal{U}}_{-}^{\nu}$ that was excluded from the procedure.

Therefore, all elements of collection $\bar{\mathcal{U}}^{\nu}$ are subsets that contain optimal solutions. The values of all upper bounds are the same, equal to the value of the lower bound of the collection, and equal to the optimal value of the criterion.

Hence,

$$\bar{\mathcal{U}}^{\nu} = \{\{u(\cdot)\} \mid u(\cdot) \in \mathcal{U}_f\} \Rightarrow (\bar{\mathcal{U}}^{\nu} = \bar{\mathcal{U}}^*). \tag{11.60}$$

This collection has the property:

$$\bar{B}(\bar{\mathcal{U}}^{\nu}) = \beta(\bar{\mathcal{U}}^{\nu}) \setminus \beta(\bar{\mathcal{U}}^{\nu})_{-} = \bar{\mathcal{U}}^{\nu} \setminus \varnothing = \bar{\mathcal{U}}^{\nu+1} = \bar{\mathcal{U}}^{\nu} = \bar{\mathcal{U}}^*. \tag{11.61}$$

The termination condition can be stated in the following way: When $\bar{B}(\bar{\mathcal{U}}^{n}) = \bar{\mathcal{U}}^{n+1} = \bar{\mathcal{U}}^{n}$, the procedure should terminate because collection $\bar{\mathcal{U}}^{n}$ contains subsets whose elements are optimal solutions. The values of upper and lower bounds are then the same and equal to the optimal criterion value:

$$\bar{B}(\{\{u^{j}*(\cdot)\}\}) = b(\bar{\mathcal{U}}^*) = J_c'(u^{j}*(\cdot)) = J_c^* \\ (\{u^{j}*(\cdot)\} \in \bar{\mathcal{U}}^*). \tag{11.62}$$

The computer program STECSOT (STructurE and Cycle Split Optimization Technique) is developed for application of the algorithm, described in this part. The program is described in Appendix VII.

Part IV

DETERMINATION OF OPTIMAL CONTROL (SIGNAL PLAN)

In this part, the method for optimal signal plan determination, presented in Part III, is applied for solving several different problems of optimal traffic control.

All elements of the method are fully described in the procedure for determination of the optimal signal plan by which intersection capacity is maximized.

The method is then applied to several other problems, indirectly related to the intersection capacity. These are the problems of determining the signal plan that equalizes saturation degrees of vehicle traffic streams, and the signal plan that is optimal in the sense of capacity factor.

Another problem solved in this part is determination of the signal plan by which time losses, i.e., the total delay of vehicles on intersection approaches is minimized.

Problems of determining extreme values of some signal plan parameters are also solved here, using the method described in Part III. These parameters are: the cycle time—its minimal and maximal value, the number of control vectors in the signal plan—minimal and maximal number, etc.

12. CAPACITY OPTIMIZATION

The criterion function used for signal plan determination, as mentioned in Section 9.3, can be theoretical or practical capacity of one traffic stream, one signal group, or the whole intersection. Of course, instead of the capacity, the criterion function can be the capacity per cycle, i.e., the maximal number of vehicles of one traffic stream, signal group, or the whole intersection that can pass through the intersection during a cycle.

It means that there exist several capacity maximization problems. The optimization criteria in problem statements have to be expressed by variables **u** and τ, i.e., expression (4.19) should be used for green times of signal groups:

$$g_p = \sum_{k=1}^{K} u_p^k \tau^k = \mathbf{u}_p \cdot \tau^T , \quad (p \in \mathscr{P}) .$$

This is the expression for green time allocated to traffic stream σ_i, where $\sigma_i \in D'_p, (i \in \mathscr{J}, p \in \mathscr{P})$.

It has, also, to be taken into consideration that saturation flows of traffic streams depend on the decision of giving or not giving the simultaneous right-of-way to opposing and opposed traffic streams, i.e., whether the "filtering" is permitted or not. If filtering is permitted, then the saturation flow s_b of the opposed traffic stream, in interval *k*, depends on the control vector $\mathbf{u}^k$ and on the volume of the opposing traffic stream σ_a, i.e.,

$$s_b^k = s_b(\mathbf{u}^k, q_a) .$$

All expressions for capacity, given in Section 9.3, become functions of **u** and τ when substituting g_p $(p \in \mathscr{P}')$ in them with expression (4.19). Since all constraints are expressed as functions of **u** and τ, the capacity maximization problems become the problems of mathematical programming, in which the structure **u** and the cycle time split τ have to be determined so as to maximize the optimality criterion, i.e., the capacity.

When solving the problem of capacity maximization, for traffic stream, signal group, or whole intersection, the flow balance constraints α.4 are omitted from the problem statement. This means that in capacity maximization problems there do not exist the constraints ensuring that all vehicles coming to the intersection during a cycle can leave it in the same cycle.

12.1. The capacity per cycle

The expressions for capacity per cycle given here are explicit functions of **u** and τ variables.

- The capacity per cycle of a signal group D'_p (9.14):

$$J_c(u_p(\cdot)) = \overline{\omega}_p^g = s_p^g \sum_{k=1}^{\mathrm{K}} u_p^k \tau^k = s_p^g(\mathbf{u}_p \cdot \tau^T), \quad (p \in \mathscr{P}'), \tag{12.1}$$

where

$$s_p^g = \sum_{e=1}^{\mathrm{E}(p)} s_{pe}, \quad (p \in P').$$

If each traffic stream makes a signal group, i.e., if $\mathscr{I}' = \mathscr{P}'$, then the capacity per cycle of a signal group is in fact the capacity per cycle of the associated traffic stream (9.13). In this case:

$$\begin{aligned} J_c(u_i(\cdot)) &= \overline{\omega}_i^s = s_i \sum_{k=1}^{\mathrm{K}} u_p^k \tau^k = \\ &= s_i(\mathbf{u}_p \tau^T), \quad (p = i, i \in \mathscr{I}', \mathscr{I}' = \mathscr{P}'). \end{aligned} \tag{12.2}$$

- The capacity per cycle of an opposed traffic stream σ_b filtering through the opposing traffic stream σ_a:

$$J_c(u_b(\cdot)) = \sum_{k=1}^{\mathrm{K}} \left(u_a^k \cdot u_b^k \frac{q_a e^{-\alpha' q_a}}{1 - e^{-\beta_b q_a}} + u_b^k (1 - u_a^k) s_b \right) \tau^k. \tag{12.3}$$

- The intersection capacity per cycle (9.16):

$$J_{c1} = \overline{\Omega}^g = \sum_{p=1}^{\mathrm{P}'} s_p^g \cdot \sum_{k=1}^{\mathrm{K}} u_p^k \, \tau^k = \sum_{k=1}^{\mathrm{K}} \psi_{c1}^k \, \tau^k = \psi_{c1} \tau^T, \tag{12.4}$$

where

$$\psi_{c1} = [\psi_{c1}^1, \psi_{c1}^2, \ldots, \psi_{c1}^{\mathrm{K}}],$$

$$\psi_{c1}^k = \sum_{p=1}^{\mathrm{P}'} s_p^g u_p^k = s^g \mathbf{u}^k,$$

$$s^g = [s_1^g, s_2^g, \ldots, s_p^g, \ldots, s_{\mathrm{P}'}^g].$$

If each traffic stream makes a signal group, the intersection capacity per cycle is (9.15):

$$J_{c2} = \overline{\Omega}^s = \sum_{i=1}^{\mathrm{I}'} s_i g_p = \sum_{i=1}^{\mathrm{I}'} s_i \sum_{k=1}^{\mathrm{K}} u_i^k \tau^k = \sum_{k=1}^{\mathrm{K}} \psi_{c2}^k \tau^k = \psi_{c2} \tau^T , \qquad (12.5)$$

where:

$$\psi_{c2} = [\psi_{c2}^1, \psi_{c2}^2, \ldots, \psi_{c2}^{\mathrm{K}}] ;$$

$$\psi_{c2}^k = \sum_{i=1}^{\mathrm{I}'} s_i u_i^k = s\,\mathbf{u}^k , \quad (k \in \mathscr{K}),$$

$$s = [s_1, s_2, \ldots, s_i, \ldots, s_{\mathrm{I}'}] .$$

The optimal signal plan, maximizing the capacity per cycle, is obtained by solving the optimization problem with the criterion defined by one of expressions (12.1) to (12.5), subject to the constraints formulated in Section 8.9.

The problem of maximizing the capacity per cycle of one traffic stream, $J_c(u_p(\cdot))$, can be formulated in another way. The maximal number of vehicles of stream σ_i ($\mathscr{I}' = \mathscr{P}', i = p$) that can pass through the intersection during a cycle can be determined under the assumption that the flow balance constraints α.4 are satisfied for all other vehicle traffic streams.

Intersection traffic control by the optimal signal plan determined by solving the stated problem of maximizing the capacity $J_c(u_i(\cdot))$ of traffic stream σ_i makes sense only if the average volume of the stream is greater or equal to the capacity, i.e.,

$$q_i \geq J_{c2}^*(u_i(\cdot)) ,$$

where:

$$J_{c2}^*(u_i(\cdot)) = \max_{u(\cdot) \in \mathscr{U}_f} J_{c2}(u_i(\cdot)) .$$

If the average volume q_i of traffic stream σ_i is greater than $J_{c2}^*(u_i(\cdot))$, then stream σ_i is *saturated*.

It is possible, also, that this problem does not have any solution because constraints α.4 might not be satisfied for some other traffic streams. It means that queues on the approaches used by these streams will not discharge until the end of their green time. These streams are saturated, also.

The *intersection is saturated* if no queue can discharge until the end of its associated green interval. In this case, the maximal capacity value can be obtained by solving the optimization problem without flow balance constraints α.4. The effective green times in this case have only to satisfy minimal effective green time constraints, α.2, maximal effective red time constraints, α.3, and the constraints on the sum of control vector durations, ε.1 or ε.2.

The maximal capacity per cycle of a signal group, similarly to the maximal capacity of a traffic stream, can be determined as the solution of the optimization problem with constraints α.4 included or omitted.

12.2. Maximization of the intersection capacity per cycle

Application of the signal plan obtained by solving the capacity per cycle maximization problem makes sense only in the case when the intersection is saturated.

The number of vehicles that pass through an intersection during a cycle can be maximized only if some approach or the whole intersection is saturated. Therefore, the plan obtained as a solution of the capacity per cycle maximization problem should not be implemented in the cases when the average number of vehicles arriving to the intersection during one cycle can leave it during the same cycle.

The intersection capacity maximization problem, in the case when each traffic stream makes a signal group, can be formulated as follows:

Determine the signal plan

$$u(\cdot) = (\mathbf{u}, \tau)^T$$

that maximizes the function (12.5)

$$J_{c2} = \sum_{i=1}^{I'} s_i\, g_i = \sum_{i=1}^{I'} s_i \sum_{k=1}^{K} u_i^k\, \tau^k = \sum_{k=1}^{K} \psi_{c2}^k\, \tau^T ,$$

subject to constraints α.1, α.2, α.3, γ.1, γ.2, ε.2. (This means that this problem considers a saturated intersection, with no filtering permitted and no negative minimal effective intergreen times, so that constraints α.4.2 and δ are not included.)

In order to apply the described algorithm for solving this problem, it is necessary to determine the precise form of the algorithm elements introduced in Part III. These elements are defined as follows:

a) Relaxation

a1) The superset of the set of feasible solutions

Superset $\mathcal{U}^s \supset \mathcal{U}_f$ is defined by constraints α'.1, α'.2, α'.3, γ'.1, γ'.2, ε'.2 (Subsection 11.1.1).

a2) The bounded extension of function J_{c2}

The bounded extension of function J_c, as defined in Subsection 11.1.1, is defined by the mapping (11.11):

$$J'_c : \mathscr{U}^s \to \mathrm{R},$$

with the property

$$u(\cdot) \in \mathscr{U}_f \Rightarrow J'_c(u(\cdot)) = J_c(u(\cdot)).$$

In the problem of intersection capacity maximization,

$$J_c(u(\cdot)) = J_{c2}(u(\cdot)),$$

and the bounded extension of the criterion function $J_{c2}(u(\cdot))$ is:

$$J'_{c2}(u(\cdot)) = \sum_{i=1}^{\mathrm{I}'} s_i \sum_{k=1}^{\kappa} u_i^k \, \tau^k = \sum_{k=1}^{\kappa} \psi_{c2}^k \, \tau^k = \psi'_{c2} \tau'^T, \tag{12.6}$$

where:

$$\kappa \leq \mathrm{K},$$

$$\psi'_{c2} = [\psi_{c2}^1, \psi_{c2}^2, \ldots, \psi_{c2}^{\kappa}],$$

$\tau' = [\tau^1, \tau^2, \ldots, \tau^{\kappa}]$, and

$$\psi_{c2}^k = \sum_{i=1}^{\mathrm{I}'} s_i u_i^k = s\,\mathbf{u}^k, \quad (k \in \mathscr{K}),$$

where

$$s = [s_1, s_2, \ldots, s_i, \ldots, s_{\mathrm{I}'}],$$

$$\mathbf{u}^{kT} = (u_1^k, u_2^k, \ldots, u_{\mathrm{I}'}^k).$$

The component ψ'^k_{c2} of vector ψ'_{c2} represents the number of vehicles per second that would pass through the intersection if control vector $\mathbf{u}^k$ were applied. The volumes of all traffic streams gaining the right-of-way by control vector $\mathbf{u}^k$ are equal to saturation flow volumes.

b) The branching rule

Branching rule β, i.e., the function

$$\beta : \overline{\mathscr{U}} \to \overline{\mathscr{U}},$$

is defined in Subsection 11.1.3, and it is the same in all problems of optimal signal plan determination, and hence for capacity optimization problem as well. After the branching of subset $\mathscr{U}_\gamma$, i.e., after determining $\beta(\mathscr{U}_\gamma)$, bounds have to be calculated.

c) The bounding function

Two types of bounds are used in the branch-and-bound procedure (Subsection 11.1.4), upper and lower bound.

c1) Upper bound

The upper bound is defined as the function:

$$B : \xi \to R$$

with properties:

$$J'_{c2}(u(\cdot)) \le B(\mathscr{U}_\alpha), \quad (u(\cdot) \in \mathscr{U}_\alpha,\ \mathscr{U}_\alpha \in \beta(\mathscr{U}_\gamma)), \tag{12.7}$$

$$B(\{u(\cdot)\}) = J'_{c2}(u(\cdot)). \tag{12.8}$$

$J'_{c2}(u(\cdot))$ is the volume per cycle, i.e., the number of vehicles that pass through the intersection when traffic is controlled by signal plan $u(\cdot)$.

On the basis of property (12.7), $B(\mathscr{U}_\alpha)$ can be defined as:

$$B(\mathscr{U}_\alpha) = \max\{J'_{c2}(u(\cdot)) \mid u(\cdot) \in \mathscr{U}_\alpha\}, \tag{12.9}$$

which means that $B(\mathscr{U}_\alpha)$ is the maximal volume attainable if signal plan $u(\cdot)$ belongs to subset $\mathscr{U}_\alpha$.

At the beginning of the procedure $\mathscr{U}_\alpha = \mathscr{U}^s$, and expression (12.9) becomes:

$$B(\mathscr{U}^s) = \max\{J'_{c2}(u(\cdot)) \mid u(\cdot) \in \mathscr{U}^s\}. \tag{12.10}$$

The signal plans that satisfy given constraints belong to set $\mathscr{U}^s$. Structures of these signal plans have various lengths. The structures can be denoted as:

$$\mathbf{u} = [\underline{\mathbf{u}}', \underline{\mathbf{u}}''],$$

where

$$\underline{\mathbf{u}}' = [\mathbf{u}^1, \mathbf{u}^2, \ldots, \mathbf{u}^\kappa]$$

is the part of the structure that is the same for all signal plans belonging to one subset.

At the beginning of the procedure $\kappa = 1$ and an initial control vector $\mathbf{u}^1$ has to be chosen. In such a way the subset of signal plans is defined such that their structures have $\mathbf{u}^1$ as the initial part of the structure. A member of this subset is also the signal plan whose structure is

$$\mathbf{u} = \underline{\mathbf{u}}' = [\mathbf{u}^1].$$

It is suitable to choose as $\mathbf{u}^1$ the control vector $\widetilde{\mathbf{u}}^1$, which ensures the maximal flow volume through the intersection.

Therefore, applying the (infeasible) signal plan

$$u^1(\cdot) = (\underline{\mathbf{u}}', \tau^1)^T = ([\widetilde{\mathbf{u}}^1], c)^T \tag{12.11}$$

would provide passage of the maximal number of vehicles through the intersection during a cycle. Forming the control vector sequence by adding other control vectors after $\mathbf{u}^1$ will have as the consequence reduction of flow rate through the intersection.

The shaded area in Fig. 12.1 represents the number of vehicles that pass through the intersection if the structure of the signal plan used to control the traffic is $[\mathbf{u}^1, \mathbf{u}^2, \ldots, \mathbf{u}^\kappa]$. The maximal number of vehicles, when $\mathbf{u}^1$ is applied for control during the entire cycle time c (i.e., the upper bound $B(\mathscr{U}^s)$) is represented by the area of the rectangle whose edges are of length c and ψ^1_{c2}.

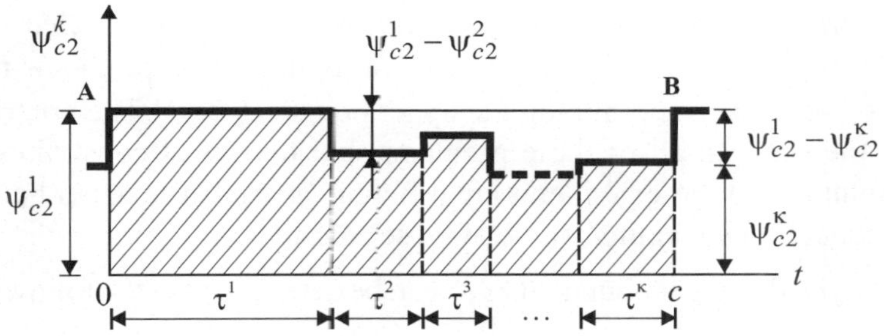

Figure 12.1

Obviously, the flow volume through the intersection per cycle is greater if the cycle duration is divided to κ control vectors than when divided to K control vectors, since $\mathrm{K} \geq \kappa$ (K is the number of control vectors in a feasible signal plan).

The maximal number of vehicles that pass through the intersection during a cycle, $c \cdot \psi^1_{c2}$, is reduced by adding control vector $\mathbf{u}^2$ to the sequence. Further extension of the control vectors sequence reduces the maximal flow rate value even more. Therefore, the maximal flow rate can be determined by optimal allocation of the cycle time c to control vectors $\mathbf{u}^1, \mathbf{u}^2, \ldots, \mathbf{u}^\kappa$.

The upper bound is determined according to expression (12.10) as follows:

$$\begin{aligned} B(\mathscr{U}^s) &= \max\{J'_{c2}(u(\cdot)) \mid u(\cdot) \in \mathscr{U}^s\} \\ &= \max\{\psi^1_{c2}(\mathbf{u}^1)\cdot c \mid \mathbf{u}^1 \in \mathbf{U}_f\} \\ &= c\cdot \max\{\psi^1_{c2}(\mathbf{u}^1) \mid \mathbf{u}^1 \in \mathbf{U}_f\} = \widetilde{\psi}^1_{c2}\cdot c\,, \end{aligned} \tag{12.12}$$

where:

$$\max\{\psi^1_{c2}(\mathbf{u}^1) \mid \mathbf{u}^1 \in \mathbf{U}_f\} = \widetilde{\psi}^1_{c2}\,.$$

The upper bound of subset $\mathscr{U}^s$ is

$$B(\mathscr{U}^s) = \widetilde{\psi}^1_{c2}\cdot c = c\cdot \max\{s\cdot \mathbf{u}^1 \mid \mathbf{u}^1 \in \mathbf{U}_f\} = c(s\cdot \widetilde{\mathbf{u}}^1)\,.$$

According to expression (12.9), $u(\cdot)\in\mathscr{U}_\alpha$ and $\mathscr{U}_\alpha$ is defined by constraints α'.1, α'.2, α'.3, γ'.1, γ'.2, ε'.2, meaning that element $u(\cdot)$ has to satisfy these constraints. In order to satisfy these constraints, it is necessary to know which control vector follows $\mathbf{u}^\kappa$. There are several control vectors $\mathbf{u}^{\kappa+1}$ and they all are elements of set $\beta_1(\mathscr{U}_\alpha)$. The choice of particular elements of set $\beta_1(\mathscr{U}_\alpha)$ enables formulation of different optimization problems that have to be solved when determining the upper bound $B(\mathscr{U}_\alpha)$. The criterion function is the same in all of these problems, but the set of constraints is different. Solutions of these problems are different, as well. The upper bound $B(\mathscr{U}_\alpha)$ is equal to the maximal value of these solutions.

Accordingly, the upper bound $B(\mathscr{U}_\alpha)$ can be determined in the following way:

$$\begin{aligned} B(\mathscr{U}_\alpha) &= \max\{J'_{c2}(u(\cdot)) \mid u(\cdot) \in \mathscr{U}_\alpha\} \\ &= \max_{\mathbf{u}^{\kappa+1}}\{\max_{\tau^1,\ldots,\tau^\kappa}\{\sum_{k=1}^{\kappa}\psi^k_{c2}\tau^k\}\} \end{aligned} \tag{12.13}$$

subject to given constraints. Besides this, there holds

$$\mathbf{u}^{\kappa(\mathrm{mod}\,K)+1} \in \Gamma_s\mathbf{u}^\kappa\,.$$

With regard to constraint ε'.2,

$$\sum_{k=1}^{\kappa}\tau^k = c\,,$$

τ^1 can be eliminated from expression (12.13) by substituting

$$\tau^1 = c - \sum_{k=1}^{\kappa}\tau^k\,.$$

Expression (12.13) for $B(\mathcal{U}_\alpha)$ then becomes

$$B(\mathcal{U}_\alpha) = \max_{\mathbf{u}^{\kappa+1}}\{\max_{\tau^2,\ldots,\tau^\kappa}\{\psi_{c2}^1 \cdot c - \sum_{k=2}^{\kappa}(\psi_{c2}^1 - \psi_{c2}^k)\tau^k\}\}. \quad (12.14)$$

When $\kappa = 1$, expression (12.13) is used rather than expression (12.14).

The maximal value of expression (12.14) is achieved, as shown in Fig. 12.1, when the nonshaded area in rectangle 0cBA is minimal.

Therefore, in determining the upper bound $B(\mathcal{U}_\alpha)$, the linear programming problem has to be solved several times (for different $\mathbf{u}^{\kappa+1}$).

c2) The lower bound

The lower bound, $b(\bar{u})$, for a collection $\bar{u}$ can be determined if the collection contains feasible solutions. In this case,

$$b(\bar{u}) = \max\{J_{c2}(u(\cdot)) \mid (\{u(\cdot)\} \in \bar{u})\}. \quad (12.15)$$

The method of maximal capacity determination is illustrated by the following examples.

Example 12.1

α) Determine the optimal signal plan maximizing the capacity of the intersection presented in Fig. 2.5, together with its compatibility graph G_s.

The cycle duration is $c = 90$ s. Saturation flow volumes and minimal effective green times for all traffic streams on this intersection are given in Table 12.1.

Table 12.1

i	1	2	3	4	5	6
s_i (veh/h)	1836	1650	1620	1650	1600	0
g_{mi} (s)	20	15	15	15	15	16

The matrix of minimal effective intergreen times (in seconds) is

$$Z = \begin{bmatrix} 0 & 0 & 0 & 4 & 0 & 4 \\ 0 & 0 & 3 & 5 & 0 & 0 \\ 0 & 3 & 0 & 3 & 5 & 2 \\ 2 & 1 & 2 & 0 & 0 & 2 \\ 0 & 0 & 1 & 0 & 0 & 0 \\ 8 & 0 & 4 & 8 & 0 & 0 \end{bmatrix}.$$

The problem has to be solved under the assumption that each traffic stream is a signal group.

The problem statement

Determine the signal plan to maximize criterion function

$$J_{c2}=\sum_{i=1}^{5} s_i \sum_{k=1}^{K} u_i^k \tau^k = \psi_{c2}\cdot\tau^T ,$$

subject to constraints α.1, α.2, γ.1, γ.2, ε.2, and

$$\psi_{c2}=[s\mathbf{u}^1 \;\; s\mathbf{u}^2 \;\; \ldots \;\; s\mathbf{u}^K], \quad (\mathbf{u}^k \in \mathbf{U}_f),$$

$$s=[s_1,s_2,\ldots,s_5]=[1850,1650,1620,1650,1600].$$

The set of feasible control vectors, $\mathbf{U}_f$, for this example is determined in Example 8.2 using CLIQ program:

$$\begin{aligned}\mathbf{U}_f &= \{\mathbf{u}(1),\mathbf{u}(2),\ldots,\mathbf{u}(r),\ldots,\mathbf{u}(16)\}\\ &= \{(0,0,0,0,0,0)^T,(1,0,0,0,0,0)^T,(0,1,0,0,0,0)^T,(0,0,1,0,0,0)^T,\\ &\quad (0,0,0,1,0,0)^T,(0,0,0,0,1,0)^T,(0,0,0,0,0,1)^T,(1,1,0,0,0,0)^T,\\ &\quad (1,0,1,0,0,0)^T,(1,0,0,0,1,0)^T,(0,1,0,0,1,0)^T,(0,1,0,0,0,1)^T,\\ &\quad (0,0,0,1,1,0)^T,(0,0,0,0,1,1)^T,(1,1,0,0,1,0)^T,(0,1,0,0,1,1)^T\}.\end{aligned}$$

The components of vector ψ_{c2} can be calculated for all feasible control vectors, i.e., elements of set $\mathbf{U}_f$. These components belong to the set

$$\begin{aligned}\overline{\psi}_{c2} &= \{s\mathbf{u}(1),\, s\mathbf{u}(2),\ldots,\, s\mathbf{u}(r),\ldots, s\mathbf{u}(R)\}\\ &= \{s\mathbf{u}(1),\, s\mathbf{u}(2),\ldots,\, s\mathbf{u}(r),\ldots, s\mathbf{u}(16)\}\\ &= \{0,\ 0.510,\ 0.458,\ 0.450,\ 0.458,\ 0.444,\ 0,\ 0.968,\\ &\quad 0.960,\ 0.902,\ 0.902,\ 0.458,\ 0.902,\ 0.444,\ 1.426,\ 0.902\},\end{aligned}$$

where:

$$\psi_{c2}^k \in \overline{\psi}_{c2}, \quad (k \in \mathcal{K}),$$

$$\mathcal{J}=\{1,2,3,4,5,6\}, \quad \mathcal{J}'=\mathcal{J}\setminus\{6\},$$

$$g_m=[g_{m1},g_{m2},\ldots,g_{m6}]=[20,15,15,15,15,16].$$

The control vector transition graph, $G_s=(\mathbf{U}_f,\Gamma_s)$, for this example is determined in Example 8.7α and shown in Fig. 8.17. The same graph is presented in Fig. 12.2 with value $s\mathbf{u}(r)$, corresponding to control vector $\mathbf{u}(r)$, written next to each node representing the control vector.

The elements of matrix Z^f, representing minimal effective intergreen times, are determined for each pair of control vectors using expression (8.61):

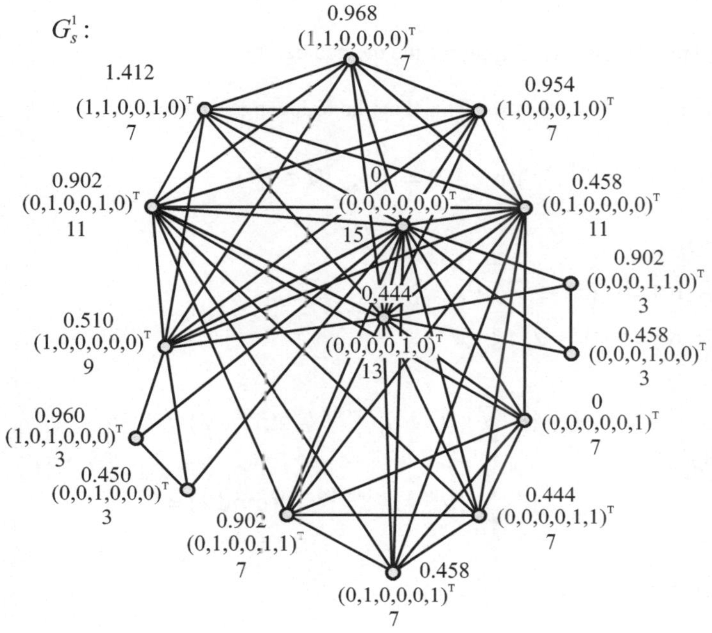

Figure 12.2

Z^f	$(0,0,0,0,0,0)^T$	$(1,0,0,0,0,0)^T$	$(0,1,0,0,0,0)^T$	$(0,0,1,0,0,0)^T$	$(0,0,0,1,0,0)^T$	$(0,0,0,0,1,0)^T$	$(0,0,0,0,0,1)^T$	$(1,1,0,0,0,0)^T$	$(1,0,1,0,0,0)^T$	$(1,0,0,0,1,0)^T$	$(0,1,0,0,1,0)^T$	$(0,1,0,0,0,1)^T$	$(0,0,0,1,1,0)^T$	$(0,0,0,0,1,1)^T$	$(1,1,0,0,1,0)^T$	$(0,1,0,0,1,1)^T$
$(0,0,0,0,0,0)^T$	0	0	0	0	0	0	0	0	0	0	0	0	0	0	0	0
$(1,0,0,0,0,0)^T$	0	0	0	0	4	0	4	0	0	0	0	4	4	4	0	4
$(0,1,0,0,0,0)^T$	0	0	0	3	5	0	0	0	3	0	0	0	5	0	0	0
$(0,0,1,0,0,0)^T$	0	0	3	0	3	5	2	3	0	5	5	3	5	5	5	5
$(0,0,0,1,0,0)^T$	0	2	1	2	0	0	2	2	2	2	1	2	0	2	2	2
$(0,0,0,0,1,0)^T$	0	0	0	1	0	0	0	0	1	0	0	0	0	0	0	0
$(0,0,0,0,0,1)^T$	0	8	0	4	8	0	0	8	8	8	0	0	8	0	8	0
$(1,1,0,0,0,0)^T$	0	0	0	3	5	0	4	0	3	0	0	4	5	4	0	4
$(1,0,1,0,0,0)^T$	0	0	3	0	4	5	4	3	0	5	5	4	5	5	5	5
$(1,0,0,0,1,0)^T$	0	0	0	1	4	0	4	0	1	0	0	4	4	4	0	4
$(0,1,0,0,1,0)^T$	0	0	0	3	5	0	0	0	3	0	0	0	5	0	0	0
$(0,1,0,0,0,1)^T$	0	8	0	4	8	0	0	8	8	8	0	0	8	0	8	0
$(0,0,0,1,1,0)^T$	0	2	1	2	0	0	2	2	2	2	1	2	0	2	2	2
$(0,0,0,0,1,1)^T$	0	8	0	4	8	0	0	8	8	8	0	0	8	0	8	0
$(1,1,0,0,1,0)^T$	0	0	0	3	5	0	4	0	3	0	0	4	5	4	0	4
$(0,1,0,0,1,1)^T$	0	8	0	4	8	0	0	8	8	8	0	0	8	0	8	0

The solution

The stated problem is solved applying the method described in Part III.

Relaxation: Superset $\mathscr{U}^s \supset \mathscr{U}_f$ is defined by the following constraints:

α'.1. $$\sum_{k=1}^{\kappa}(u_i^k + u_i^{k(\text{mod}K)+1})(\text{mod}\,2) = \theta_p\,, \quad (i \in \mathscr{J})$$

where

$$(\kappa < \text{K}) \Rightarrow ((\exists i \in \mathscr{J}), \theta_i < 2)$$

$$(\kappa = \text{K}) \Rightarrow (\theta_i = 2,\, i \in \mathscr{J})$$

α'.2. $$\sum_{k=1}^{\kappa} u_i^k \tau^k \geq \begin{cases} 0, & \text{if } \sum_{k=1}^{\kappa}(u_i^k + u_i^{k(\text{mod}K)+1})(\text{mod}\,2) < 2 \\ g_{mi}, & \text{if } \sum_{k=1}^{\kappa}(u_i^k + u_i^{k(\text{mod}K)+1})(\text{mod}\,2) = 2 \end{cases}, \quad (i \in \mathscr{J})$$

γ'.1. $$\mathbf{u}^{k(\text{mod}K)+1} \in \Gamma_s \mathbf{u}^k\,, \quad (k < \kappa)$$

γ'.2. $$\sum_{l=0}^{\omega-1} \tau^{\kappa-(\kappa+l-k)(\text{mod}K)} \geq \max\{z^f(\mathbf{u}^{\kappa-(\kappa+\omega-k)(\text{mod}K)}, \mathbf{u}^{k(\text{mod}K)})$$

$$(k \leq \kappa\,, \omega \in \Omega'(\mathbf{u}^k, \mathbf{u}^{k(\text{mod}K)+1})\}$$

ε'.2. $$\sum_{k=1}^{\kappa} \tau^k = c$$

Bounding rules

<u>The upper bound</u>

The upper bound $B(\mathscr{U}_\alpha)$ is determined by expression (12.14), i.e.,

$$B(\mathscr{U}_\alpha) = \max_{\mathbf{u}^{\kappa+1}}\{\max_{\tau^2,\ldots,\tau^\kappa}\{\psi_{c2}^1 \cdot c - \sum_{k=2}^{\kappa}(\psi_{c2}^1 - \psi_{c2}^k)\tau^k\}\}.$$

For $\kappa = 1$ the following expression is used:

$$B(\mathscr{U}^s) = \tilde{\psi}_{c2}^1 \cdot c\,.$$

In this example $c = 90\text{ s}$, $\tilde{\psi}_{c2}^1 = 1.412\text{ veh/s}$, and $\tilde{\mathbf{u}}^1 = (1,1,0,0,1,0)^T$, so that the value of the upper bound is

$$B(\mathscr{U}^s) = 90 \cdot 1.426 = 128.34\text{ veh}\,.$$

The upper bound, $B(\mathcal{U}_\alpha)$, is the solution of the optimization problem, which is obtained by solving several linear programming problems. As an example, the upper bound $B(\mathcal{U}_\alpha)$ is determined for subset $\mathcal{U}_\alpha$, the elements of which have the property that the initial part of their structure is the same and equal to

$$\underline{\mathbf{u}}'(\mathcal{U}_\alpha) = [\mathbf{u}^1, \mathbf{u}^2, \ldots, \mathbf{u}^8] = \begin{bmatrix} 1 & 1 & 1 & 1 & 0 & 0 & 0 & 0 \\ 1 & 0 & 0 & 0 & 0 & 0 & 0 & 0 \\ 0 & 0 & 0 & 1 & 1 & 0 & 0 & 0 \\ 0 & 0 & 0 & 0 & 0 & 0 & 1 & 0 \\ 1 & 1 & 0 & 0 & 0 & 0 & 0 & 1 \\ 0 & 0 & 0 & 0 & 0 & 0 & 0 & 0 \end{bmatrix}.$$

The initial part of the structure $\underline{\mathbf{u}}'(\mathcal{U}_\alpha)$ of subset $\mathcal{U}_\alpha$ elements is marked by the bold line on the control vector transition graph G_s in Fig. 11.4.

The value of coefficients ψ_{c2}^k, ($\mathbf{u}^k \in \mathbf{U}_f$) is assigned to each node of graph G_s in Fig. 12.2.

The expression for the upper bound $B(\mathcal{U}_\alpha)$ in this case is:

$$\begin{aligned} B(\mathcal{U}_\alpha) &= \max_{\mathbf{u}^9}\{\max_{\tau^2,\ldots,\tau^8}\{\psi_{c2}^1 \cdot c - \sum_{k=2}^{8}(\psi_{c2}^1 - \psi_{c2}^k)\tau^k\}\} \\ &= \max_{\mathbf{u}^9}\{\max_{\tau^2,\ldots,\tau^8}\{127.08 - (0.458\tau^2 + 0.902\tau^3 + 0.452\tau^4 + \\ &\qquad 0.962\tau^5 + 1.412\tau^6 + 0.954\tau^7 + 0.968\tau^8)\}\}. \end{aligned}$$

The set to which control vector $\mathbf{u}^9$ belongs is determined in Example 11.2:

$$\mathbf{u}^9 \in \{(0,0,0,0,1,1)^T, (0,1,0,0,1,1)^T, (0,1,0,0,1,0)^T\}.$$

The elements of this set are marked in Fig. 11.4 by circles around the nodes representing them in graph G_s.

The optimization problems can now be formulated in the following way:
Find the values of variables $\tau^2, \tau^3, \ldots, \tau^8$ so as to maximize expression

$$\{127.08 - (0.458\tau^2 + 0.902\tau^3 + 0.452\tau^4 + 0.962\tau^5 + 1.412\tau^6 + 0.954\tau^7 + 0.968\tau^8)\}.$$

The constraints the variables $\tau^2, \tau^3, \ldots, \tau^8$ have to satisfy depend on $\mathbf{u}^9$. The constraints are defined and the optimization problems solved for particular control vectors $\mathbf{u}^9$:

$\mathbf{u}^9 = (0,0,0,0,1,1)^T$

The constraints:

$$\begin{array}{llllllllll}
a) & \tau^1 + \tau^2 + \tau^3 + \tau^4 + \tau^5 + \tau^6 + \tau^7 + \tau^8 & = 90 \\
b) & \tau^2 + \tau^3 & \geq 3 \\
c) & \tau^3 & \geq 1 \\
d) & \tau^4 + \tau^5 & \geq 15 \\
e) & \tau^5 + \tau^6 & \geq 4 \\
f) & \tau^6 & \geq 3 \\
g) & \tau^7 & \geq 15 \\
h) & \tau^8 & \geq 2
\end{array}$$

The solution:

$$\tau^1 = 52\text{s},\ \tau^2 = 2\text{ s},\ \tau^3 = 1\text{ s},\ \tau^4 = 14\text{ s},$$

$$\tau^5 = 1\text{ s},\ \tau^6 = 3\text{ s},\ \tau^7 = 15\text{ s},\ \tau^8 = 2\text{ s},$$

and according to expression (12.13),

$$\max_{\tau^1,\ldots,\tau^8}\{\sum_{k=1}^{8}\psi_{c2}^k\tau^k\} = 97.49 \text{ veh}.$$

$\mathbf{u}^9 = (0,1,0,0,1,1)^T$

Constraints a) to g) are the same as for $\mathbf{u}^9 = (0,0,0,0,1,1)^T$; hence, the solution is the same.

$\mathbf{u}^9 = (0,1,0,0,1,0)^T$

The constraints a) to g) in this case are the same like in the previous two cases and the constraint h) is $\tau^8 \geq 1$.

The solution is:

$$\tau^1 = 53\text{ s},\ \tau^2 = 2\text{ s},\ \tau^3 = 1\text{ s},\ \tau^4 = 14\text{ s},$$

$$\tau^5 = 1\text{ s},\ \tau^6 = 3\text{ s},\ \tau^7 = 15\text{ s},\ \tau^8 = 1\text{ s}.$$

$$\max_{\tau^1,\ldots,\tau^8}\{\sum_{k=1}^{8}\psi_{c2}^k\tau^k\} = 99.096 \text{ veh}.$$

Thus,

$$B(\mathcal{U}_\alpha) = \max\{98.114; 98.114; 99.096\} = 99.096 \text{ veh}.$$

Optimal solution of this problem, $u^* = (\mathbf{u}^*_s\ \tau^*)^T$, is obtained using the program STECSOT (Appendix VII), which successively performs operations of branching, determining upper and lower bounds, and branch-and-bound recursive operation. The optimal structure and the optimal cycle time split are given by the following expressions:

$$\mathbf{u}^* = \begin{bmatrix} 1&1&1&1&0&0&0&0&0&0&0&0 \\ 1&0&0&0&0&0&0&0&0&1&1&1 \\ 0&0&0&1&1&0&0&0&0&0&0&0 \\ 0&0&0&0&0&0&1&1&0&0&0&0 \\ 1&1&0&0&0&0&0&1&1&1&1&1 \\ 0&0&0&0&0&0&0&0&0&0&1&0 \end{bmatrix}.$$

$$\tau^* = \begin{bmatrix} 28 & 2 & 1 & 14 & 1 & 3 & 2 & 13 & 1 & 1 & 16 & 8 \end{bmatrix}$$

The structure $\mathbf{u}^*$ of this solution is marked by the bold line on control vector transition graph G_s in Fig. 12.3.

The optimal criterion value is:

$$J_{c2}^* = J_{c2}(u^*(\cdot)) = J_{c2}(\mathbf{u}^*, \tau^*)^T = 91.48 \text{ veh}.$$

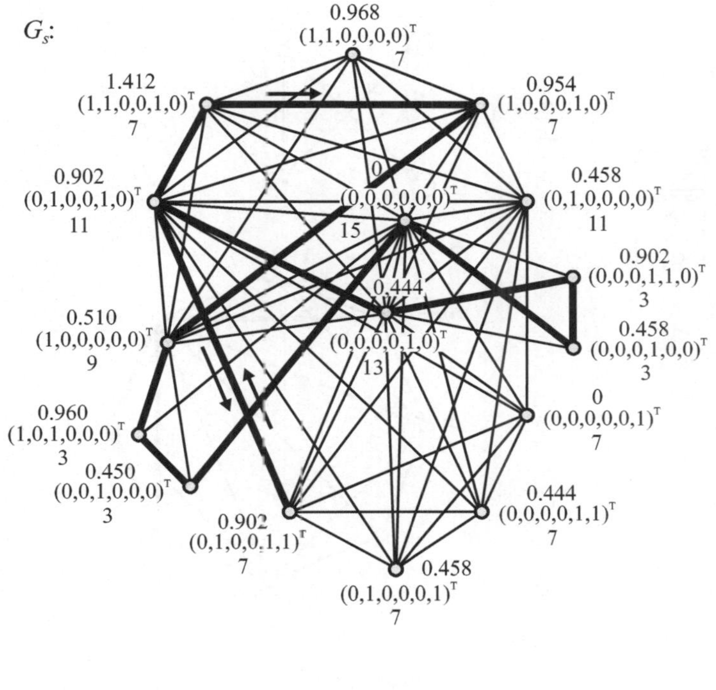

Figure 12.3

β) Determine the signal plan maximizing the capacity of the intersection presented in Fig. 12.4.

The intersection has four traffic streams which are, at the same time, the signal groups. Its compatibility graph G_g, control vector transition graph G_s, and minimal effective intergreen matrix are given in the same figure.

The cycle duration is 90 s.

The minimal effective green times and saturation flows are given in Table 12.2.

Table 12.2

i	1	2	3	4
s_i (veh/h)	0	1800	0	1600
g_{mi} (s)	10	15	10	10

One step of the solution procedure is graphically presented in Fig. 12.5.

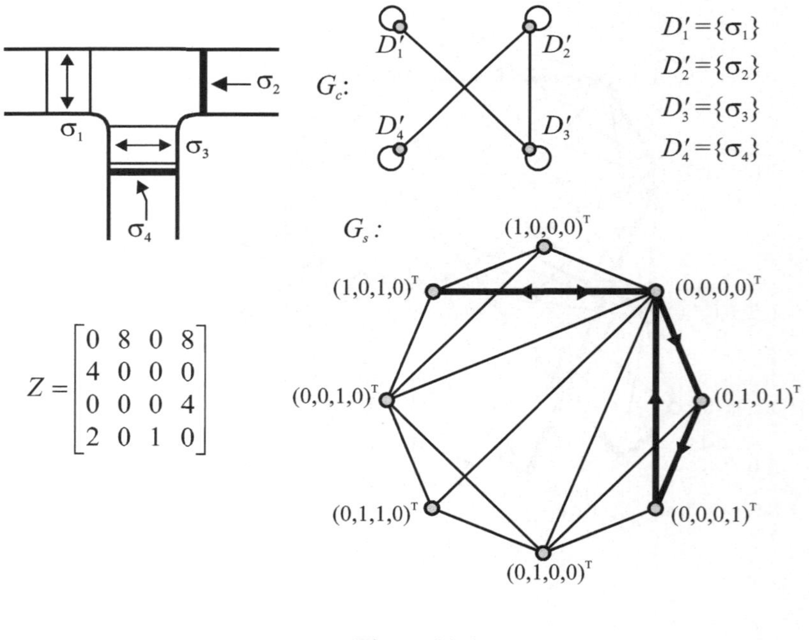

Figure 12.4

The procedure can be represented by a graph having the tree structure. The nodes of the graph represent subsets of set $\mathscr{U}^s$. The root of the tree is the subset with the property that all its members have the control vector ensuring maximal flow rate, $\tilde{\mathbf{u}}^1 = (0,1,0,1)^T$, as the initial part of their structure. Every node is connected to the root by a single path. Each subset $\mathscr{U}_\alpha$, represented by a node, has the property that all its members have the same initial part of the structure, $[\mathbf{u}^1,\mathbf{u}^2,\ldots,\mathbf{u}^\kappa]$, ($\mathbf{u}^1 = \tilde{\mathbf{u}}^1$). This sequence of control vectors is defined by the path from the root of the tree to the node that represents subset $\mathscr{U}_\alpha$. Next to each node, the mark of the last control vector in the sequence, $\mathbf{u}^\kappa$, is written.

The edges of this graph connected the nodes representing subsets $\mathscr{U}_\alpha$ with the nodes that are obtained by applying operation β to these subsets. This means that the node representing subset $\mathscr{U}_\alpha$ is connected to the node that "precedes" it, which represents the subset by whose branching subset $\mathscr{U}_\alpha$ evolved, and also to the nodes that "succeed" it, i.e., the nodes representing subsets obtained by subset $\mathscr{U}_\alpha$ branching. Thus, each node is connected to one "predecessor" node and one or more "successor" nodes. The exceptions, when considering collection $\overline{\mathscr{U}}^n$, are the following nodes:

a) The root of the tree, for which no predecessor exists.

b) "Leaves" of the tree, having no successor nodes. These leaves represent:
   - b1) The subsets that belong to collections $\overline{\mathscr{U}}_-^1$, $\overline{\mathscr{U}}_-^2$,..., $\overline{\mathscr{U}}_-^n$ that were excluded during the procedure because their elements were not candidates for branching in the next step of the procedure.
   - b2) The subsets that are candidates for further branching, i.e., elements of collection $\overline{\mathscr{U}}^n$.
   - b3) The singleton subsets, which are elements of collection $\overline{\mathscr{U}}^*$.

In the node, i.e., the ellipse representing subset $\mathscr{U}_\alpha$ in the graph (Fig. 12.5), the value of its upper bound $B(\mathscr{U}_\alpha)$ is written. This value was calculated in the step when this subset was a leaf of the tree.

The nodes marked by "▲" represent the subsets for which no further branching will be performed because their upper bound is less than or equal to the lower bound of the collection $\overline{\mathscr{U}}^{25}$. The value of the lower bound of the collection $\overline{\mathscr{U}}^{25}$ is

$$b(\overline{\mathscr{U}}^{25}) = 65.08 \text{ veh}.$$

The node that represents the subset containing one feasible solution, and for which the upper bound is equal to the value of optimality criterion chosen as the lower bound of collection $\overline{\mathscr{U}}^{25}$, is marked by bold line.

The nodes marked by "•" represent the subsets for which no further branching will be performed because the duration of some control vector in the part of the structure that is the same for all members of the subset represented by this node is equal to zero.

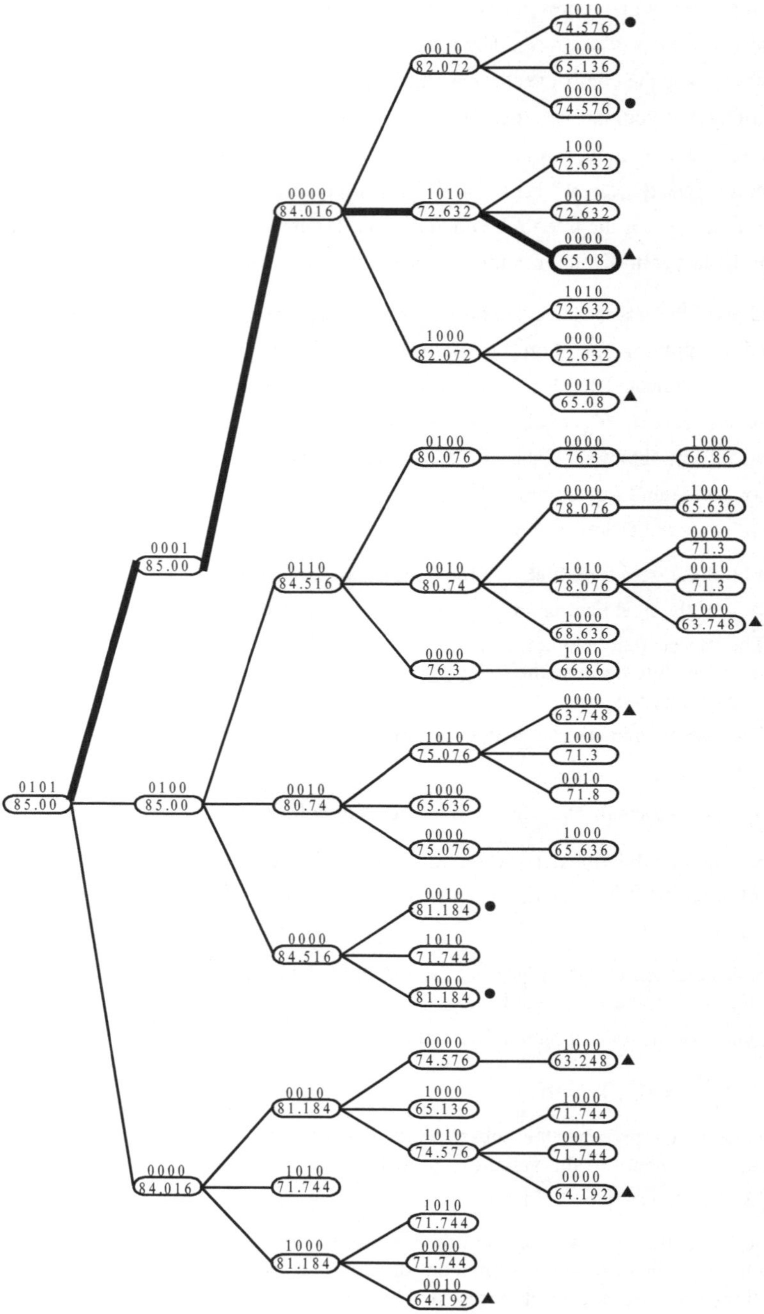
0101
85.00
0001
85.00
0100
85.00
0000
84.016
0000
84.016
0110
84.516
0010
80.74
0000
84.516
0010
82.072
1010
72.632
1000
82.072
1010
74.576
1000
65.136
0000
74.576
1000
72.632
0010
72.632
0000
65.08
1010
72.632
0000
72.632
0010
65.08
0100
80.076
0010
80.74
0000
76.3
0000
76.3
1000
66.86
0000
78.076
1000
65.636
1010
78.076
0000
71.3
0010
71.3
1000
63.748
1000
68.636
1000
66.86
1010
75.076
0000
63.748
1000
71.3
0010
71.8
1000
65.636
0000
75.076
1000
65.636
0010
81.184
1010
71.744
1000
81.184
0010
81.184
0000
74.576
1000
63.248
1000
65.136
1010
74.576
1000
71.744
0010
71.744
0000
64.192
1010
71.744
1000
81.184
1010
71.744
0000
71.744
0010
64.192

Figure 12.5

The optimal criterion value is

$$J_{c2}^* = J_{c2}(u^*(\cdot)) = J_{c2}(\mathbf{u}^*, \tau^*)^T = 65.08 \text{ veh}.$$

The optimal structure and cycle split in this solution is:

$$\mathbf{u}^* = \begin{bmatrix} 0 & 0 & 0 & 1 & 0 \\ 1 & 0 & 0 & 0 & 0 \\ 0 & 0 & 0 & 1 & 0 \\ 1 & 1 & 0 & 0 & 0 \end{bmatrix}.$$
$$\tau^* = \begin{bmatrix} 68 & 2 & 2 & 10 & 8 \end{bmatrix}$$

The structure of the optimal solution is shown on graph G_s (Fig. 12.4).

12.3. Equalizing saturation degrees of vehicle traffic streams

According to a standard heuristic rule, often used by traffic engineers, green time should be allocated to each vehicle traffic stream to ensure approximately the same saturation degrees for all vehicle traffic streams on the intersection. Webster [89] points out, and it is also suggested in some manuals [24], that such a way of green times determination leads, approximately, to the minimal delay of vehicles on the intersection.

The values of effective green times ensuring approximately the same saturation degrees of vehicle traffic streams, and satisfying all necessary constraints, can be determined by solving certain suitably stated optimization problems.

These optimization problems differ by the criterion that should be optimized. The suitable optimization criteria are: the mean square of differences between saturation degrees of traffic streams (9.19), the mean square of differences between reciprocal values of saturation degrees of traffic streams (9.20), the maximal saturation degree value (9.21). All these criteria should be minimized [35].

The capacity factor maximization (9.22) also contributes to the saturation degrees equalization. This problem, however, can be deduced to the problem of minimizing the maximal saturation degree.

In next subsections the optimization problems with the mentioned criteria are stated, together with examples of optimal signal plan determination.

12.3.1. Minimization of the sum of square differences between saturation degrees of traffic streams

The expression for this criterion (9.19) is given in Subsection 9.3.4. The problem of optimization of this criterion can be stated as follows: Determine the signal plan $u(\cdot) = (\mathbf{u}, \tau)^T$, so as to minimize the function:

$$\text{a)}\quad J_c = \sum_{i=1}^{I'-1} \sum_{j=i+1}^{I'} (\rho_i - \rho_j)^2 = c^2 \sum_{i=1}^{I'-1} \sum_{j=i+1}^{I'} \left(\frac{q_i}{s_i \sum_{k=1}^{K} u_p^k \tau^k} - \frac{q_j}{s_j \sum_{k=1}^{K} u_q^k \tau^k} \right)^2,$$

subject to constraints α.1, α.2, α.3, α.4, γ.1, γ.2, ε (ε.1 or ε.2).

In problems of practical capacity optimization these constraints are extended by constraints (9.18):

$$\rho_i \le \overline{\rho}_i, \quad (i \in \mathscr{I}'),$$

where $\overline{\rho}_i$ is the maximal acceptable saturation degree of traffic stream σ_i.

Equalization of saturation degrees can be achieved, also, by minimizing the criterion function (9.20):

$$\text{b)}\quad J_c = \sum_{i=1}^{I'-1} \sum_{j=i+1}^{I'} \left(\frac{1}{\rho_i} - \frac{1}{\rho_j} \right)^2 = \frac{1}{c^2} \sum_{i=1}^{I'-1} \sum_{j=i+1}^{I'} \left(\frac{s_i}{q_i} \sum_{k=1}^{K} u_p^k \tau^k - \frac{s_j}{q_j} \sum_{k=1}^{K} u_q^k \tau^k \right)^2,$$

subject to the given constraints.

The constraints α'.1, α'.2, α'.3, α'.4, γ'.1, γ'.2, ε'.1, or ε'.2, by which the superset $\mathscr{U}^s$ is defined, have to be used when calculating upper bounds (in this problem, the optimization means minimization).

The lower bound is obtained by minimizing function J'_c, which represents extension of criterion function J_c. This extension is obtained by substituting K in the criterion function by $\kappa \le K$, and the sum of square differences in this case contains only the elements with the property:

$$sign\left(\sum_{k=1}^{\kappa} u_p^k \cdot \sum_{k=1}^{\kappa} u_q^k \right) = 1.$$

The squares of differences of saturation degrees ρ_i and ρ_j are included in the expression for criterion function if $i < j$. The number of such elements is $\binom{I'}{2}$.

The following example illustrates determination of optimal signal plans by minimizing these two criteria.

Example 12.2.

α) Determine the signal plan for intersection presented in Fig. 2.5, with its transition graph given in Fig 8.18, so as to minimize the optimality criterion defined by expression (9.19). The signal plan structure is given, shown in Fig. 12.6, and defined by the following expression:

$$\mathbf{u} = \begin{bmatrix} 0 & 0 & 0 & 1 & 1 & 1 & 0 & 0 & 0 & 0 \\ 1 & 1 & 1 & 1 & 0 & 0 & 0 & 0 & 0 & 0 \\ 0 & 0 & 0 & 0 & 0 & 1 & 1 & 0 & 0 & 0 \\ 0 & 0 & 0 & 0 & 0 & 0 & 0 & 0 & 1 & 0 \\ 1 & 1 & 1 & 1 & 0 & 0 & 0 & 0 & 0 & 1 \\ 0 & 1 & 0 & 0 & 0 & 0 & 0 & 0 & 0 & 0 \end{bmatrix}.$$

Signal groups comprise single traffic streams ($p = i,\ \mathscr{P}' = \mathscr{J}'$).

Saturation flows, minimal effective green times, and traffic streams volumes are given in Table 12.3.

The values of minimal effective intergreen times are elements of matrix Z.

Cycle time is 90 s.

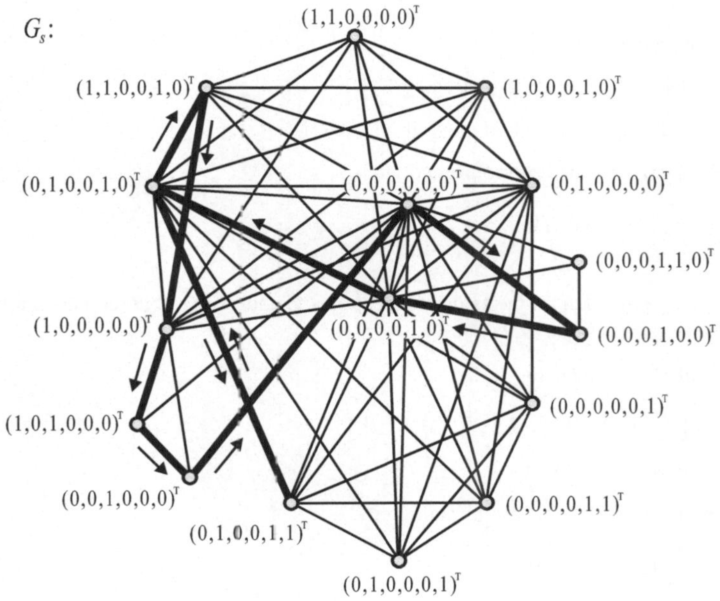

Figure 12.6

Table 12.3

i	1	2	3	4	5	6
s_i (veh/h)	1850	1650	1620	1650	1600	0
q_i (veh/h)	92.5	82.5	81	82.5	80	0
g_{mi} (s)	25	15	15	15	15	16
r_{Mi} (s)	70	75	75	70	85	–

$$Z = [z_{ij}]_{I \times I} = \begin{bmatrix} 0 & 0 & 0 & 4 & 0 & 4 \\ 0 & 0 & 3 & 5 & 0 & 0 \\ 0 & 3 & 0 & 3 & 5 & 2 \\ 2 & 1 & 2 & 0 & 0 & 2 \\ 0 & 0 & 1 & 0 & 0 & 0 \\ 6 & 0 & 8 & 6 & 0 & 0 \end{bmatrix}$$

The optimal cycle time split

$$\tau^* = [1, 17, 9, 0, 4, 23, 4, 4, 27, 1]$$

is obtained by applying the described method.

The optimal value of the criterion function is

$$J_c^* = 1.417234 \cdot 10^{-4}.$$

β) Determine the signal plan for the same intersection and data as in α, but with the optimality criterion defined by expression (9.21).

The optimal cycle time split

$$\tau^* = [1, 16, 8, 2, 3, 22, 5, 5, 27, 1]$$

is obtained in the same way as in α.

The optimal value of the criterion function is

$$J_c^* = 1.417235 \cdot 10^{-4}.$$

12.3.2. Minimization of the maximal saturation degree

As already pointed out in Subsection 9.3.4, minimization of the maximal saturation degree value leads to equalization of saturation degrees.

If g_p in expression (9.20), defining the optimization criterion, is substituted by:

$$g_p = \sum_{k=1}^{K} u_p^k \tau^k , \quad (p \in \mathscr{P}'),$$

the criterion function becomes:

$$\begin{aligned} J_c &= \max\{\rho_i \mid i \in \mathscr{I}'\} \\ &= \max\left\{ \frac{q_i\, c}{s_i \sum_{k=1}^{K} u_p^k \tau^k} \mid \sigma_i \in D'_p, i \in \mathscr{I}', p \in \mathscr{P}' \right\}. \end{aligned} \tag{12.16}$$

When signal groups are singleton sets, then $i = p$ ($i \in \mathscr{I}'$, $p \in \mathscr{P}'$, $\mathscr{I}' = \mathscr{P}'$).

The problem of minimizing the maximal saturation degree can be stated as follows: Determine the signal plan so as to minimize criterion (12.16), subject to constraints $\alpha.1$, $\alpha.2$, $\alpha.3$, $\alpha.4$, $\gamma.1$, $\gamma.2$, ε ($\varepsilon.1$ or $\varepsilon.2$).

This problem can be stated in another way by introducing variable ρ, defined as:

$$\rho = \max\{\rho_i \mid i \in \mathscr{I}'\}. \tag{12.17}$$

In this case the following inequalities hold:

$$\rho - \rho_i \geq 0 , \quad (i \in \mathscr{I}') \tag{12.18}$$

or

$$\rho \sum_{k=1}^{K} u_i^k \tau^k - \frac{q_i\, c}{s_i} \geq 0 , \quad (i \in \mathscr{I}'),$$

i.e.,

$$\sum_{k=1}^{K} u_i^k \tau^k - \frac{1}{\rho} \frac{q_i\, c}{s_i} \geq 0 , \quad (i \in \mathscr{I}'). \tag{12.19}$$

In the case of undersaturated intersection $\rho < 1$ or $\rho_i \leq \bar{\rho}_i$, ($i \in \mathscr{I}'$).

The problem of minimizing the maximal saturation degree can now be stated as follows: Determine the signal plan so as to minimize optimality criterion

$$J_c = \rho,$$

subject to constraints

$$\sum_{k=1}^{K} u_i^k \tau^k - \frac{1}{\rho}\gamma_i \geq 0, \quad (i \in \mathscr{I}'),$$

and α.1, α.2, α.3, γ.1, γ.2, and ε (ε.1 or ε.2).

If the notation

$$\frac{1}{\rho} = \mu \tag{12.20}$$

is introduced, the optimality criterion, which has to be minimized, becomes:

$$J_c = \frac{1}{\mu}.$$

Instead of (12.20), another criterion can be introduced:

$$\widetilde{J}_c = \mu, \tag{12.21}$$

which has to be maximized.

Substituting (12.20) in (12.19), the following expression is obtained:

$$\sum_{k=1}^{K} u_i^k \tau^k - \mu\gamma_i \geq 0, \quad (i \in \mathscr{I}'). \tag{12.22}$$

The problem of optimizing criterion (12.21), subject to constraints (12.22) and other given constraints, represents actually the problem of maximizing the capacity factor μ, which is considered in the next paragraph.

Expression (12.20) points out the fact that *the capacity factor is equal to the reciprocal value of maximal saturation degree.*

12.3.3. Capacity factor maximization

The capacity factor is one of generally accepted criteria, particularly suitable for determination of "long-term" signal plans for isolated intersections (Subsection 9.3.5), i.e., signal plans that will not be frequently changed. It means that the signal plan obtained by optimizing this criterion can be used in the conditions when flow volumes change [45].

There exist several variants of the capacity factor maximization problem. In some problems, the cycle time is given, and in others not. Also, problem formulations are different if "filtering" of some traffic stream through others is permitted or not.

In some cases, the problem of capacity maximization includes the maximal effective red constraints. These constraints are usually introduced with the intention to limit the length of certain queues that form during red signal indications. One consequence of maximal effective red constraints is that there exists a maximal value of cycle time duration. When these constraints are present, the maximal capacity value does not monotonously change with cycle changes from minimal to maximal cycle time [45]. Rather than that, the maximal capacity value increases with cycle time until a certain, optimal cycle time value is attained, and after that, until the maximal cycle time value, it decreases.

The problem statements and solved examples in the following text refer to two cases:

- Determination of the optimal signal plan when filtering of one traffic stream through others is not permitted.
- Determination of the optimal signal plan when filtering is permitted.

a) The capacity factor maximization if the filtering is not permitted

The problem of capacity factor maximization in the case when filtering is not permitted, for a given signal plan structure, becomes a linear programming problem. This problem, as already mentioned, can be solved with the cycle time given in advance, or the value of the cycle time is obtained as the solution of the problem. Both cases are formulated and illustrative examples are given below.

a1) The capacity factor maximization when the cycle time is not given

The optimal signal plan obtained as the solution of this problem will contain, also, the information on the optimal cycle time value.

This problem can be stated as follows: Find the signal plan $u(\cdot) = (\mathbf{u}, \tau)^T$ so as to minimize the capacity factor

$$J_c = \mu ,$$

subject to constraints $\alpha.1$, $\alpha.2$, $\alpha.3$, $\alpha.4.1$ (modified), β, $\gamma.1$, $\gamma.2$, $\varepsilon.2$, and the constraints defined by expression (9.18).

Constraints α.4, defined as:

$$\sum_{k=1}^{K} u_p^k \tau^k \geq \gamma_p, \quad (i \in \mathscr{P}'),$$

where, according to expression (8.14):

$$\gamma_p = \max\left\{\frac{c q_i}{\bar{\rho}_i s_i} \mid \sigma_i \in D'_p\right\} = c\gamma'_p =$$

$$= c \max\left\{\frac{q_i}{\bar{\rho}_i s_i} \mid \sigma_i \in D'_p\right\} \quad (p \in \mathscr{P}'), \tag{12.23}$$

are modified by multiplying each γ_p by μ ($p \in \mathscr{P}'$). Hence, the flow balance constraints in this case are defined by the following expression:

$$\sum_{k=1}^{K} u_p^k \tau^k - \mu\gamma_p \geq 0, \quad (p \in \mathscr{P}').$$

This inequality is obviously identical to inequality (12.22) in the case when each traffic stream makes a signal group.

After substituting

$$c = \sum_{k=1}^{K} \tau^k,$$

this constraint becomes:

$$\sum_{k=1}^{K} u_p^k \tau^k - \gamma'_p \cdot \mu \sum_{k=1}^{K} \tau^k \geq 0, \quad (p \in \mathscr{P}'). \tag{12.24}$$

The constraints formed by expression (12.24) are, obviously, nonlinear, and in this case the problem of optimal signal plan determination and determination of upper bounds are problems of nonlinear mathematical programming.

For solving this problem, the method described in Part III is used.

- *Superset* $\mathscr{U}^s$ of the set of feasible signal plans $\mathscr{U}_f$ is defined by constraints (Part III): α'.1, α'.2, α'.3, α'.4 (mod.), β, γ'.1, γ'.2, ε'.2, and constraint (9.18).
- *The extension of criterion* function J is defined in the same way as the optimality criterion, i.e.,

  $$J' = \mu.$$

- *The branching* rule is determined as described in Subsection 11.1.3.

- *The upper bound* is determined by solving the following optimization problem: Maximize function

  $$J' = \mu ,$$

  subject to constraints α'.1, α'.2, α'.3, α'.4 (mod.), β, γ'.1, γ'.2, ε'.1, and constraint (9.18).

- The choice of initial control vector $\mathbf{u}^1$

  When solving this problem, the initial control vector should be the vector that will surely be present in the optimal solution. However, in the majority of real problems, usually there does not exist a control vector, which must be a part of each feasible signal plan structure.

  The fact that each signal group has to gain its right-of-way once during the cycle (constraint α.1) can be used to determine the rule for choosing the initial vector $\mathbf{u}^1$.

  In the set of feasible control vectors, $\mathbf{U}_f$, there may exist several control vectors with the value of control variable $u_l(\cdot)$ equal to 1. One of these control vectors has to be an element of any feasible signal plan structure.

  The problem, thus, has to be solved several times, each time taking as the initial control vector $\mathbf{u}^1$ one of the control vectors having some control variable $u_l(\cdot)$ with value 1. In order to apply this method the minimal number of times possible, it is rational to choose the variable $u_l(\cdot)$ which has value 1 in the minimal number of control vectors, compared to other signal groups.

  If set $\mathbf{U}_{fl}$ is defined as:

  $$\mathbf{U}_{fl} = \{\mathbf{u}(r) \mid u_l(r) = 1\} , \tag{12.25}$$

  then the initial control vectors $\mathbf{u}^1$ will be chosen from set $\mathbf{U}_{fl^a}$, which has the following property:

  $$card\,\mathbf{U}_{fl^a} = \inf\left\{ card\,\mathbf{U}_{fl} \mid \mathbf{U}_{fl} \subset \mathbf{U}_f, \left(\bigcup_{l=1}^{I} \mathbf{U}_{fl} \right) \bigcup \{(0,0,\ldots,0)^T\} = \mathbf{U}_f \right\} .$$

For the intersection with four signal groups, presented in Fig. 12.4 in Example 12.2, the cardinal numbers of sets $\mathbf{U}_{fl}$ for particular values of l are given in Table IV.4.

Table 12.4

l	1	2	3	4
$card\ \mathbf{U}_{fl}$	2	3	3	2

Hence $l^a \in \{1,4\}$. Any of these two values can be chosen for l^a. In the case $l^a = 4$, the initial phase $\mathbf{u}^1$ takes first the value of the first, and then of the second control vector from set

$$\mathbf{U}_{f4} = \{(0,1,0,1)^T, (0,0,0,1)^T\}.$$

Example 12.3

Determine signal plan for the intersection presented in Fig. 2.5 to maximize its capacity factor.

The graph of control vectors transition, G_s, for this intersection is given in Fig. 8.17.

Each signal group comprises a single traffic stream ($\mathscr{P} = \mathscr{J}$).

The data on saturation flow volumes, minimal effective green times, maximal effective red times, and flow volumes of vehicle traffic streams are given in Table 12.5.

Minimal effective intergreen times are given as elements of matrix Z.

Table 12.5

i	1	2	3	4	5	6
s_i (veh/h)	1850	1650	1620	1650	1600	0
q_i (veh/h)	185	330	162	165	160	0
g_{mi} (s)	25	15	15	15	15	16
r_{Mi} (s)	60	65	65	60	75	–

$$Z = [z_{pq}]_{P\times P} = \begin{array}{c|cccccc} p\backslash q & 1 & 2 & 3 & 4 & 5 & 6 \\ \hline 1 & 0 & 0 & 0 & 4 & 0 & 4 \\ 2 & 0 & 0 & 3 & 5 & 0 & 0 \\ 3 & 0 & 3 & 0 & 3 & 5 & 2 \\ 4 & 2 & 1 & 2 & 0 & 0 & 2 \\ 5 & 0 & 0 & 1 & 0 & 0 & 0 \\ 6 & 8 & 0 & 4 & 8 & 0 & 0 \end{array}$$

The maximal acceptable saturation degrees are:

$$\overline{\rho}_i = 0.9\,, \quad (i \in \mathscr{Z}' = \mathscr{P}')\,.$$

Applying the described method the following optimal solution is obtained:

$$\begin{array}{c} \\ \mathbf{u}^* = \end{array} \begin{array}{c} \begin{array}{cccccccccc} \mathbf{u}^1 & \mathbf{u}^2 & \mathbf{u}^3 & \mathbf{u}^4 & \mathbf{u}^5 & \mathbf{u}^6 & \mathbf{u}^7 & \mathbf{u}^8 & \mathbf{u}^9 & \mathbf{u}^{10} \end{array} \\ \begin{bmatrix} 1 & 0 & 0 & 0 & 0 & 0 & 0 & 0 & 0 & 0 \\ 0 & 0 & 0 & 0 & 0 & 1 & 1 & 1 & 0 & 0 \\ 1 & 1 & 0 & 0 & 0 & 0 & 0 & 0 & 0 & 1 \\ 0 & 0 & 0 & 1 & 0 & 0 & 0 & 0 & 0 & 0 \\ 0 & 0 & 0 & 0 & 0 & 1 & 1 & 1 & 0 & 0 \\ 0 & 0 & 0 & 0 & 0 & 0 & 1 & 0 & 0 & 0 \end{bmatrix} . \end{array}$$

$$\tau^* = \begin{bmatrix} 25 & 6 & 3 & 18 & 1 & 1 & 16 & 0.5 & 3.5 & 4 \end{bmatrix}$$

The value of index l^a belongs to the set $\{3,4\}$. For both values of l^a the same optimal solution was obtained.

The graph of control vectors transition and the optimal signal plan structure are shown in Fig. 12.7.

The optimal criterion value and cycle time are:

$$\mu^* = 2.019 \text{ and } c^* = 78\text{ s}\,.$$

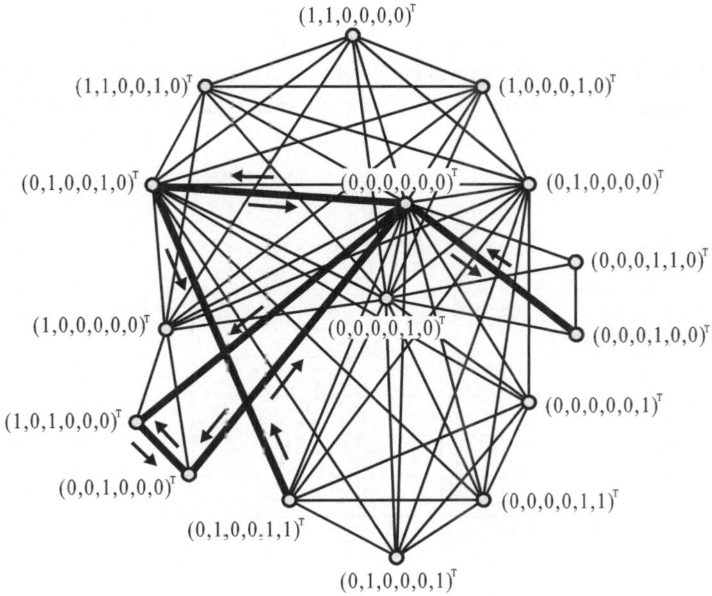

Figure 12.7

a2) The capacity factor maximization when the cycle time is given

The difference between the previous problem and this one is in the constraints. Instead of ε.2, constraint ε.1 is added to constraints α.1, α.2, α.3, α.4 (mod.), β, γ.1, γ.2.

The constraints defining superset $\mathcal{U}^s$ have to be extended by constraint ε.1.

The method applied to solve this problem is the same as in the previous case. The problems of determining the optimal signal plan and upper bounds in this problem are problems of linear programming.

Example 12.4

Find the optimal capacity factor values for the same intersection and data as in Example 12.3, with the cycle time assuming different values, from the minimal, $c_{\min} = 70$ s, to the maximal $c_{\max} = 86$ s value. The minimal and maximal cycle times were obtained as solutions of optimization problems formulated in Chapter 14.

The problem is solved for the sequence of cycle time values indicated in Fig. 12.8. This figure shows how the capacity factor changes as the function of cycle duration. This function, $\mu^*(c)$, has the maximal value

$$\mu^*_{\max} = 2.019\,,$$

for cycle time

$$c^* = 78 \text{ s}\,.$$

These values are the same as obtained in Example 12.3 when cycle time was not given in advance. The optimal signal plan for $c^* = 78$ s is, of course, equal to the optimal signal plan obtained in Example 12.3.

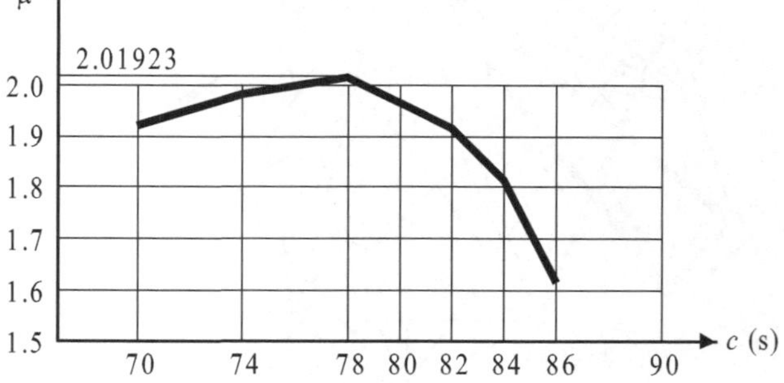

Figure 12.8

b) The capacity factor maximization in the case filtering is permitted

The problem of signal plan determination to maximize the capacity factor when filtering is permitted, for a known signal plan structure represents the problem of nonlinear mathematical programming [79].

This problem can be stated as follows: Find signal plan $u(\cdot) = (\mathbf{u}, \tau)^T$ to maximize the capacity factor

$$J_c = \mu ,$$

subject to constraints α.1, α.2, α.3, α.4.1 (mod.), α.4.2.I (mod.), α.4.2.II (mod.), α.4.2.III (mod.), β, γ.1 (mod.), γ.2, ε.1 (or ε.2).

Constraints α.4.1 (mod.) have to be satisfied by traffic streams that are neither opposed nor opposing. Constraints α.4.2.I (mod.) relate to opposing traffic streams, while constraints α.4.2.II (mod.) and α.4.2.III (mod.) refer to opposed traffic streams. Constraints γ.1 (mod.) point out that the control vectors transition graph, G_s, has to be modified in accordance with specific requirements posed in this problem, using the method described in Section 8.6.

The modification of constraints α.4.2.I, α.4.2.II, α.4.2.III, and α.4.1, means that all given average flow volumes, q_i, in the expressions that define these constraints have to be multiplied by the capacity factor, μ.

In Subsection 8.2.4 it was suggested that when filtering is permitted, the sequence of control vectors should be such that the opposing stream first gets its right-of-way. After its queue is discharged, the right-of-way should be given to both the opposing and the opposed traffic stream. Finally, the opposed traffic stream should keep its right-of-way alone. According to these requirements, the graph of control vectors transition, G_s, has to be extended including new nodes and new oriented edges.

Example 12.5

Determine the maximal capacity factor values for the intersection presented in Fig. 12.9, together with its compatibility graph and control vectors transition graph. The problem should be solved for two cases: when filtering of traffic stream σ_2 through traffic stream σ_3 is permitted, and when not. Two signal plan structures are given: $\mathbf{u}(u^r(\cdot))$ and $\mathbf{u}(u^s(\cdot))$.

The cycle time is $c = 100$ s.

The values of saturation flow volumes, average values of traffic flow volumes, minimal effective green times, and maximal effective red times are given in Table 12.6.

Table 12.6

i	1	2	3	4	5	6
s_i (veh/h)	1850	1650	1620	1650	1600	0
q_i (veh/h)	185	165	162	165	160	0
g_{mi} (s)	25	15	15	15	15	16
r_{Mi} (s)	70	75	75	70	85	–

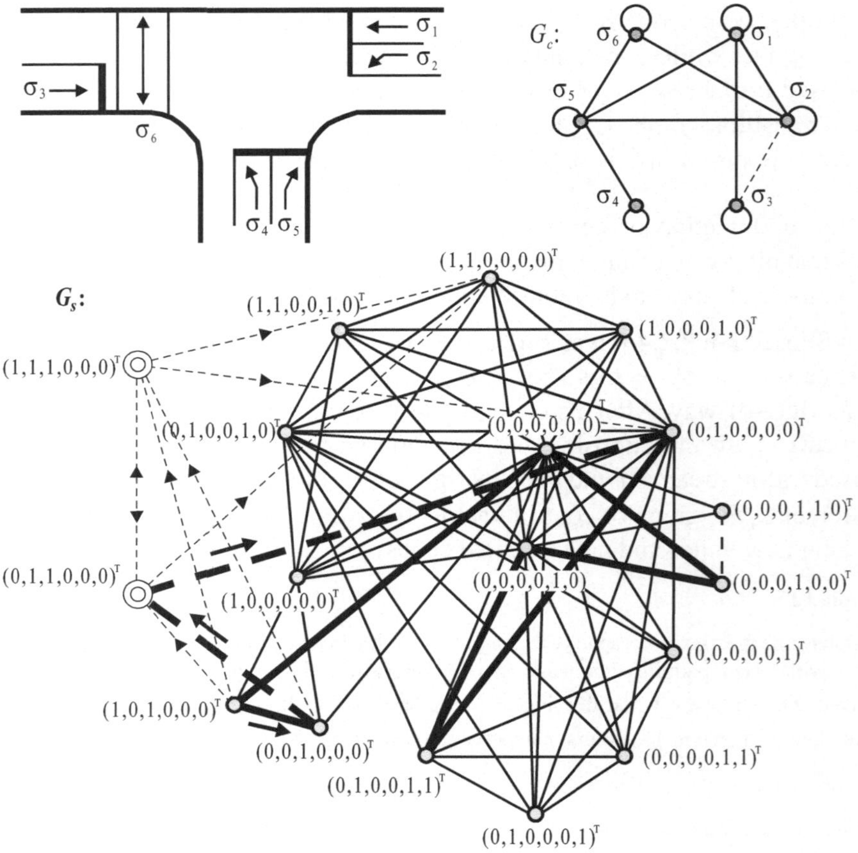

Figure 12.9

The minimal effective intergreen times are given as elements of matrix Z:

$$Z = [z_{pq}]_{P\times P} = \begin{array}{c} {}_p\backslash^q \\ 1 \\ 2 \\ 3 \\ 4 \\ 5 \\ 6 \end{array} \begin{array}{c} \begin{matrix} 1 & 2 & 3 & 4 & 5 & 6 \end{matrix} \\ \begin{bmatrix} 0 & 0 & 0 & 4 & 0 & 4 \\ 0 & 0 & 3 & 5 & 0 & 0 \\ 0 & 3 & 0 & 3 & 5 & 2 \\ 2 & 1 & 2 & 0 & 0 & 2 \\ 0 & 0 & 1 & 0 & 0 & 0 \\ 6 & 0 & 8 & 6 & 0 & 0 \end{bmatrix} \end{array}.$$

The chosen signal plan structures, $\mathbf{u}(u^r(\cdot))$ and $\mathbf{u}(u^s(\cdot))$ are:

$$\mathbf{u}(u^r(\cdot)) = \begin{bmatrix} 1 & 0 & 0 & 0 & 0 & 0 & 0 & 0 \\ 0 & 0 & 0 & 1 & 1 & 0 & 0 & 0 \\ 1 & 1 & 0 & 0 & 0 & 0 & 0 & 0 \\ 0 & 0 & 0 & 0 & 0 & 0 & 1 & 0 \\ 0 & 0 & 0 & 0 & 1 & 1 & 0 & 0 \\ 0 & 0 & 0 & 0 & 1 & 0 & 0 & 0 \end{bmatrix} \quad \mathbf{u}(u^s(\cdot)) = \begin{bmatrix} 1 & 0 & 0 & 0 & 0 & 0 & 0 & 0 \\ 0 & 0 & 1 & 1 & 1 & 0 & 0 & 0 \\ 1 & 1 & 1 & 0 & 0 & 0 & 0 & 0 \\ 0 & 0 & 0 & 0 & 0 & 0 & 1 & 0 \\ 0 & 0 & 0 & 0 & 1 & 1 & 0 & 0 \\ 0 & 0 & 0 & 0 & 1 & 0 & 0 & 0 \end{bmatrix}.$$

The problem will be solved for the sequence of average flow volumes q_3 of traffic stream σ_3, which is given in Table 12.7 for the case when filtering is not permitted, and in Table 12.8 for the case when filtering is permitted.

Table 12.7

q_3 (veh/h)	μ*
302	1.6400
402	1.2330
482	1.0284
492	1.0070
495	1.0015

Table 12.8

q_3 (veh/h)	μ*	t^3 (s)
162	2.7000	0
252	2.1407	0
262	2.0789	1
272	2.0237	2
302	1.8607	4
472	1.1905	4
562	0.9999	4

Since the structure and cycle time are given, the problem of capacity maximization becomes the problem of linear programming when filtering is not permitted, or the problem of nonlinear programming when filtering is permitted.

The problem is to determine the values of variables $\tau^1, \tau^2, \ldots, \tau^8$ and μ. All constraints are linear except constraints α.4.2.I (mod), α.4.2.II (mod), and α.4.2.III (mod). The expressions defining these constraints are as follows:

$$\sum_{k=1}^{8} (s_3 u_3^k (1-u_2^k) + \mu q_3 u_3^k u_2^k)\tau^k \geq c\mu q_3$$

$$\sum_{k=1}^{8} ((1-u_3^k u_2^k)\mu q_3 - u_3^k (1-u_2^k) s_3)\tau^k \geq 0$$

$$\sum_{k=1}^{8} \left(u_3^k u_2^k \cdot \frac{\mu q_3 e^{-\alpha' \mu q_3}}{1-e^{-\beta_2 \mu q_3}} + u_2^k (1-u_3^k) s_2(0) \right) \tau^k \geq c\mu q_2 .$$

The coefficients α' and β_2, explained in Subsection 8.2.4, in this case have the following values:

$$\alpha' = 4.5 \text{ s}, \quad \beta_2 = \frac{1}{s_2(0)} = \frac{1}{1650} = 2.2 \text{ s}.$$

The solution

When no filtering is permitted, the set of feasible control vectors is the same as in Example 8.7α. This set is presented on graph G_s, in Fig. 12.9, by the set of nodes that are not circled.

When filtering is permitted, the set of feasible control vectors has to be extended. In the case the filtering of traffic stream σ_2 through σ_3 is permitted, the set of feasible control vectors, $\mathbf{U}_f$, should be extended by the set:

$$\mathbf{U}'_f = \{(1,1,1,0,0,0)^T, (0,1,1,0,0,0)^T\}.$$

The elements of this set are represented by the circled nodes on graph G_s in Fig. IV.9.

The structural constraints are such that traffic stream σ_3 first gets its right-of-way, after that, both σ_2 and σ_3 simultaneously, and at last σ_3 alone. Therefore, new edges of graph G_s are defined as follows:

$$\Gamma'_s(0,0,1,0,0,0)^T = \Gamma_s(0,0,1,0,0,0)^T \cup \{(0,1,1,0,0,0)^T, (1,1,1,0,0,0)^T\},$$

$$\Gamma'_s(1,0,1,0,0,0)^T = \Gamma_s(1,0,1,0,0,0)^T \cup \{(0,1,1,0,0,0)^T, (1,1,1,0,0,0)^T\},$$

$$\Gamma'_s(0,1,1,0,0,0)^T = \{(0,1,0,0,0,0)^T, (1,1,0,0,0,0)^T, (1,1,1,0,0,0)^T\},$$

$$\Gamma'_s(1,1,1,0,0,0)^T = \{(0,1,0,0,0,0)^T, (1,1,0,0,0,0)^T, (0,1,1,0,0,0)\},$$

where $G_s = (\mathbf{U}_f, \Gamma_s)$ is the control vectors transition graph in the case the filtering is not permitted.

G'_s is defined as:

$$G'_s = ((\mathbf{U}_f \cup \mathbf{U}'_f), \Gamma''_s),$$

where

$$\Gamma''_s(\mathbf{u}(r)) = \Gamma_s(\mathbf{u}(r)) \cup \Gamma'_s(\mathbf{u}(r)).$$

Program LINGO was used in solving this problem.

For the case when no filtering is permitted, the maximal capacity factor values, determined for various values of q_3, are given in Table 12.7. The values obtained in the case the filtering is permitted are given in Table 12.8. Comparing the results in Tables 12.7 and 12.8, it can be concluded that higher maximal capacity values can be achieved if filtering is permitted.

The structure $\mathbf{u}(u^s(\cdot))$ is presented in Fig. 12.9.

13. DELAY MINIMIZATION

The total vehicle delay on intersection approaches is one of the most important criteria for signal plan determination, as noted in Section 9.4. The method described in Part III, based on the branch-and-bound principle, is used for minimization of this criterion.

For calculation of delay Webster's formula is used. Formula (9.29) is used when the cycle time is given, and formula (9.31) if not. The problem statement and illustrative examples, for both cases, are given below.

13.1. Delay minimization in the case the cycle time is known

The problem of optimal signal plan determination when the cycle time is given, in the case when filtering is not permitted, can be stated as follows: Find the signal plan $u(\cdot) = (\mathbf{u}, \tau)^T$ to minimize the mathematical expectation of vehicle delay on the intersection (9.29):

$$J_c = 0.9 \sum_{p=1}^{P'} \sum_{e=1}^{e=E(p)} \left(a_{pe}^1 \left(\sum_{k=1}^{K} (1 - u_p^k) \tau^k \right)^2 + \frac{a_{pe}^2}{\left(\sum_{k=1}^{K} u_p^k \tau^k \right)^2 - a_{pe}^3 \left(\sum_{k=1}^{K} u_p^k \tau^k \right)} \right),$$

subject to constraints α.1, α.2, α.3, α.4.1, β, γ.1, γ.2, ε.1. Constants a_{pe}^1, a_{pe}^2, a_{pe}^3 are defined in Section 9.4.

The criterion function in this problem is nonlinear, and the constraints, for a known structure, are linear functions of variables $\tau^1, \tau^2, \ldots, \tau^K$.

The solution of the stated nonlinear programming problem will be the global minimum if all constraints and criterion function are convex with respect to variables $\tau^1, \tau^2, \ldots, \tau^K$. Since the constraints are linear, it is necessary to check only the convexity of the criterion function (9.29). Appendix VIII presents the proof of convexity of this function.

The method described in Part III is applied for solving this problem. Superset $\mathscr{U}^s$ of the set of feasible solutions $\mathscr{U}_f$ is defined by constraints α'.1, α'.2, α'.3, α'.4.1, β', γ'.1, γ'.2, ε.1.

The extension of criterion function J_c is defined as follows:

$$J'_c = 0.9\sum_{p=1}^{P'}\sum_{e=1}^{e=E(p)}\left(a_{pe}^1\left(\sum_{k=1}^{K}(1-u_p^k)\tau^k\right)^2 + \frac{a_{pe}^2}{\left(\sum_{k=1}^{K}u_p^k\tau^k\right)^2 - a_{pe}^3\left(\sum_{k=1}^{K}u_p^k\tau^k\right)}\right). \quad (13.1)$$

The branching rule used in this case is the same as defined in Subsection 11.1.3.

The bounding rules

The lower bound

The lower bound $B(\mathcal{U}_\alpha)$ is defined as follows:

$$J'_c(u(\cdot)) \geq B(\mathcal{U}_\alpha), \quad (u(\cdot) \in \mathcal{U}_\alpha, \mathcal{U}_\alpha \in \beta(\mathcal{U}_\gamma)),$$

$$B(\{u(\cdot)\}) = J'_c(u(\cdot)).$$

According to property (11.37), the lower bound is defined as:

$$B(\mathcal{U}_\alpha) = \min\{J'_c(u(\cdot)) \mid u(\cdot) \in \mathcal{U}_\alpha\},$$

i.e., the lower bound is determined as the solution of the following optimization problem: Find $u(\cdot) \in \mathcal{U}_\alpha$ so as to minimize $J'_c(u(\cdot))$, subject to constraints α'.1, α'.2, α'.3, α'.4.1, β', γ'.1, γ'.2, ε.1.

The problem of lower bound determination, in this case, is the problem of nonlinear programming with nonlinear criterion function and linear constraints.

The upper bound

The upper bound $b(\bar{\mathcal{u}})$ of collection $\bar{\mathcal{u}}$ can be determined if the collection $\bar{\mathcal{u}}$ contains feasible solutions. In this case:

$$b(\bar{\mathcal{u}}) = \min\{J'_c(u(\cdot)) \mid (\{u(\cdot)\} \in \bar{\mathcal{u}}) \wedge (u(\cdot) \in \mathcal{U}_f)\}$$
$$= \min\{J_c(u(\cdot)) \mid (\{u(\cdot)\} \in \bar{\mathcal{u}})\}.$$

Example 13.1

Find the signal plan that minimizes delay on the intersection shown in Fig. 13.1, together with its compatibility graph G_C, control vectors transition graph G_S, and signal plan structure.

Each signal group contains only one traffic stream. The saturation flows, arrival flow volumes, and minimal effective green times are given in Table 13.1.

The cycle time ranges from 70 s to 120 s.

Table 13.1

i	1	2	3	4	5	6
s_i (veh/h)	1850	1650	1620	1650	1600	0
q_i (veh/h)	370	330	324	330	320	0
g_{mi} (s)	25	15	15	15	15	16

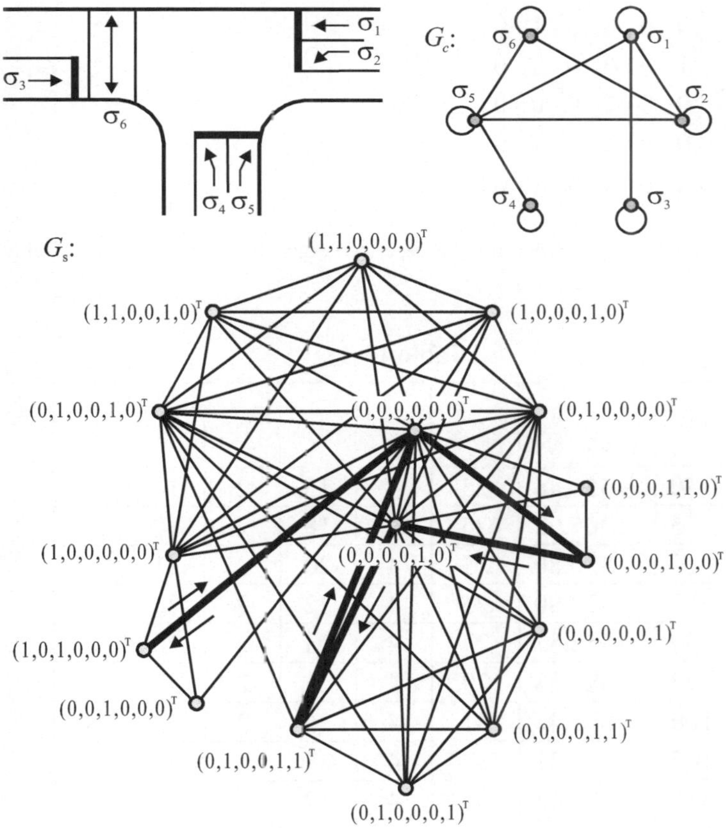

Figure 13.1

Minimal effective intergreen times are presented as elements of matrix Z.

$$Z = \begin{array}{c|cccccc} {}_p\backslash^q & 1 & 2 & 3 & 4 & 5 & 6 \\ \hline 1 & 0 & 0 & 0 & 4 & 0 & 4 \\ 2 & 0 & 0 & 3 & 5 & 0 & 0 \\ 3 & 0 & 3 & 0 & 3 & 5 & 2 \\ 4 & 2 & 1 & 2 & 0 & 0 & 2 \\ 5 & 0 & 0 & 1 & 0 & 0 & 0 \\ 6 & 6 & 0 & 8 & 6 & 0 & 0 \end{array}$$

The signal plan structure shown in Fig. 13.1 is:

$$\mathbf{u} = \begin{bmatrix} 1 & 0 & 0 & 0 & 0 & 0 \\ 0 & 0 & 0 & 0 & 1 & 0 \\ 1 & 0 & 0 & 0 & 0 & 0 \\ 0 & 0 & 1 & 0 & 0 & 0 \\ 0 & 0 & 0 & 1 & 1 & 0 \\ 0 & 0 & 0 & 0 & 1 & 0 \end{bmatrix}.$$

The results, obtained by applying the described algorithm, are presented in Table 13.2 and Fig. 13.2.

Table 13.2

c (s)	J_c (veh s/c)	τ_1	τ_2	τ_3	τ_4	τ_5	τ_6
70	1551.46	25	4	15	2	16	8
75	1305.92	25	4	18	2	18	8
80	1330.80	25	4	20	2	21	8
85	1427.80	25	4	22	2	24	8
90	1552.57	27	4	23	2	26	8
95	1683.58	29	4	25	2	27	8
100	1821.46	31	4	26	2	29	8
105	1966.31	32	4	28	2	31	8
110	2121.43	34	4	29	2	33	8
115	2276.54	36	4	30	2	35	8
120	2441.51	38	4	32	2	36	8

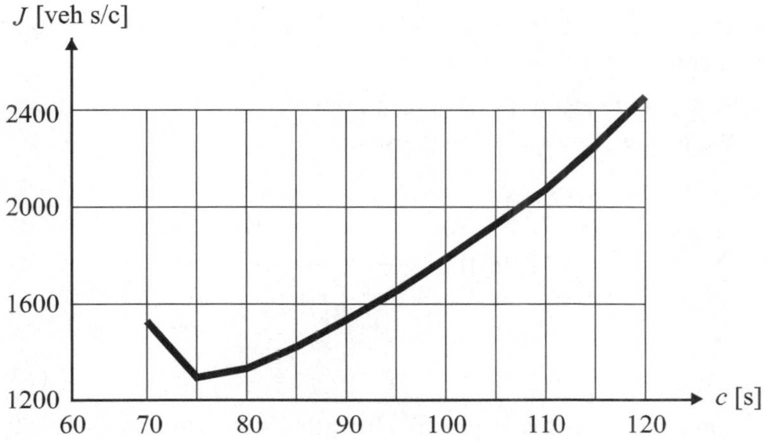

Figure 13.2

13.2. Delay minimization when cycle time is not given

The problem of optimal signal plan determination when cycle time is not given, in the case when filtering is not permitted, can be stated as follows: Find the signal plan $u(\cdot) = (\mathbf{u}, \tau)^T$ so as to minimize the mathematical expectation of vehicle delay on the intersection (9.31):

$$J_c = 0.9 \sum_{p=1}^{P'} \sum_{e=1}^{e=E(p)} \left(a_{pe}^1 \left(\sum_{k=1}^{K} (1 - u_p^k) \tau^k \right)^2 + \frac{\bar{a}_{pe}^2 \left(\sum_{k=1}^{K} \tau^k \right)^3}{\left(\sum_{k=1}^{K} u_p^k \tau^k \right)^2 - \bar{a}_{pe}^3 \left(\sum_{k=1}^{K} \tau^k \right) \left(\sum_{k=1}^{K} u_p^k \tau^k \right)} \right),$$

subject to constraints α.1, α.2, α.3, α.4.1, β, γ.1, γ.2.

The solution of the stated nonlinear programming problem will be the global minimum if all constraints and criterion function are convex with respect to variables $\tau^1, \tau^2, \ldots, \tau^K$. Since the constraints are linear, for a known structure, it is necessary to check only the convexity of the criterion function (9.31). Appendix VIII presents the proof of convexity of this function.

Like in the previous case, the method described in Part III is applied for solving this problem. The differences, in respect to the previous case, are as follows:

- When defining the superset $\mathscr{U}^s$ of the self of feasible signal plans, $\mathscr{U}_f$, constraint ε.1 is not used.
- Instead of expression (13.11), as the extension of criterion function J_c the following expression is used:

$$J_c' = 0.9\sum_{p=1}^{P'}\sum_{e=1}^{e=E(p)}\left(a_{pe}^1\left(\sum_{k=1}^{K}(1-u_p^k)\tau^k\right)^2 + \frac{\bar{a}_{pe}^2\left(\sum_{k=1}^{K}\tau^k\right)^3}{\left(\sum_{k=1}^{K}u_p^k\tau^k\right)^2 - \bar{a}_{pe}^3\left(\sum_{k=1}^{K}\tau^k\right)\left(\sum_{k=1}^{K}u_p^k\tau^k\right)}\right). \tag{13.2}$$

The proof of the convexity of this function is the same as for function $J_c(u)$ (9.31), which is given in Appendix VIII.

Example 13.2

For the same intersection as in Example 13.1, determine the optimal signal plan for the given signal plan structure, in the case the cycle time is not given. Other data are the same as in Example 13.1.

The stated problem is the problem of nonlinear programming, and its solution is:

$\tau^1 = 25\text{ s}$, $\tau^2 = 4\text{ s}$, $\tau^3 = 18\text{ s}$, $\tau^4 = 2\text{ s}$, $\tau^5 = 18\text{ s}$, $\tau^6 = 8\text{ s}$,

$c^* = 75\text{ s}$.

The optimal criterion function value is 1305.92 veh s/c .

It is evident that the solution of this problem is the same as in the preceding example. The curve in Fig. 13.2 does have the minimum value for $c = 75$ s .

14. EXTREME VALUES OF SIGNAL PLAN PARAMETERS

The majority of optimization criteria used in traffic control problems for isolated intersections are the criteria that are significant in traffic, economical, or ecological sense, such as the capacity, the delay, etc. Optimization of signal plan parameters would not make any sense by itself; however, for more precise analysis of traffic control problems on isolated intersections, it is often suitable to determine extreme values of some signal plan parameters. Such signal plan parameters are: effective green times of some or all signal groups, the cycle time, the number of control vectors in the signal plan (the length of the structure), etc.

14.1. Maximization of effective green times

The sum of green times is chosen as the optimization criterion in one of the first papers stating the problem of traffic control on an isolated intersection as the problem of linear programming [22]. The criterion is to be maximized by the optimal signal plan. Another similar problem statement can be found in the paper by R. Camus et al., presented on the II IAESTED Symposium [18]. Their paper is concerned with traffic control on complex intersections, and they state the problem as the mathematical programming problem in which the criterion function is the sum of green times of all signal groups or only of those that contain arrival traffic streams.

These criterion functions can be expressed as follows:

- Effective green time allocated to signal group D'_p:

$$J_p^1 = \sum_{k=1}^{K} u_p^k \tau^k , \quad (p \in \mathscr{P}). \tag{14.1}$$

- The sum of effective green times allocated to subset $\mathscr{P}_c$ of signal groups set $\mathscr{P}$:

$$J^2 = \sum_{k=1}^{K} \sum_{p \in \mathscr{P}_c} u_p^k \tau^k , \quad (\mathscr{P}_c \subset \mathscr{P}). \tag{14.2}$$

- The total green time, i.e., the sum of effective green times allocated to all signal groups:

$$J^3 = \sum_{k=1}^{K} \sum_{p=1}^{P'} u_p^k \tau^k . \tag{14.3}$$

The listed optimization criteria, together with appropriate constraints, define the combinatorial optimization problems, i.e., the problems of finding the optimal closed path on control vectors transition graph, and the optimal cycle time split. For a given signal plan structure, the problems of signal plan optimization reduce to determination of the optimal cycle time split.

14.2. Cycle time minimization

The information about minimal cycle duration is very important when the intersection is to be included in a centralized traffic control system. In this case a single cycle time value has to be chosen for all intersections in the network. This value has to be greater or equal to the maximal value of minimal cycle times of all intersections.

The presence of minimal effective green constraints and minimal effective intergreen constraints points to the conclusion that, for a given signal plan structure **u**, there has to exist a cycle time value such that these constraints cannot be satisfied for shorter cycle time values. The cycle time of any feasible signal plan has to be equal or greater than this minimal value. The determination of the signal plan with minimum cycle time is also a combinatorial optimization problem defined on the graph of control vector transitions G_s.

The minimal cycle time values are different for different signal plan structures. The minimal among them, $c_{\min}$, is called *critical cycle time value.* There exists no feasible signal plan with the cycle time value less than the critical one. The critical cycle time value may correspond to more than one signal plan structure, i.e., there can exist one or more closed paths on graph G_s, which all have to be determined when finding the signal plan having the minimal cycle time.

The statement of the optimization problem, in this case, is: Determine the signal plan $u(\cdot) = (\mathbf{u}, \tau)^T$ so as to minimize the function:

$$J_c = \sum_{k=1}^{K} \tau^k , \tag{14.4}$$

subject to constraints α.1, α.2, α.3, α.4, β, γ.1, γ.2. Constraint ε.1 is not included because it states that the sum of all control vector durations is equal to the cycle time, which is the optimization criterion in this case.

Besides the problem of minimal cycle time duration, solved by searching in the set $\mathscr{U}_f$ of all feasible signal plans, it is sometimes interesting to find the minimal cycle time for a given signal plan structure. In this case, this problem is transformed to the problem in which control vectors $\mathbf{u}^1,\mathbf{u}^2,\ldots,\mathbf{u}^{\mathrm{K}}$ are known, the structural constraints, α.1 and γ.1, are satisfied, and there remain only the constraints with time variables (constraints α.2, α.3, α.4, β, and γ.2). The optimization criterion, which has to be minimized is, again:

$$J_c = \sum_{k=1}^{\mathrm{K}} \tau^k .$$

The method described in Part III is used for minimizing function (14.4).

Superset $\mathscr{U}^s$ of the set of feasible signal plans, $\mathscr{U}_f$, is defined by constraints α'.1, α'.2, α'.3, α'.4, β, γ'.1, γ'.2.

The extension of criterion function J_c is defined by the expression

$$J'_c = \sum_{k=1}^{\kappa} \tau^k , \quad (\kappa \le \mathrm{K}) . \tag{14.5}$$

The branching rule is defined according to the procedure described in Subsection 11.1.3.

The lower bound is determined as the solution of the following optimization problem: Minimize function (14.5) subject to constraints α'.1, α'.2, α'.3, α'.4, β, γ'.1, γ'.2.

The initial control vector $\mathbf{u}^1$ is chosen as described in Subsection 12.3.3.

Example 14.1

Determine all signal plans with the minimal cycle time for the intersection and data from Example 12.1β.

The solution:

Applying the method described in Part III, with the set of initial control vectors:

$$\mathbf{u}^1 \in \mathbf{U}_{f4} = \{(0,1,0,1)^T,(0,0,0,1)^T\},$$

the set of problem solutions is obtained. In the case the initial vector is $\mathbf{u}^1 = (0,1,0,1)^T$, 26 optimal solutions are obtained, and for $\mathbf{u}^1 = (0,0,0,1)^T$ the number of optimal solutions is 11. In both sets 5 optimal solutions are the same; thus the total number of optimal solutions is 32.

The minimal cycle time is 37 s.

One of the optimal solutions is:

$$\mathbf{u}^* = \begin{bmatrix} 0 & 0 & 0 & 1 & 0 & 0 & 0 \\ 0 & 0 & 0 & 0 & 0 & 1 & 1 \\ 0 & 0 & 1 & 1 & 1 & 0 & 0 \\ 1 & 0 & 0 & 0 & 0 & 0 & 1 \end{bmatrix}.$$

$$\tau^* = \begin{bmatrix} 2 & 1 & 1 & 10 & 8 & 7 & 8 \end{bmatrix}$$

This solution is marked by a bold line on control vectors transition graph G_s in Fig. 14.1.

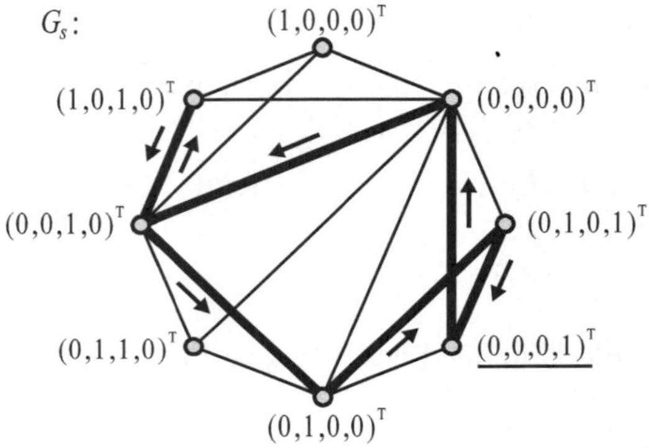

Figure 14.1

14.3. Cycle time maximization

The maximal cycle time in many countries is fixed by traffic regulations, usually with the value of 120 s. However, in some traffic control problems, it might be necessary to introduce the maximal red time constraints, in order to prevent forming of long queues on some intersection approaches. Due to these constraints, there can exist the maximal cycle time that is less than the one defined by regulations.

The cycle time maximization problem can be stated as follows: Determine the signal plan $u(\cdot) = (\mathbf{u}, \tau)^T$ so as to minimize the function given by expression (14.4):

$$J_c = \sum_{k=1}^{K} \tau^k ,$$

subject to constraints $\alpha.1$, $\alpha.2$, $\alpha.3$, $\alpha.4$, β, $\gamma.1$, $\gamma.2$.

This problem can be solved in the same way as the cycle minimization problem if the criterion function, which now has to be minimized, is expressed as

$$J_c^1 = -J_c \,.$$

The minimal cycle time value, $c_{\min} = 70$ s, and the maximal cycle time value, $c_{\max} = 86$ s, are determined in Example 12.4 using the procedure described in Section 14.2.

14.4. Extreme values of the length of signal plan structure

The number of control vectors (the length of signal plan structure) is different for different signal plans that belong to the set of feasible signal plans. Solutions of optimization problems on the graph of control vectors transition that have been considered here contain the optimal number of control vectors, as well, in the sense of the adopted optimization criterion. Therefore, solutions of different optimization problems will contain different optimal number of control vectors.

An interesting question is whether, for given constraints, there exist a minimal and a maximal number of control vectors, such that the structure length of any feasible signal plan lies between these two "extreme" values.

This question was once interesting because there existed two traffic control approaches [2] regarding implementation of control vector sequences in traffic signal controllers. These approaches resulted in construction of two types of traffic controllers: phase-oriented, and signal–group-oriented. The main features of these approaches are as follows:

- *Phase (control vector) based control.* In applying this approach, the constraints (minimal effective green times, minimal effective intergreen times, etc.) have to be defined for each control vector (phase), and they refer to all the components of the vector. The intention, when using this approach, was to minimize the number of phases. This reasoning was based on the commonly accepted assumption that the lost time on an intersection would be reduced by decreasing the number of phases used to control the intersection.
- *Signal–group control.* In applying this approach, control variables are associated to traffic streams, i.e., signal groups. The constraints (minimal effective green times, minimal effective intergreen times, etc.) are here defined for each signal group. The intention, when using this approach, is to "maximize overlaps" because this will reduce the total time necessary to fulfill the capacity constraints.

This question is not so significant nowadays. Having modern, flexible, microprocessor-based traffic controllers makes any control approach rather easy to implement.

Observing the control problem statement, it can be noted that among the constraints that define signal plan feasibility (Section 8.9), there exist constraints related to control vectors (phases) and their sequences, as well as constraints related to signal groups. Therefore, a correct problem statement includes elements of both control approaches. As the solution of any optimization problem the optimal signal plan will be obtained with the optimal structure length K and other signal plan parameters optimal in the sense of the chosen optimization criterion.

14.4.1. Determining the maximal length of signal plan structure

Determination of the maximal number of control vectors in the signal plan structure (the structure length) can be stated as the problem of finding the best closed path on the graph of control vectors transition.

The set of feasible signal plans is defined by constraints listed in Section 8.9.

Taking into consideration that the number of control vectors in a signal plan is the feature of the signal plan structure, it is also interesting to solve this problem respecting only the structural constraints, $\alpha.1$ and $\gamma.1$.

Two variants of the problem of maximizing the length of signal plan structure will be considered here: (a) with constraints $\alpha.1$ and $\gamma.1$ only, (b) with all constraints listed in Section 8.9.

a) Maximizing the length of signal plan structure with only structural constraints included in the problem statement

The maximal number of control vectors in a signal plan structure is determined in the doctoral thesis of Isabel Tully [85]. There, it is shown that the maximal number of control vectors in a signal plan equals twice the number of signal groups. This result is valid in the case when only structural constraints $\alpha.1$ and $\gamma.1$ are included in the problem statement. Other constraints, including constraints on control vector durations ($\tau^1, \tau^2, \ldots, \tau^K$), are not part of the problem statement.

This problem can be formulated as the optimization problem on the graph of control vectors transition: Find the signal plan to maximize function

$$J_c : \mathcal{U}_f \rightarrow \mathrm{R}$$

given by expression

$$J_c = \mathrm{K} \tag{14.6}$$

subject to constraints α.1 and γ.1, i.e.,

$$\alpha.1)\quad \mathbf{u}^{k(\mathrm{mod}\,\mathrm{K})+1} \in \Gamma_s \mathbf{u}^k\,,\quad (k \in \mathscr{K})$$

$$\gamma.1)\quad \sum_{k=1}^{\mathrm{K}} (u_p^k + u_p^{k(\mathrm{mod}\,\mathrm{K})+1})(\mathrm{mod}\,2) = 2\,,\quad (p \in \mathscr{P})\,.$$

The total number of changes of signal indications (from effective red to effective green, and vice versa) for one control variable (assigned to a signal group) is 2, and, obviously, the total number of changes in a signal plan is:

$$\sum_{p=1}^{\mathrm{P}} \sum_{k=1}^{\mathrm{K}} (u_p^k + u_p^{k(\mathrm{mod}\,\mathrm{K})+1})(\mathrm{mod}\,2) = 2\mathrm{P}\,. \tag{14.7}$$

Since the total number of changes of signal indications in a signal plan equals 2P, it is obvious that the signal plan will contain the maximal number of control vectors if the sum of changes of particular components in adjacent control vectors is 1, i.e.,

$$\sum_{p=1}^{\mathrm{P}} (u{*}_p^k + u{*}_p^{k(\mathrm{mod}\,\mathrm{K})+1})(\mathrm{mod}\,2) = 1\,,\quad (k \in \mathscr{K})\,. \tag{14.8}$$

In this case,

$$\sum_{k=1}^{\mathrm{K}} \left(\sum_{p=1}^{\mathrm{P}} (u{*}_p^k + u{*}_p^{k(\mathrm{mod}\,\mathrm{K})+1})(\mathrm{mod}\,2) \right) = \mathrm{K}_{\max} = 2\mathrm{P}\,. \tag{14.9}$$

It means that

$$J_c = J_c^* = \mathrm{K}_{\max} = 2\mathrm{P}\,. \tag{14.10}$$

When determining $\mathrm{K}_{\max}$, only structural constraints α.1 и γ.1 were considered.

It would be interesting to know whether it is possible that adjacent control vectors in a signal plan differ by the value of one component only. The existence of such a signal plan is easy to prove when the graph of control vectors transition, G_s, does not have oriented edges. In this case, the signal plan can be constructed in such a way that each control vector gives the right-of-way to a single signal group, i.e., with only one component having the value of 1, and these control vectors are separated from one another by the "all red" control vector. Namely, one of the properties of graph G_s is that the node representing control vector $(0,0,\ldots,0)^T$ is connected to all other nodes.

For the intersection presented in Fig. 12.4, containing four signal groups, the structure of such a signal plan is:

$$\mathbf{u} = \begin{bmatrix} 1 & 0 & 0 & 0 & 0 & 0 & 0 & 0 \\ 0 & 0 & 1 & 0 & 0 & 0 & 0 & 0 \\ 0 & 0 & 0 & 0 & 1 & 0 & 0 & 0 \\ 0 & 0 & 0 & 0 & 0 & 0 & 1 & 0 \end{bmatrix}.$$

This structure has the property (14.8), so that $K_{max} = 2P = 2 \cdot 4 = 8$.

The structure with maximal length ($K_{max} = 2 \cdot 6 = 12$) is presented in Fig. 12.3 (Example 12.1). It is the structure of the optimal signal plan that maximizes the total traffic volume in that example.

b) Maximizing the length of signal plan structure with all constraints included in the problem statement

The maximal length of signal plan structure can be less than 2P in the case its maximization is performed subject to all constraints, rather than the structural constraints only. Also, for some given cycle time values, it might be impossible to find a signal plan with the structure length equal to 2P.

In order to determine the maximal number of control vectors, the optimization problem has to be solved with all constraints included. The problem can be stated as follows: Determine the signal plan to maximize (14.6):

$$J_c = K,$$

subject to constraints α.1, α.2, α.3, α.4.1, β, γ.1, γ.2, ε.1.

This problem can be solved, also, by the branch-and-bound method described in Part III.

Superset $\mathcal{U}^s$ of the set of feasible signal plans $\mathcal{U}_f$ is defined by constraints (Part III) α'.1, α'.2, α'.3, α'.4, β, ε'.1.

The extension of criterion function J_c is defined by expression

$$J'_c = \kappa, \tag{14.11}$$

for $\kappa = 2,3,\ldots,K$.

The branching rule is determined in the same way as in solving other problems, as described in Subsection 11.1.3.

The upper bound in this case is defined as:

$$B(\mathcal{U}_\alpha) = \kappa - 1 + \left[\frac{\kappa}{K}\right] + \left(2P - \sum_{k=1}^{\kappa-1+[\kappa/K]} \sum_{p=1}^{P} (u_p^k + u_p^{k(\mathrm{mod}K)+1})\,(\mathrm{mod}\,2)\right), \tag{14.12}$$

for $\kappa = 2,3,\ldots,K$.

If $\mathscr{U}_\alpha = \mathscr{U}^s$, then

$$B(\mathscr{U}^s) = 2P. \tag{14.13}$$

The initial control vector should be chosen in the same way as in Subsection 12.3.3.

The choice of the subset for branching. Subset $\mathscr{U}_\gamma$ with the property

$$B(\mathscr{U}_\gamma) = \max B(\mathscr{U}_\alpha), \quad (\mathscr{U}_\alpha \in \bar{u})$$

should be chosen for branching, where $\bar{u}$ is the collection whose elements are candidates for branching.

14.4.2. Determining the minimal length of signal plan structure

Similar to the problem of determining the maximal number of control vectors, the problem of determining the minimal number of control vectors in signal plan structure can be stated. The optimization criterion is the same as in the previous problem, but instead of maximization, this is the problem of minimization.

This problem can be stated as follows: Determine the signal plan by which function

$$J_c = K$$

is minimized, subject to constraints α.1, α.2, α.3, α.4, γ.1, γ.2, ε.1.

This is also a problem of finding the optimal closed path on graph G_s. The solution method is the same as in solving the previous problem.

In branching operation, the subset with the least lower bound is chosen for further branching. Lower bounds are calculated using expression (14.12). The value of the first lower bound, which is calculated for set $\mathscr{U}^s$, is equal to the number of independent cliques of graph G_g—the graph of signal groups compatibility.

Example 14.2

Determine the signal plans having the minimal and the maximal number of control vectors for the intersection presented in Fig. 12.4, together with its graphs G_c and G_s. In this case $G_c = G_g$. Other data, on saturation flow volumes and minimal effective green times, are the same as in Example 12.1β.

The minimal effective intergreen times are elements of matrix Z:

$$Z = \begin{bmatrix} 0 & 9 & 0 & 5 \\ 4 & 0 & 0 & 0 \\ 0 & 0 & 0 & 4 \\ 6 & 0 & 1 & 0 \end{bmatrix}.$$

The cycle time is 38 s (the minimal cycle time obtained as the solution of the problem of cycle time minimization).

The signal plan with the maximal structure length ($\mathrm{K}_{\max} = 8$) is:

$$u^*(\cdot) = (\mathbf{u}^*, \tau^*)^T$$

$$\mathbf{u}^* = \begin{bmatrix} 0 & 0 & 0 & 0 & 1 & 0 & 0 & 0 \\ 1 & 1 & 1 & 0 & 0 & 0 & 0 & 1 \\ 0 & 0 & 1 & 1 & 1 & 1 & 0 & 0 \\ 1 & 0 & 0 & 0 & 0 & 0 & 0 & 0 \end{bmatrix}.$$

$$\tau^* = [10 \;\; 1 \;\; 1 \;\; 4 \;\; 10 \;\; 8 \;\; 2 \;\; 2]$$

The signal plan with the minimal structure length ($\mathrm{K}_{\min} = 5$) for the same cycle time is:

$$u^*(\cdot) = (\mathbf{u}^*, \tau^*)^T$$

$$\mathbf{u}^* = \begin{bmatrix} 0 & 0 & 0 & 1 & 0 \\ 1 & 1 & 0 & 0 & 0 \\ 0 & 0 & 1 & 1 & 0 \\ 1 & 0 & 0 & 0 & 0 \end{bmatrix}.$$

$$\tau^* = [10 \;\; 5 \;\; 4 \;\; 10 \;\; 9]$$

The signal plan with the minimal structure length, for cycle time 40 s, is:

$$u^*(\cdot) = (\mathbf{u}^*, \tau^*)^T$$

$$\mathbf{u}^* = \begin{bmatrix} 1 & 0 & 0 & 0 \\ 0 & 0 & 1 & 0 \\ 1 & 0 & 0 & 0 \\ 0 & 0 & 1 & 0 \end{bmatrix}.$$

$$\tau^* = [10 \;\; 9 \;\; 15 \;\; 6]$$

It is interesting to note that for signal plans with the minimal structure length $\mathrm{K}_{\min} = 4$, a longer cycle time is "necessary" ($c = 40$ s) than for the signal plan with the minimal structure length $\mathrm{K}_{\min} = 5$ ($c = 38$ s). It is an example of the surprises brought to us by Combinatorics.

Part V

EFFECTS OF THE CHOICE OF THE COMPLETE SET OF SIGNAL GROUPS ON INTERSECTION PERFORMANCE

The quality of the choice of a complete set of signal groups can be assessed by the value of function (3.29):

$$J_G : \mathfrak{D}_b \rightarrow R,$$

where

$$J_G(\mathfrak{D}_a^m) = \text{opt}\ \{J_c(u(\cdot)) \mid u(\cdot) \in \mathcal{U}_f^m\}, \quad (\mathfrak{D}_a^m \in \mathfrak{D}_b), \tag{V.1}$$

and $\mathcal{U}_f^m$ represents the set of feasible controls (signal plans) if the complete set of signal groups $\mathfrak{D}_a^m$ is chosen.

Expression (V.1) can be used for defining a relation of total ordering in collection $\mathfrak{D}_b$. For that, it is necessary to determine the value of J_G for each complete set of signal groups $\mathfrak{D}_a^m \in \mathfrak{D}_b$. However, in the collection of complete sets of signal groups, $\mathfrak{D}_b$, there exists the relation of partial ordering R_p (3.31), which can be used to reduce the number of J_G value calculations.

15. THE RELATION OF PARTIAL ORDERING (REFINEMENT) AND THE SET OF FEASIBLE CONTROLS

A better quality of traffic process control will be achieved, under assumption $\mathfrak{D}_a^r R_p \mathfrak{D}_a^q$, if the complete set of signal groups $\mathfrak{D}_a^r$ is chosen rather than set $\mathfrak{D}_a^q$. This claim is based on the following considerations. The problems of optimal traffic control on an intersection, in these two cases, are defined by pairs $(J_c, \mathcal{U}_f^r)$ and $(J_c, \mathcal{U}_f^q)$, where J_c is the optimization criterion, and $\mathcal{U}_f^r$ and $\mathcal{U}_f^q$ are the sets of feasible controls if the corresponding complete sets of signal groups $\mathfrak{D}_a^r$ and $\mathfrak{D}_a^q$ are chosen. Taking into consideration the definition of relation R_p (3.31), the following can be claimed:

If $\mathfrak{D}_a^r R_p \mathfrak{D}_a^q$, then

$$(\exists D_h^q \in \mathfrak{D}_a^q) \;\; D_h^q = \bigcup_{\alpha=1}^{d} D_{n_\alpha}^r , \quad (D_{n_1}^r, D_{n_2}^r, \ldots, D_{n_d}^r \in \mathfrak{D}_a^r) \tag{15.1}$$

with $d \geq 2$. This means that some signal groups, elements of complete set of signal groups $\mathfrak{D}_a^q$, represent unions of some signal groups that belong to another complete set of signal groups, $\mathfrak{D}_a^r$. Since a single control variable is assigned to each signal group, then, if complete set of signal groups $\mathfrak{D}_a^q$ is chosen, this means that one control variable is used to control all signal groups $D_{n_1}^r, D_{n_2}^r, \ldots, D_{n_d}^r \in \mathfrak{D}_a^r$, the union of which constitutes signal group D_h^q. Therefore, the following equalities hold:

$$\begin{aligned} u_{n_1}^k &= u_{n_2}^k , \quad (k \in \mathcal{K}) \\ u_{n_1}^k &= u_{n_3}^k , \quad (k \in \mathcal{K}) \\ &\ldots \\ u_{n_1}^k &= u_{n_d}^k , \quad (k \in \mathcal{K}) . \end{aligned} \tag{15.2}$$

If in a complete set of signal groups $\mathfrak{D}_a^q$ there exist several signal groups that are unions of some signal groups that belong to a complete set of signal groups $\mathcal{D}_a^r$, then for each such signal group the set of equalities can be formulated, similar to (15.2).

The set of constraints defining the set of feasible controls, $\mathcal{U}_f^q$, therefore, besides the constraints defining set $\mathcal{U}_f^r$, contains additional constraints similar to (15.2). Hence, the following expression is valid:

$$(\mathcal{U}_f^q \subset \mathcal{U}_f^r) \Rightarrow \min\{J_c(u(\cdot) \mid u(\cdot) \in \mathcal{U}_f^r\} \le \min\{J_c(u(\cdot) \mid u(\cdot) \in \mathcal{U}_f^q\}$$

$$\Rightarrow (J_G(\mathcal{D}_a^r) \le J_G(\mathcal{D}_a^q) \Rightarrow \mathcal{D}_a^r \prec \mathcal{D}_a^q . \qquad (15.3)$$

Here it is assumed that the optimal value of criterion J_c is its minimal value. The relation $\prec$ indicates that $\mathcal{D}_a^r$ is "better" than $\mathcal{D}_a^q$, in the sense of criterion J_G. Complete set of signal groups $\mathcal{D}_a^r$ is better than $\mathcal{D}_a^q$ in the sense of any criterion related to intersection performances.

The consequence of the presence of constraints (15.2) is that the graph of control vector transitions

$$G_s^q = (\mathbf{U}_f^q, \Gamma_s^q) = (\mathbf{U}_f^q, R_s^q)$$

will be a subgraph of graph

$$G_s^r = (\mathbf{U}_f^r, \Gamma_s^r) = (\mathbf{U}_f^r, R_s^r),$$

where $\mathbf{U}_f^q$ and $\mathbf{U}_f^r$ are the sets of feasible control vectors corresponding to complete sets of signal groups $\mathcal{D}_a^q$ or $\mathcal{D}_a^r$, respectively. Relations R_s^q and R_s^r are the control vectors transition relations in sets $\mathcal{D}_a^q$ and $\mathcal{D}_a^r$, respectively. Namely, it is obvious that set $\mathbf{U}_f^q$ contains only the control vectors whose components $u_{n_1}^k$ and $u_{n_2}^k$, as well as $u_{n_1}^k$ and $u_{n_3}^k$, etc., have the same values for any k, while $\mathbf{U}_f^r$ contains also the vectors with different values of these components. Since G_s^q is a subgraph of G_s^r, the following is evident:

$$\mathbf{U}_f^q \subset \mathbf{U}_f^r , \qquad (15.4)$$

$$R_s^q = R_s^r \cap (\mathbf{U}_f^q \times \mathbf{U}_f^q) . \qquad (15.5)$$

The combinatorial nature of the problem of choosing the complete set of signal groups is reflected in fast growth of the number of complete signal groups with the number of traffic streams, as presented in Table 3.1.

16. THE HEURISTICS FOR THE CHOICE OF THE COMPLETE SET OF SIGNAL GROUPS

The process of choosing the complete set of signal groups that will be used for traffic control on an intersection can be simplified by using properties of the "refinement" relation, R_p, in the collection of complete sets of signal groups, $\mathfrak{D}_b$.

The heuristics selects a subset $\mathfrak{D}'_b$ of set $\mathfrak{D}_b$ and performs searching for a suitable complete set of signal groups only in subset $\mathfrak{D}'_b$, which has a smaller number of elements than $\mathfrak{D}_b$. The procedure for selecting $\mathfrak{D}'_b$ consists of the following steps. First, optimal values of the chosen criterion are determined for subsets of $\mathfrak{D}_b$ that have a minimal number of elements. Among them there exists a subset, $\mathfrak{D}^t_a$, with the best criterion value. Then, from the graph by which Hasse diagram is represented, the subgraph

$$G_H = (\mathfrak{D}'_b, R'_p) \tag{16.1}$$

is extracted.

Set $\mathfrak{D}'_b \subset \mathfrak{D}_b$ is defined by the expression:

$$\mathfrak{D}'_b = \{\mathfrak{D}^\delta_a \mid (\mathfrak{D}^\delta_a, \mathfrak{D}^t_a) \in R_p, \quad \mathfrak{D}^\delta_a, \mathfrak{D}^t_a \in \mathfrak{D}_b\} \bigcup \{\mathfrak{D}^t_a\}, \tag{16.2}$$

where R_p is the refinement relation.

Hence, relation R'_p is:

$$R'_p = R_p \bigcap (\mathfrak{D}^\delta_a \times \mathfrak{D}^\delta_a). \tag{16.3}$$

The search for the complete set of signal groups that will be chosen for control is then performed over elements of set $\mathfrak{D}'_b$.

The effects of the choice of a complete set of signal groups on intersection capacity are discussed in papers [33], [34], [36]. In the example given in [34] it is shown that the change in intersection capacity value, as the function of the choice of the complete set of signal groups, can be even 42% (the difference between the greatest and the smallest value of the criterion, expressed in percentages).

The effect of the choice of the complete set of signal groups on the optimal capacity factor value is analyzed in the following example.

Example 16.1

Determine the effects of the choice of the complete set of signal groups on the optimal capacity factor value for the intersection given in Fig. 13.1. Form the Hasse diagram and show the rationality of the choice achieved by the proposed heuristics.

Minimal effective intergreen times, in the case each signal group controls a single traffic stream, are given by matrix Z.

The values of saturation flow volumes, minimal effective green times, maximal effective red times, and average flow volumes, are given in Table 16.1.

The cycle time is $c = 90$ s.

$$Z = \begin{bmatrix} 0 & 0 & 0 & 4 & 0 & 4 \\ 0 & 0 & 3 & 5 & 0 & 0 \\ 0 & 3 & 0 & 3 & 5 & 2 \\ 2 & 1 & 2 & 0 & 0 & 2 \\ 0 & 0 & 1 & 0 & 0 & 0 \\ 8 & 0 & 4 & 8 & 0 & 0 \end{bmatrix}$$

Table 16.1

i	1	2	3	4	5	6
s_i (veh/h)	1850	1650	1620	1650	1600	0
q_i (veh/h)	185	330	162	165	160	0
g_{mi} (s)	25	15	15	15	15	16
r_{Mi} (s)	70	75	75	70	85	–

The results obtained by applying the method described in Subsection 12.3.3 are presented in Table 16.2 and in Figs. 16.1 to 16.10.

Besides the maximal capacity values for each complete set of signal groups, Table 16.2 contains the percentage of differences between these values and the maximal one in collection $\mathcal{D}_b$ of all complete sets of signal groups. $\mathcal{D}_b$, i.e.,

$$\nu^r = \frac{\mu^*_{\max} - \mu^r_{\max}}{\mu^*_{\max}}, \tag{16.4}$$

where:

$$\mu^*_{\max} = \max\{\mu^r_{\max} \mid \mathcal{D}^r_a \in \mathcal{D}_b\}, \tag{16.5}$$

$$\mu^r_{\max} = \max\{\mu^r \mid u(\cdot) \in \mathcal{U}^r_f,\ r \in \{1,2,\ldots,card\,\mathcal{D}_b\}\}. \tag{16.6}$$

Parameters ν^r_β, given in Table 16.2, are defined as follows:

$$\nu^r_\beta = \frac{\mu^{r*}_{\max} - \mu^r_{\max}}{\mu^{r*}_{\max}}, \tag{16.7}$$

where:

$$\mu^{r*}_{\max} = \max\{\mu^{\nu}_{\max} \mid \mathfrak{D}^{\nu}_{a} \in \mathfrak{D}^{s}_{b}\}, \tag{16.8}$$

$$\mathfrak{D}^{s}_{b} = \{\mathfrak{D}^{\nu}_{a} \mid card\, \mathfrak{D}^{\nu}_{a} = card\, \mathfrak{D}^{r*}_{a}\}. \tag{16.9}$$

Members $\mathfrak{D}^{\nu}_{a}$ of collection $\mathfrak{D}^{s}_{b}$ have the same cardinality as element $\mathfrak{D}^{r}_{a}$, i.e., the element corresponding to ν^{r}_{β}.

Table 16.2

r	The complete set of signal groups $\mathfrak{D}^{r}_{a}$	$\mu^{r}_{\max}$	ν^{r} [%]	ν^{r}_{β} [%]
1	$\mathfrak{D}^{1}_{a} = \{D_1, D_2, D_3, D_4, D_5, D_6\} = \{\{\sigma_1\},\{\sigma_2\},\{\sigma_3\},\{\sigma_4\},\{\sigma_5\},\{\sigma_6\}\}$	2.075	0	0
2	$\mathfrak{D}^{2}_{a} = \{D_3, D_4, D_5, D_6, D_7\} = \{\{\sigma_3\},\{\sigma_4\},\{\sigma_5\},\{\sigma_6\},\{\sigma_1,\sigma_2\}\}$	1.30	37.35	37.35
3	$\mathfrak{D}^{3}_{a} = \{D_2, D_4, D_5, D_6, D_8\} = \{\{\sigma_2\},\{\sigma_4\},\{\sigma_5\},\{\sigma_6\},\{\sigma_1,\sigma_3\}\}$	1.85	10.84	10.84
4	$\mathfrak{D}^{4}_{a} = \{D_2, D_3, D_4, D_6, D_9\} = \{\{\sigma_2\},\{\sigma_3\},\{\sigma_4\},\{\sigma_6\},\{\sigma_1,\sigma_5\}\}$	1.60	22.89	22.89
5	$\mathfrak{D}^{5}_{a} = \{D_1, D_3, D_4, D_6, D_{10}\} = \{\{\sigma_1\},\{\sigma_3\},\{\sigma_4\},\{\sigma_6\},\{\sigma_2,\sigma_5\}\}$	2.075	0	0
6	$\mathfrak{D}^{6}_{a} = \{D_1, D_2, D_3, D_6, D_{11}\} = \{\{\sigma_1\},\{\sigma_2\},\{\sigma_3\},\{\sigma_6\},\{\sigma_4,\sigma_5\}\}$	2.025	2.41	2.41
7	$\mathfrak{D}^{7}_{a} = \{D_3, D_4, D_6, D_{12}\} = \{\{\sigma_3\},\{\sigma_4\},\{\sigma_6\},\{\sigma_1,\sigma_2,\sigma_5\}\}$	1.30	37.35	29.73
8	$\mathfrak{D}^{8}_{a} = \{D_3, D_6, D_7, D_{11}\} = \{\{\sigma_3\},\{\sigma_6\},\{\sigma_1,\sigma_2\},\{\sigma_4,\sigma_5\}\}$	1.25	39.76	32.43
9	$\mathfrak{D}^{9}_{a} = \{D_4, D_6, D_8, D_{10}\} = \{\{\sigma_4\},\{\sigma_6\},\{\sigma_1,\sigma_3\},\{\sigma_2,\sigma_5\}\}$	1.85	10.84	0
10	$\mathfrak{D}^{10}_{a} = \{D_2, D_6, D_8, D_{11}\} = \{\{\sigma_2\},\{\sigma_6\},\{\sigma_1,\sigma_3\},\{\sigma_4,\sigma_5\}\}$	1.80	13.25	2.70

Figs. 16.1 to 16.10 present, for each complete set of signal groups $\mathcal{D}_a^r \in \mathcal{D}_b$, the control vector transitions graph, G_s^r, the compatibility graph, G_g^r, the matrix of minimal effective intergreen times, and the optimal solution.

The optimal signal plan is represented by the bold line on the graph of control vector transitions. The optimal structure, $\mathbf{u}^*$, cycle time split, τ^*, and the maximal capacity factor value, $\mu^* = \mu_{\max}$, are also given in the figures.

The edges of graphs G_s^r, which do not belong to graphs G_s^q, are marked by dashed lines so as to indicate the fact that $G_s^q = (\mathbf{U}_f^q, \Gamma_s^q)$ is a subgraph of $G_s^r = (\mathbf{U}_f^r, \Gamma_s^r)$ if $(\mathcal{D}_a^r, \mathcal{D}_a^q) \in R_p$. The edges that belong to both graphs or only graph G_s^q are marked by continuous lines.

The Hasse diagram, $(\mathcal{D}_a, R_p)$, of the partially ordered set $\mathcal{D}_b$, by relation R_p [19], is presented in Fig. 16.11. The nodes of the diagram represent the complete sets of signal groups. The maximal capacity factor, which can be obtained if the corresponding complete set of signal groups is chosen, is written next to each node.

From the Hasse diagram, it can be concluded that the highest value of the maximal capacity factor is obtained if the complete set with four signal groups is chosen. This maximal value is 1.85.

According to the proposed heuristics, it is necessary to extract graph $G_H = (\mathcal{D}_b', R_p')$ from the Hasse diagram (see Appendix IV). From Fig. 16.11 it can be noted that:

$$\mathcal{D}_b' = \{\mathcal{D}_a^1, \mathcal{D}_a^3, \mathcal{D}_a^5, \mathcal{D}_a^9\},$$

and $R_p' \subset R_p$ is defined by the set of ordered pairs:

$$R_p' = \{(\mathcal{D}_a^9, \mathcal{D}_a^3), (\mathcal{D}_a^9, \mathcal{D}_a^5), (\mathcal{D}_a^9, \mathcal{D}_a^1), (\mathcal{D}_a^3, \mathcal{D}_a^1), (\mathcal{D}_a^5, \mathcal{D}_a^1)\}.$$

The edges corresponding to elements of relation R_p' are marked by bold lines in the Hasse diagram. The complete set of signal groups, according to the proposed heuristics, should be chosen from set $\mathcal{D}_b'$. In this case, set $\mathcal{D}_b'$, besides the best complete set of signal groups, $\mathcal{D}_a^9$, with four signal groups, contains also the best complete set with five signal groups, $\mathcal{D}_a^5$.

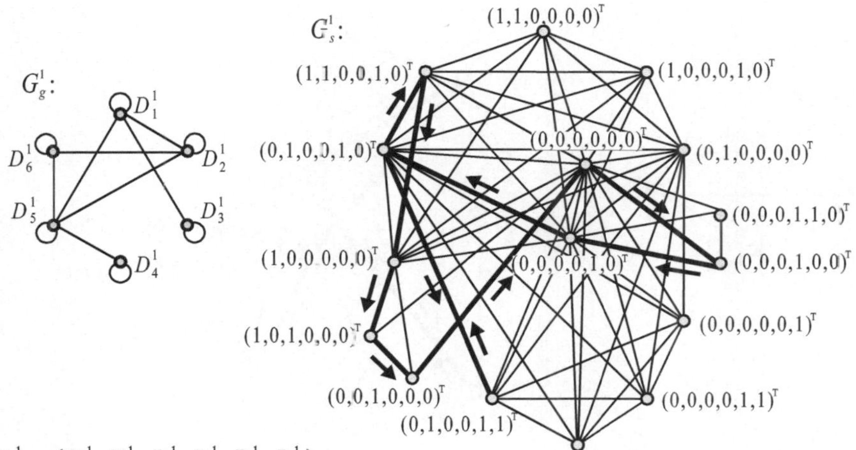

$$\mathfrak{D}_a^1 = \{D_1^1, D_2^1, D_3^1, D_4^1, D_5^1, D_6^1\}$$
$$= \{D_1, D_2, D_3, D_4, D_5, D_6\}$$
$$= \{\{\sigma_1\}, \{\sigma_2\}, \{\sigma_3\}, \{\sigma_4\}, \{\sigma_5\}, \{\sigma_6\}\}$$

$$Z = [z_{pq}]_{6\times 6} = \begin{bmatrix} 0 & 0 & 0 & 4 & 0 & 4 \\ 0 & 0 & 3 & 5 & 0 & 0 \\ 0 & 3 & 0 & 3 & 5 & 2 \\ 2 & 1 & 2 & 0 & 0 & 2 \\ 0 & 0 & 1 & 0 & 0 & 0 \\ 8 & 0 & 4 & 8 & 0 & 0 \end{bmatrix}$$

The optimal solution:

$$\mathbf{u}^1* = \begin{bmatrix} 0 & 0 & 0 & 1 & 1 & 1 & 0 & 0 & 0 & 0 \\ 1 & 1 & 1 & 1 & 0 & 0 & 0 & 0 & 0 & 0 \\ 0 & 0 & 0 & 0 & 0 & 1 & 1 & 0 & 0 & 0 \\ 0 & 0 & 0 & 0 & 0 & 0 & 0 & 0 & 1 & 0 \\ 1 & 1 & 1 & 1 & 0 & 0 & 0 & 0 & 0 & 1 \\ 0 & 1 & 0 & 0 & 0 & 0 & 0 & 0 & 0 & 0 \end{bmatrix}$$

$$\tau^1* = \begin{bmatrix} 1 & 16 & 8 & 16.5 & 3 & 5.5 & 15.25 & 3 & 20.75 & 1 \end{bmatrix}$$

$$\mu^1_{\max} = 2.075$$

Figure 16.1

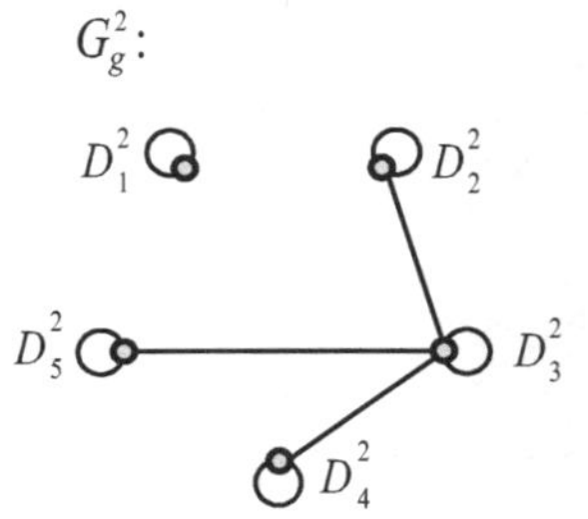

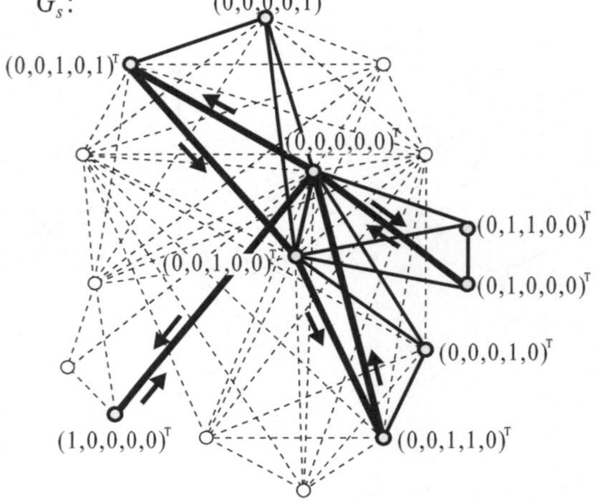

$$\mathfrak{D}_a^2 = \{D_1^2, D_2^2, D_3^2, D_4^2, D_5^2\}$$
$$= \{D_3, D_4, D_5, D_6, D_7\}$$
$$= \{\{\sigma_3\}, \{\sigma_4\}, \{\sigma_5\}, \{\sigma_6\}, \{\sigma_1, \sigma_2\}\}$$

$$Z = [z_{pq}]_{5\times 5} = \begin{bmatrix} 0 & 3 & 5 & 2 & 3 \\ 2 & 0 & 0 & 2 & 2 \\ 1 & 0 & 0 & 0 & 0 \\ 4 & 8 & 0 & 0 & 8 \\ 3 & 5 & 0 & 4 & 0 \end{bmatrix}$$

The optimal solution:

$$\mathbf{u}^2{*} = \begin{bmatrix} 1 & 0 & 0 & 0 & 0 & 0 & 0 & 0 \\ 0 & 0 & 1 & 0 & 0 & 0 & 0 & 0 \\ 0 & 0 & 0 & 0 & 1 & 1 & 1 & 0 \\ 0 & 0 & 0 & 0 & 0 & 0 & 1 & 0 \\ 0 & 0 & 0 & 0 & 1 & 0 & 0 & 0 \end{bmatrix}$$

$$\tau^2{*} = \begin{bmatrix} 15 & 3 & 20 & 2 & 26 & 4 & 16 & 4 \end{bmatrix}$$

$$\mu_{\max}^2 = 1.30$$

Figure 16.2

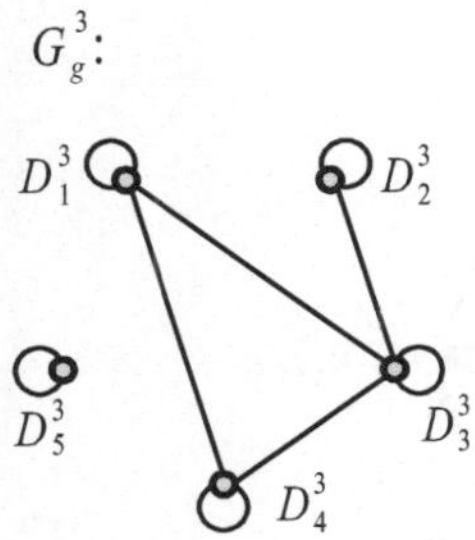

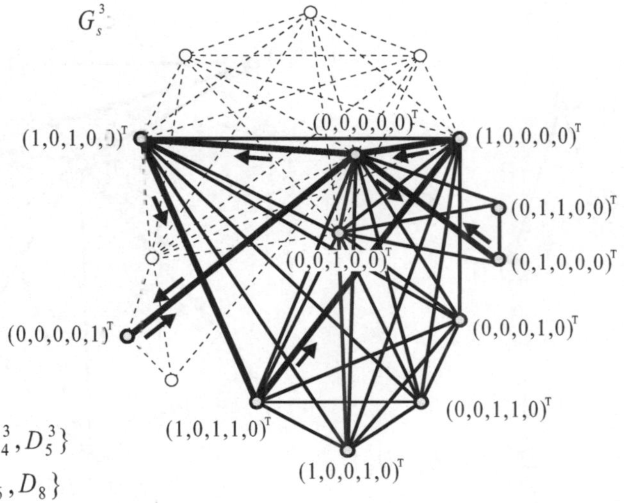

$$\mathfrak{D}_a^3 = \{D_1^3, D_2^3, D_3^3, D_4^3, D_5^3\}$$
$$= \{D_2, D_4, D_5, D_6, D_8\}$$
$$= \{\{\sigma_2\}, \{\sigma_4\}, \{\sigma_5\}, \{\sigma_6\}, \{\sigma_1, \sigma_3\}\}$$

$$Z = [z_{pq}]_{5\times 5} = \begin{bmatrix} 0 & 5 & 0 & 0 & 3 \\ 1 & 0 & 0 & 2 & 2 \\ 0 & 0 & 0 & 0 & 2 \\ 0 & 8 & 0 & 0 & 8 \\ 3 & 4 & 5 & 4 & 0 \end{bmatrix}$$

The optimal solution is:

$$\mathbf{u}^3* = \begin{bmatrix} 0 & 0 & 0 & 0 & 1 & 1 & 1 & 0 \\ 0 & 0 & 1 & 0 & 0 & 0 & 0 & 0 \\ 0 & 0 & 0 & 0 & 1 & 1 & 0 & 0 \\ 0 & 0 & 0 & 0 & 0 & 1 & 0 & 0 \\ 1 & 0 & 0 & 0 & 0 & 0 & 0 & 0 \end{bmatrix}$$

$$\tau^3* = \begin{bmatrix} 25 & 4 & 20 & 1 & 16 & 5 & 3 & 4 \end{bmatrix}$$

$$\mu_{\max}^3 = 1.85$$

Figure 16.3

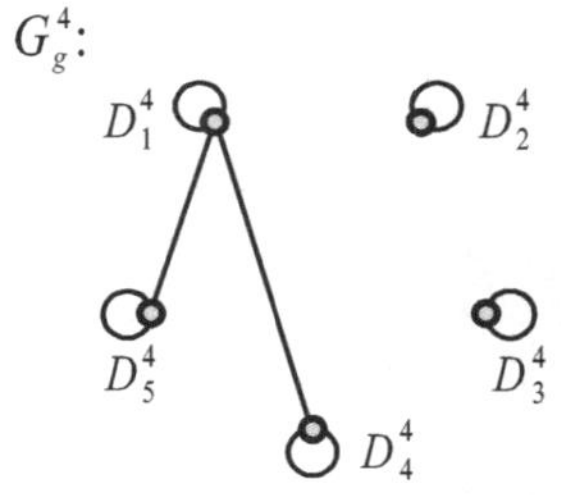

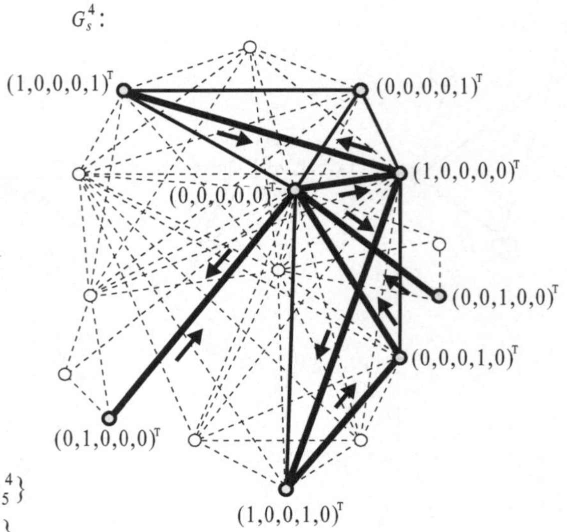

$$
\begin{aligned}
\mathfrak{D}_a^4 &= \{D_1^4, D_2^4, D_3^4, D_4^4, D_5^4\} \\
&= \{D_2, D_3, D_4, D_6, D_9\} \\
&= \{\{\sigma_2\}, \{\sigma_3\}, \{\sigma_4\}, \{\sigma_6\}, \{\sigma_1, \sigma_5\}\}
\end{aligned}
$$

$$
Z = [z_{pq}]_{5\times 5} = \begin{bmatrix} 2 & 3 & 5 & 0 & 0 \\ 3 & 0 & 3 & 2 & 5 \\ 1 & 2 & 0 & 2 & 2 \\ 0 & 4 & 8 & 0 & 8 \\ 0 & 1 & 4 & 4 & 0 \end{bmatrix}
$$

The optimal solution is:

$$
\mathbf{u}^{4*} = \begin{bmatrix} 0 & 0 & 1 & 1 & 1 & 1 & 0 & 0 & 0 & 0 \\ 0 & 0 & 0 & 0 & 0 & 0 & 0 & 0 & 1 & 0 \\ 1 & 0 & 0 & 0 & 0 & 0 & 0 & 0 & 0 & 0 \\ 0 & 0 & 0 & 0 & 0 & 1 & 1 & 0 & 0 & 0 \\ 0 & 0 & 0 & 1 & 0 & 0 & 0 & 0 & 0 & 0 \end{bmatrix}
$$

$$
\tau^{4*} = \begin{bmatrix} 20 & 1 & 1 & 25 & 4 & 2 & 14 & 4 & 16 & 3 \end{bmatrix}
$$

$$
\mu_{\max}^4 = 1.60
$$

Figure 16.4

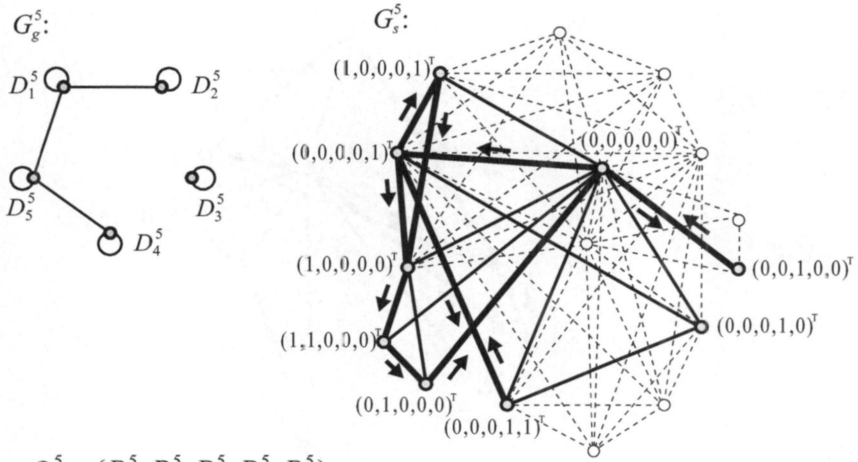

$$\mathcal{D}_a^5 = \{D_1^5, D_2^5, D_3^5, D_4^5, D_5^5\}$$
$$= \{D_1, D_3, D_4, D_6, D_{10}\}$$
$$= \{\{\sigma_1\}, \{\sigma_3\}, \{\sigma_4\}, \{\sigma_6\}, \{\sigma_2, \sigma_5\}\}$$

$$Z = [z_{pq}]_{5\times 5} = \begin{bmatrix} 0 & 0 & 4 & 4 & 0 \\ 0 & 0 & 3 & 2 & 5 \\ 2 & 2 & 0 & 2 & 1 \\ 8 & 4 & 5 & 0 & 0 \\ 0 & 3 & 5 & 0 & 0 \end{bmatrix}$$

The optimal solution is:

$$\mathbf{u}^5{*} = \begin{bmatrix} 0 & 0 & 0 & 0 & 0 & 1 & 1 & 1 & 0 & 0 \\ 0 & 0 & 0 & 0 & 0 & 0 & 0 & 1 & 1 & 0 \\ 1 & 0 & 0 & 0 & 0 & 0 & 0 & 0 & 0 & 0 \\ 0 & 0 & 0 & 1 & 0 & 0 & 0 & 0 & 0 & 0 \\ 0 & 0 & 1 & 1 & 1 & 1 & 0 & 0 & 0 & 0 \end{bmatrix}$$

$$\tau^5{*} = [20.75 \;\; 1 \;\; 15.25 \;\; 16 \;\; 8 \;\; 2.25 \;\; 3 \;\; 19.75 \;\; 1 \;\; 3]$$

$$\mu_{\max}^5 = 2.075$$

Figure 16.5

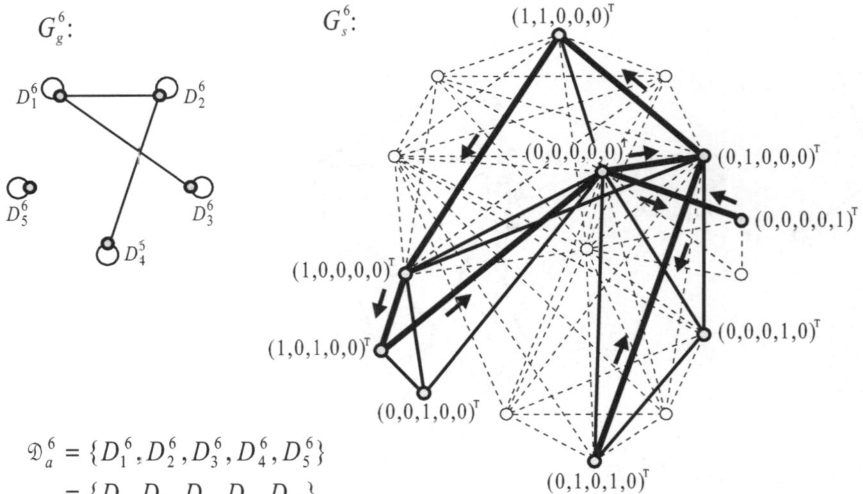

$$\mathfrak{D}_a^6 = \{D_1^6, D_2^6, D_3^6, D_4^6, D_5^6\}$$
$$= \{D_1, D_2, D_3, D_6, D_{11}\}$$
$$= \{\{\sigma_1\}, \{\sigma_2\}, \{\sigma_3\}, \{\sigma_6\}, \{\sigma_4, \sigma_5\}\}$$

$$Z = [z_{pq}]_{5\times5} = \begin{bmatrix} 0 & 0 & 0 & 4 & 4 \\ 0 & 0 & 3 & 0 & 5 \\ 0 & 3 & 0 & 2 & 5 \\ 8 & 0 & 4 & 0 & 8 \\ 2 & 1 & 2 & 2 & 0 \end{bmatrix}$$

The optimal solution is:

$$\mathbf{u}^6* = \begin{bmatrix} 0 & 0 & 0 & 0 & 0 & 1 & 1 & 1 & 0 \\ 0 & 0 & 1 & 1 & 1 & 1 & 0 & 0 & 0 \\ 0 & 0 & 0 & 0 & 0 & 0 & 0 & 1 & 0 \\ 0 & 0 & 0 & 1 & 0 & 0 & 0 & 0 & 0 \\ 1 & 0 & 0 & 0 & 0 & 0 & 0 & 0 & 0 \end{bmatrix}$$

$$\tau^6* = [20.25 \quad 1 \quad 1 \quad 16 \quad 21.75 \quad 1.75 \quad 3 \quad 20.25 \quad 5]$$

$$\mu_{\max}^6 = 2.025$$

Figure 16.6

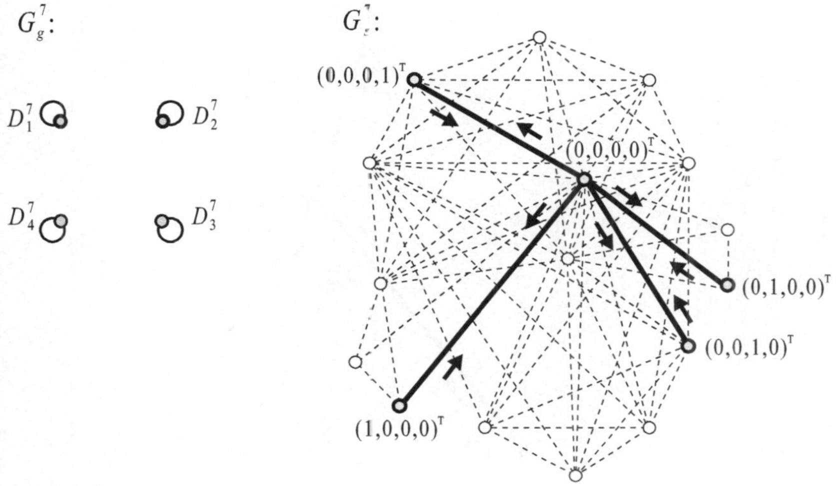

$$\mathfrak{D}_a^7 = \{D_1^7, D_2^7, D_3^7, D_4^7\}$$
$$= \{D_3, D_4, D_6, D_{12}\}$$
$$= \{\{\sigma_3\}, \{\sigma_4\}, \{\sigma_6\}, \{\sigma_1, \sigma_2, \sigma_5\}\}$$

$$Z = [z_{pq}]_{4\times 4} = \begin{bmatrix} 0 & 3 & 2 & 5 \\ 2 & 0 & 2 & 2 \\ 4 & 8 & 0 & 8 \\ 3 & 5 & 4 & 0 \end{bmatrix}$$

The optimal solution is:

$$\mathbf{u}^7 * = \begin{bmatrix} 0 & 0 & 0 & 0 & 0 & 0 & 1 & 0 \\ 1 & 0 & 0 & 0 & 0 & 0 & 0 & 0 \\ 0 & 0 & 0 & 0 & 1 & 0 & 0 & 0 \\ 0 & 0 & 1 & 0 & 0 & 0 & 0 & 0 \end{bmatrix}$$
$$\tau^7 * = \begin{bmatrix} 20 & 2 & 26 & 4 & 16 & 4 & 14 & 3 \end{bmatrix}$$
$$\mu_{\max}^7 = 1.30$$

Figure 16.7

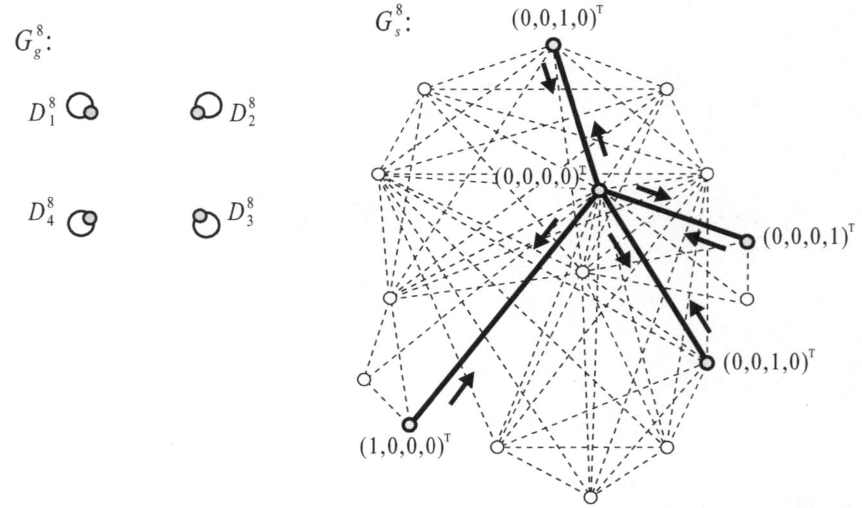

$$\mathcal{D}_a^8 = \{D_1^8, D_2^8, D_3^8, D_4^8\} = \{D_3, D_7, D_7, D_{11}\} = \{\{\sigma_3\}, \{\sigma_6\}, \{\sigma_1, \sigma_2\}, \{\sigma_4, \sigma_5\}\}$$

$$Z = [z_{pq}]_{4\times 4} = \begin{bmatrix} 0 & 2 & 3 & 5 \\ 4 & 0 & 8 & 8 \\ 3 & 4 & 0 & 5 \\ 2 & 2 & 2 & 0 \end{bmatrix}$$

The cptimal solution is:

$$\mathbf{u}^8 * = \begin{bmatrix} 0 & 0 & 0 & 0 & 0 & 0 & 1 & 0 \\ 0 & 0 & 0 & 0 & 1 & 0 & 0 & 0 \\ 1 & 0 & 0 & 0 & 0 & 0 & 0 & 0 \\ 0 & 0 & 1 & 0 & 0 & 0 & 0 & 0 \end{bmatrix}$$

$$\tau^8 * = [25 \quad 5 \quad 20 \quad 2 \quad 16 \quad 4 \quad 15 \quad 3]$$

$$\mu_{\max}^8 = 1.25$$

Figure 16.8

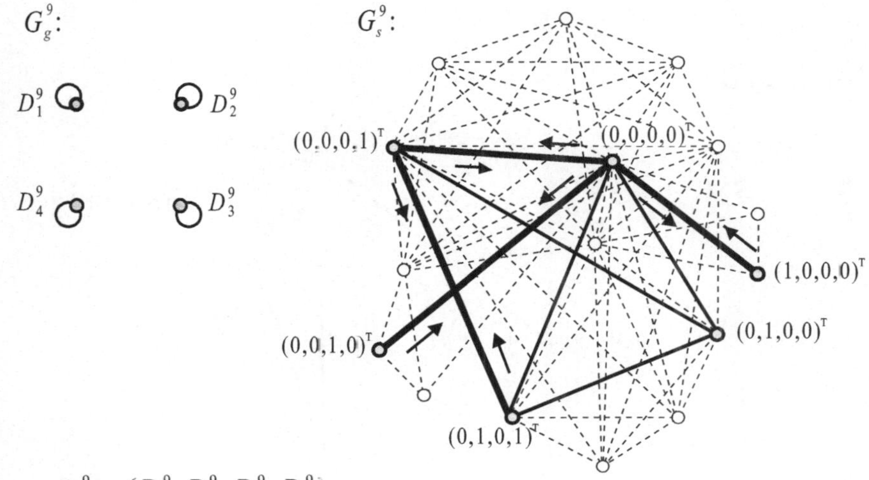

$$\mathfrak{D}_a^9 = \{D_1^9, D_2^9, D_3^9, D_4^9\}$$
$$= \{D_4, D_6, D_8, D_{10}\}$$
$$= \{\{\sigma_4\}, \{\sigma_4\}, \{\sigma_1, \sigma_3\}, \{\sigma_2, \sigma_5\}\}$$

$$Z = [z_{pq}]_{4\times 4} = \begin{bmatrix} 0 & 2 & 2 & 1 \\ 8 & 0 & 8 & 0 \\ 4 & 4 & 0 & 5 \\ 5 & 0 & 3 & 0 \end{bmatrix}$$

The optimal solution is:

$$\mathbf{u}^9{*} = \begin{bmatrix} 1 & 0 & 0 & 0 & 0 & 0 & 0 & 0 \\ 0 & 0 & 0 & 1 & 0 & 0 & 0 & 0 \\ 0 & 0 & 0 & 0 & 0 & 0 & 1 & 0 \\ 0 & 0 & 1 & 1 & 1 & 0 & 0 & 0 \end{bmatrix}$$

$$\tau^9{*} = \begin{bmatrix} 20 & 1 & 16 & 16 & 5 & 3 & 25 & 4 \end{bmatrix}$$

$$\mu_{\max}^9 = 1.85$$

Figure 16.9

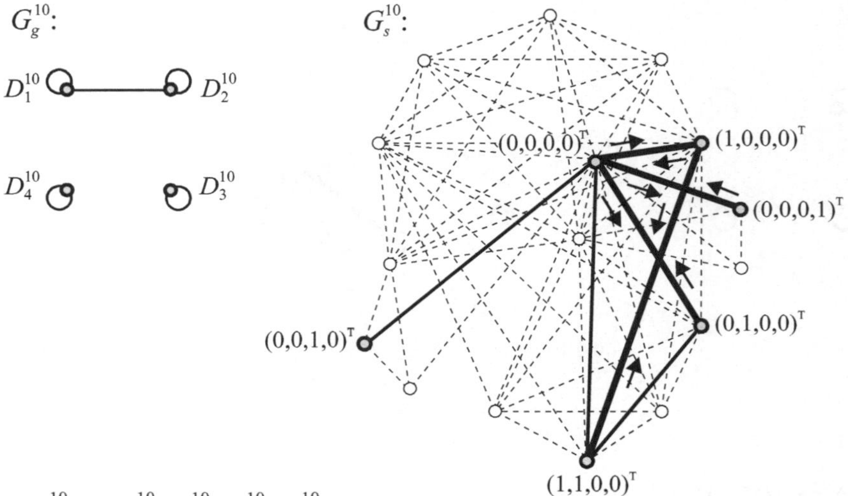

$$\mathcal{D}_a^{10} = \{D_1^{10}, D_2^{10}, D_3^{10}, D_4^{10}\}$$
$$= \{D_2, D_6, D_8, D_{11}\}$$
$$= \{\{\sigma_2\}, \{\sigma_6\}, \{\sigma_1, \sigma_3\}, \{\sigma_4, \sigma_5\}\}$$

$$Z = [z_{pq}]_{4\times 4} = \begin{bmatrix} 0 & 0 & 3 & 5 \\ 0 & 0 & 8 & 8 \\ 3 & 4 & 0 & 5 \\ 1 & 2 & 2 & 0 \end{bmatrix}$$

The optimal solution is:

$$\mathbf{u}^{10}* = \begin{bmatrix} 0 & 0 & 1 & 1 & 1 & 0 & 0 & 0 \\ 0 & 0 & 0 & 1 & 0 & 0 & 1 & 0 \\ 0 & 0 & 0 & 0 & 0 & 0 & 0 & 0 \\ 1 & 0 & 0 & 0 & 0 & 0 & 0 & 0 \end{bmatrix}$$
$$\tau^{10}* = \begin{bmatrix} 20 & 1 & 15 & 16 & 5 & 3 & 25 & 5 \end{bmatrix}$$
$$\mu_{\max}^{10} = 1.80$$

Figure 16.10

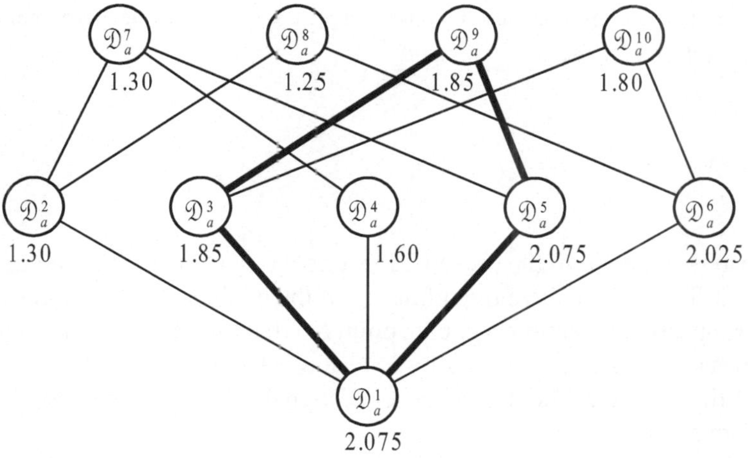

Figure 16.11

The obtained results can be summarized as follows:

- The difference between the maximal capacity factor value achieved when the complete signal group set $\mathcal{D}_a^1$ is chosen and the values obtained for all other elements of $\mathcal{D}_b$ range up to 40% of this maximal value.
- In the subsets of $\mathcal{D}_b$ containing a same number of signal groups, the maximal capacity factors differ:
  - 37.35 % in subset $\mathcal{D}_b^2$ with elements comprising 5 signal groups,
  - 29.73 % in subset $\mathcal{D}_b^3$ with elements comprising 4 signal groups.
- The greatest maximal capacity values are:
  - 2.075 for the complete set whose elements contain 6 signal groups,
  - 2.075 for the complete set whose elements contain 5 signal groups,
  - 1.85 for the complete set whose elements contain 4 signal groups.

  The greatest maximal capacity factor value in subset $\mathcal{D}_b^3$ with elements $\mathcal{D}_a^7, \mathcal{D}_a^8, \mathcal{D}_a^9, \mathcal{D}_a^{10}$ containing 4 signal groups each is less than or equal to the greatest value of the maximal capacity factor for elements in subset $\mathcal{D}_b^2$ containing complete sets $\mathcal{D}_a^2, \mathcal{D}_a^3, \mathcal{D}_a^4, \mathcal{D}_a^5, \mathcal{D}_a^6$, each with 5 signal groups.
- From the capacity factor values in Table 16.2, and the Hasse diagram, it can be seen that there exist some complete sets of signal groups, $\mathcal{D}_a^d$, with greater maximal capacity

factor value than the value obtained for other complete sets, $\mathcal{D}_a^e$, having greater number of elements than $\mathcal{D}_a^d$, i.e.,

$$card\ \mathcal{D}_a^e > card\ \mathcal{D}_a^d,$$

but in this case $(\mathcal{D}_a^e, \mathcal{D}_a^d) \notin R_p$.

The results obtained in the presented example (16.1), as well as in papers [33], [34], [36], which analyze the influence of the choice of the complete set of signal groups to intersection capacity, point out that this choice significantly affects optimal values of the optimality criterion. Also, here is shown that the problem of the choice of the complete set of signal groups can be simplified by the proposed heuristics.

It is also evident that the intentions of practitioners to choose the complete set with the least number of signal groups might lead to significant deterioration of intersection performance indices compared to choosing the set with a greater number of signal groups.

Appendix I

GRAPHS, CLIQUES

Cartesian product of sets *X* and *Y*

Cartesian product of sets X and Y is the set of ordered pairs defined by the following relation:

$$X \times Y = \{(x, y) \mid x \in X, y \in Y\}.$$

If $Y = X$, the set $X \times X$ is denoted by X^2.

Binary relation ρ in set *X*

Binary relation ρ in set X is any subset of set X^2, i.e., $\rho \subset X^2$.

Graph

Two graph definitions are used in this book:

Definition I: Graph is the ordered pair $G = (X, \rho)$, where X is a nonempty set, and ρ is a binary relation in X, i.e., $\rho \subset X^2$.

Definition II: Graph is the ordered pair $G = (X, \Gamma)$, where X is a nonempty set, and Γ is a mapping of set X in the partitive set $\mathcal{P}(X)$ of set X [9], i.e.,

$$\Gamma : X \to \mathcal{P}(X).$$

Elements of set X can be represented by dots in a plane, named the *nodes* or *vertices* of the graph.

The set of ordered pairs, ρ, which represents a relation, sometimes is denoted by U. Elements of this set are *edges* of the graph. Thus, a graph can be described as:

$$G = (X, U) = (X, \Gamma).$$

An element of set U, i.e., the ordered pair $(x, y) \in U$, is represented by the line connecting nodes x and y, oriented from x to y. If $(x, y) \in U$ and $(y, x) \in U$, nodes x and y are connected by two edges having opposite orientation. These two edges are usually represented by a single nonoriented edge. If for some element x there holds $x \in \Gamma x$, this is indicated by a loop at node x. The loop is usually not oriented.

If relation ρ is symmetric, the graph is nonoriented.

If relation ρ is antisymmetric, the graph is oriented.

If relation ρ is neither symmetric nor antisymmetric, graph $G = (X, \rho)$ is neither oriented nor nonoriented.

Adjacency matrix of a graph

The adjacency matrix of graph $G = (X, \rho)$ is square matrix

$$B = [b_{ij}]_{n \times n}.$$

The degree of matrix B is n, where

$$n = card\ X.$$

Matrix elements are defined as follows:

$$b_{ij} = \begin{cases} 1, & (x_i, x_j) \in \rho \\ 0, & (x_i, x_j) \notin \rho \end{cases}.$$

The adjacency matrix of graph G presented in Fig. AI.1 is

$$B = \begin{bmatrix} 1 & 1 & 0 & 1 & 0 \\ 1 & 0 & 1 & 1 & 0 \\ 0 & 0 & 0 & 1 & 1 \\ 1 & 0 & 1 & 0 & 0 \\ 0 & 0 & 1 & 0 & 0 \end{bmatrix}.$$

If the graph is nonoriented, its adjacency matrix is symmetric, i.e., $B = B^T$.

If the graph is oriented, then:

$$b_{ij} = 1 \Rightarrow b_{ji} = 0, \quad (i, j \in \{1,2,\ldots,n\},\ i \neq j).$$

Partitive set $\mathcal{P}(X)$ of set X

Partitive set $\mathcal{P}(X)$ of set X is the set of all subsets of set X. For example, if $X = \{a,b,c\}$, then

$$\mathcal{P}(X) = \{\{a\},\{b\},\{c\},\{a,b\},\{a,c\},\{b,c\},\{a,b,c\},\varnothing\}.$$

Subgraph

Subgraph $E = (Y,V)$ of graph $G = (X,U)$ is the graph with the following properties:

$$Y \subset X, \text{ and } V = U \cap (Y \times Y).$$

A subgraph contains, thus, only the elements of set U that are generated by elements of set Y.

Partial graph

A partial graph of graph $G = (X,U)$ is any graph $H = (X,T)$ with $T \subset U$.

For example, in Fig. AI.1, graph G_1 is a subgraph of graph G, and G_2 is a partial graph of graph G.

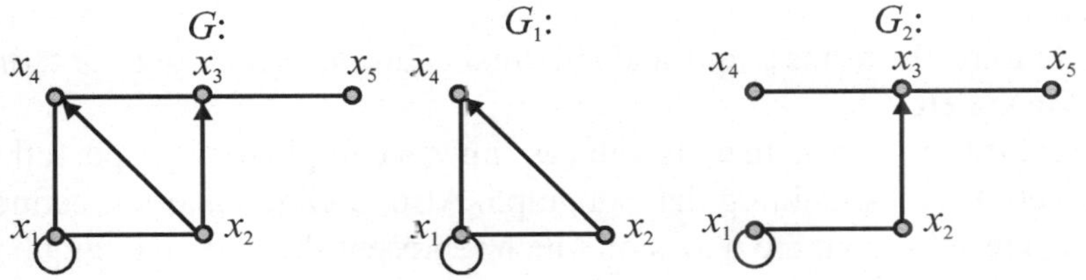

Figure AI.1

Clique (in the sense of Berge [9])

A clique of nonoriented graph $G=(X,\rho)=(X,\Gamma)$ is set $C \subset X$ with the property:

$$(x \in C, y \in C) \Rightarrow y \in \Gamma x .$$

The graph with the set of nodes C has an edge between any two nodes. Such a graph is called a *complete graph.*

The following assertions are in accordance with Berge's definition of a clique:

a) The set whose elements are the nodes at the ends of one edge (x, y) of a nonoriented graph represents a clique, i.e.,

$$C = \{x, y\} .$$

This is obvious because if (x, y) is an edge of a nonoriented graph, then:

$$y \in \Gamma x \text{ and } x \in \Gamma y .$$

b) A singleton subset containing an element x of set X, with the property $x \in \Gamma x$, represents a clique, i.e.,

$$C = \{x\} .$$

Maximal clique of graph $G=(X,\Gamma)$

Subset $C_m^k \subset X$ is the maximal clique if C_m^k is not a subset of any other clique. This means that no element can be added to subset C_m^k such that the new, extended subset forms a clique.

Remarks

In literature, the terms clique and maximal clique are often used for terms different than here.

Under the term clique, there is often assumed a complete subgraph, rather than the set of nodes defining that subgraph. Also, a clique is often defined as a complete subgraph, the nodes of which represent the maximal clique in Berge's sense.

The maximal clique is often defined as a complete subgraph with the maximal number of nodes, i.e., a subgraph of graph $G=(X,\Gamma)$, with the property

$$G_{\max} = (C_{\max}, \Gamma) ,$$

where

$$card\, C_{\max} = \max_k \{card\, C_m^k\}.$$

Path, path length, distance

A graph containing only oriented edges is called a digraph. If an edge u of a digraph connects nodes x_i and x_j, and is oriented from x_i to x_j, then it is said that edge u starts at node x_i and ends at node x_j.

The path of length k in a digraph is any sequence of edges $u_1, u_2, \ldots, u_k$, with the following properties:

α) Edge u_1 starts at an arbitrary edge of the digraph.

β) Edge u_i ($i = 2, \ldots, k$) starts at the node at which edge u_{i-1} ends.

A path connects node x_i with node x_j if the first edge of the path starts at x_i and the last edge of the path ends at x_j.

The distance between nodes x_i and x_j is equal to the length of the shortest path connecting these two nodes.

The path length of a path containing a single edge is equal to 1.

The maximal distance between any two nodes of a graph is called the *diameter* of the graph.

Appendix II

EQUIVALENCE RELATION

Equivalence relation

The equivalence relation [20] in set X is a binary relation C_e, which is reflexive, symmetric, and transitive. The relation C_e is:

- Reflexive if $x C_e x$, $(x \in X)$
- Symmetric if $x C_e y \Rightarrow y C_e x$, $(x, y \in X)$
- Transitive if $(x C_e y) \wedge (y C_e z) \Rightarrow x C_e z$, $(x, y, z \in X)$

Diagonal Δ_x, i.e., set

$$C_e = \Delta_x = \{(x,x) \mid x \in X\}$$

is a trivial example of equivalence relation.

The complete relation, i.e.,

$$C_e = X^2 = \{(x,y) \mid x, y \in X\}$$

is also an equivalence relation.

Equivalence classes

The equivalence class, $\bar{x}$, of element x in regard to equivalence relation C_e is the set defined as follows:

$$\bar{x} = \{y \mid x C_e y, y \in X\}, \quad (x \in X).$$

Two equivalent classes $\bar{x}$ and $\bar{z}$ $(x, z \in X)$ are either equivalent or disjointed (their intersection is the empty set).

Therefore, equivalence classes $\bar{x}$ ($x \in X$) define a partition of set *X*.

Every partition $\pi(X) = \{X_1, X_2, \ldots, X_n\}$ of set

$$X = X_1 \cup X_2 \cup \ldots X_{n-1} \cup X_n$$

defines an equivalence relation in set *X*. Equivalence classes are subsets X_i.

The partitions and equivalence relation in set *X* uniquely define each other and represent different expressions of a same fact [87].

Quotient set

The set of equivalent classes of all elements of set *X* is called *the quotient of X by* C_e and it is denoted by X / C_e.

An equivalence class of relation C_e in set *X*, where *X* is the set of nodes of graph $G = (X, C_e)$, represents a clique of graph G (in the Berge's sense)

Appendix III

PSEUDOCODES OF PROGRAMS CLIQ AND MINA

1. PSEUDOCODE OF PROGRAM CLIQ [34]

```
begin
class := 1
  {determination of all complete subgraphs with two
nodes
   by making all combinations of two elements from the
   number of nodes, and establishing whether
   (m_inc[i,j]=1) AND (m_inc[j,i]=1)}
  {determination of all other complete subgraphs }

  while (n_comb[class+1]<> 0) do
    class := class + 1
    for i := 1 to n_comb[class] - class do
      for j := i + 1 to n_comb[class] - class +1 do
        p = card(comb[class][j] ∩ comb[class][i])
        if (p = class - 1) then
          k :=0
          for l := j + 1 to n_comb[class] do
            p = card((comb[class][i] ∪ comb[class][j])
                        ∩comb[class][l])
            if (p = class) then k := k + 1
            if (k = class - 1) then
              n_comb[class + 1] := n_comb[class + 1] + 1
              comb[class + 1][n_comb[class + 1]] :=
                comb[class][i] ∪ comb[class][j]
            end if
          end for
        end if
      end for
```

```
    end for
  end while
end
```

2. PSEUDOCODE OF PROGRAM MINA

```
begin
  read (no_sg, no_ts, a);
  ini (no_ts, no_sg, a, q, w, t, v, zopt, ztek,
          no_posl, l, s, pregl_u_l, exists_res);
  if (not exists_res) then
    end (exists_res, no_sg, no_ts,
          zopt, no_opt_r, s, wopt);
  indl := true;
  while (true) do
  begin
    while ((t <> q) or (not indl)) do
    begin
      if (indl) then choose_list (no_ts, q, t, v,
                                  tek_l, pregl_u_l)
      else indi := true;
      parc_res (no_sg, zopt, ztek, tek_l, t, l, s,
                tek_podsk, no_posl, posl, pregl_u_l,
                exists_pr);
      if (not exists_pr) then goto 1;
      pre_testa (tek_podsk, s, ztek, w, t);
    end (while);
    if (ztek < zopt) then
      begin
        no_opt_r := 1;
        zopt := ztek;
      end
    else if (ztek = zopt) then
      no_opt_r := no_opt_r + 1;
    wopt[no_opt_r] := w;

1   if (w = [ ]) then
      end (exists_res, no_sg, no_ts,
            zopt, no_opt_r, s, wopt);
    backtr (no_sg, posl, s, l, ztek, no_posl,
            tek_l, w, t);
    indl := false;
  end(while);
end
```

Appendix IV

REFINEMENT RELATION, HASSE DIAGRAMS

Partitioning of set *X*

The family

$$\pi(X) = \{X_i \mid i \in J, X_i \subseteq X\},$$

where $X_i \neq \varnothing$ ($i \in J$), is a partition of set *X* if

$$X_i \cap X_j = \varnothing, (i, j \in J, i \neq j) \text{ and}$$

$$\bigcup_{i \in J} X_i = X,$$

where *J* is the index set of subsets of set *X*.

Refinement relation

The refinement relation [51], [25], R_p, in the set of all partitions $\Pi(X)$ of set *X* is the relation of partial order. Therefore, this relation is reflexive, antisymmetric and transitive.

$\pi_1(X)$ is refinement of $\pi_2(X)$, i.e., $\pi_1(X) R_p \pi_2(X)$ if

$$(X_1 \in \pi_1(X)) \wedge (X_2 \in \pi_2(X)) \Rightarrow X_1 \subseteq X_2.$$

Partially ordered sets can be represented by Hasse diagrams. A Hasse diagram takes the form of a nonoriented graph. According to the convention, node x is positioned below node y if $x R_p y$. Two nodes x and y are joined by an edge if $x R_p y$ and there exists no node z such that $x R_p z$ and $z R_p y$.

Example: Let $X = \{x, y, z\}$. The Hasse diagram of set $\mathcal{P}(X)$, partially ordered by the inclusion relation, is presented in Fig. AIV.1 [19].

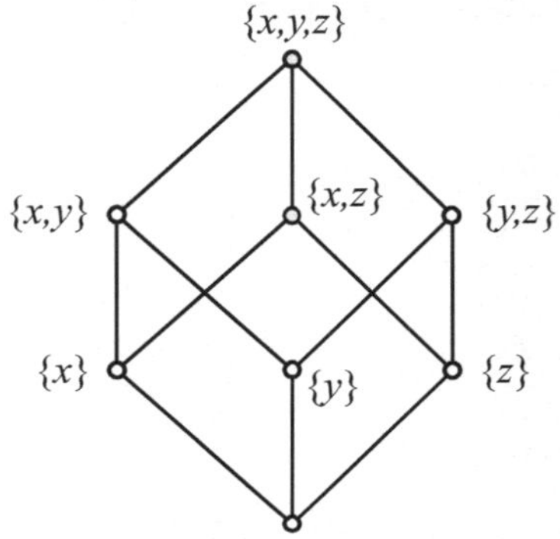

Figure AIV.1

Appendix V

EFFECTIVE VALUES OF GREEN, RED, AND INTERGREEN TIMES

1. EFFECTIVE GREEN AND RED TIMES

The real flow volume when discharging the queue of traffic stream σ_i is shown by bold line in Fig. AV.1. This function is usually transformed to the rectangular shape (ABCD), shown by dashed lines, used for defining effective green and effective red times.

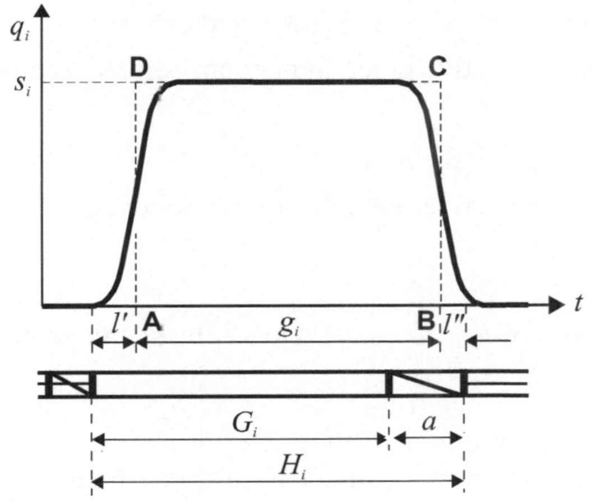

Figure AV.1

The area of rectangle ABCD is equal to the area bounded by curve $q_i(t)$ and time axis. This means that the number of vehicles that leave an intersection approach, T_i, controlled by the real sequence of signal indications, is equal to the number of vehicles that would leave the same approach if this sequence were replaced by the effective green and red time, and the volume curve transformed into the rectangle.

In this appendix it is assumed that each signal group contains a single traffic stream ($p = i$, $\mathscr{P} = \mathscr{J}$).

The following relations exist between the elements of the real sequence of signal indications and effective green and red times (Fig. AV.1):

$$H_i = G_i + a = g_i + l = g_i + l' + l'' \qquad \text{(AV.1)}$$

$$G_i + a + R_i = g_i + r_i = H_i + R_i = c \qquad \text{(AV.2)}$$

$$g_i = H_i - l \qquad \text{(AV.3)}$$

$$r_i = R_i + l \qquad \text{(AV.4)}$$

where:

G_i	–	duration of green indication,
a	–	duration of amber indication,
H_i	–	total duration of green and amber indication,
g_i	–	duration of effective green indication—effective green time,
l	–	"lost time,"
R_i	–	duration of red indication,
r_i	–	duration of effective red indication—effective red time,
c	–	cycle time.

The typical values of *lost times*, l' and l'', most frequently used [1] are $l' = l'' = 1\ \text{s}$.

The *saturation flow* volume s_i is the queue discharge rate during effective green time. The saturation flow volume has an approximately constant value that is attained after the acceleration of vehicles leaving the stop line is performed at the beginning of green time, i.e., when vehicles crossing the stop line do not accelerate any more.

2. EFFECTIVE VALUES OF INTERGREEN TIMES

Intergreen time is the minimal time that has to elapse between the end of displayed green for traffic stream σ_i until the beginning of green indication for another, incompatible traffic stream σ_j. The intergreen time is introduced due to safety reasons, and it is calculated according to the following expression (Fig. AV.2):

$$\bar{z}_{ij} = t_{ij}^{r} - t_{ij}^{e} + t^{p} + t^{g} . \qquad \text{(AV.5)}$$

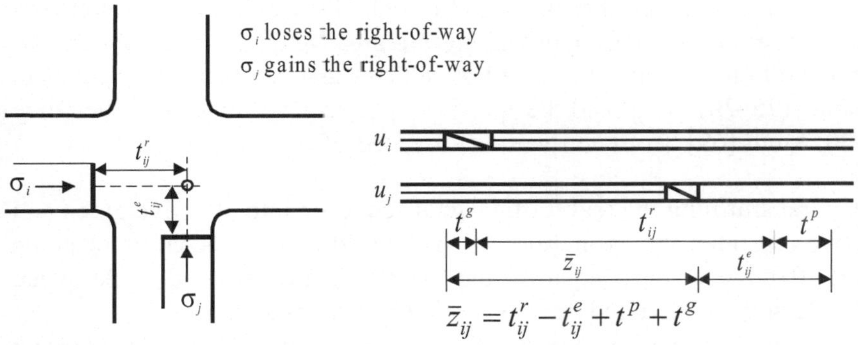

Figure AV.2

The notations in expression (AV.5) have the following meanings:

- t_{ij}^{r} – the time needed for the last vehicle of traffic stream σ_i (losing the right-of-way) to reach the conflict area with stream σ_j,
- t_{ij}^{e} – the time needed for the first vehicle of stream σ_j (gaining the right-of-way) to reach the conflict area with stream σ_i,
- t^{p} – the passing time of vehicles through the conflict area,
- t^{g} – the part of amber indication during which vehicles of stream σ_i still pass through the intersection,
- $\bar{z}_{ij}$ – the minimal intergreen time between streams σ_i and σ_j.

Expressions (AV.5) are used when conflicting traffic streams are vehicle streams. When conflicts exist between pedestrians and vehicles, the minimal intergreen times are calculated using the following expressions:

$$\bar{z}_{ij} = t_{ij}^{r} + t^{g} \tag{AV.6}$$

if vehicles (traffic stream σ_i) lose and pedestrians (traffic stream σ_j) gain the right-of-way, and

$$\bar{z}_{ij} = t_{ij}^{r} - t_{ij}^{e} \tag{AV.7}$$

if pedestrians (traffic stream σ_i) lose and vehicles (traffic stream σ_j) gain the right-of-way.

In calculating minimal intergreen times, according to expressions (AV.5) to (AV.7), it is assumed that the speed of vehicles leaving the conflict area is the lowest, and of ones approaching the area the highest possible under given conditions. Usually, the speed of vehicles leaving the intersection is in the range (25–30) km/h, and the speed of approaching vehicles in the range (40–50) km/h [69]. Common values for speed of pedestrians are 1.2 to 1.5 m/s.

By transforming the real traffic signal sequence into effective green and effective red time, the assumption is introduced that the departure volume during effective green is equal to the saturation flow volume while queue exists. During effective red time, the volume is equal to zero.

The time between the end of effective green time of one signal group and beginning of effective green time of another signal group is called the *effective intergreen time*. Its minimal value is the *minimal effective intergreen time*.

The minimal effective intergreen time between two incompatible vehicle traffic streams, σ_i and σ_j, is given by the following expression (Fig. AV.3):

$$z'_{ij} = \bar{z}_{ij} - a + l\,. \tag{AV.8}$$

In the case when pedestrians leave the conflict area, and vehicles approach, the minimal effective intergreen time is given by the following expression (Fig. AV.4):

$$z'_{ij} = \bar{z}_{ij} + l'\,. \tag{AV.9}$$

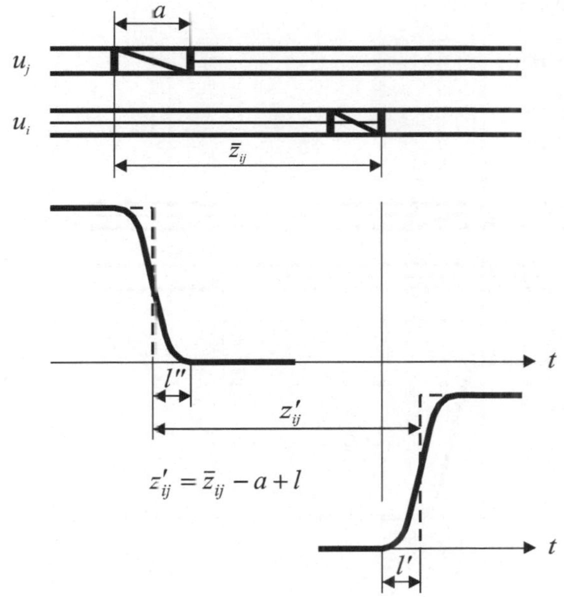

Figure AV.3

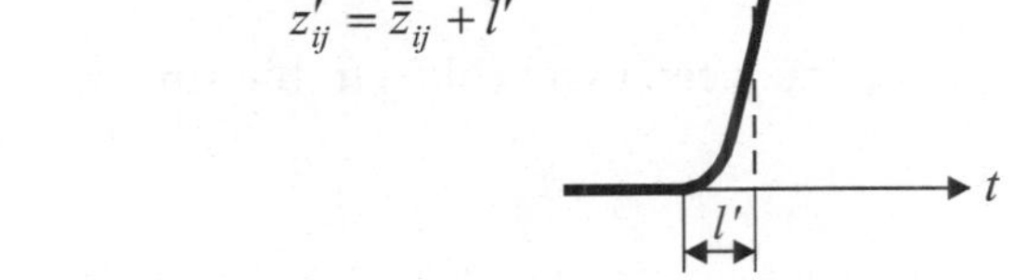

Figure AV.4

In the case when vehicles leave the conflict area, and pedestrians approach, the minimal effective intergreen time is given by the following expression (Fig. AV.5):

$$z'_{ij} = \bar{z}_{ij} - a + l''. \tag{AV.10}$$

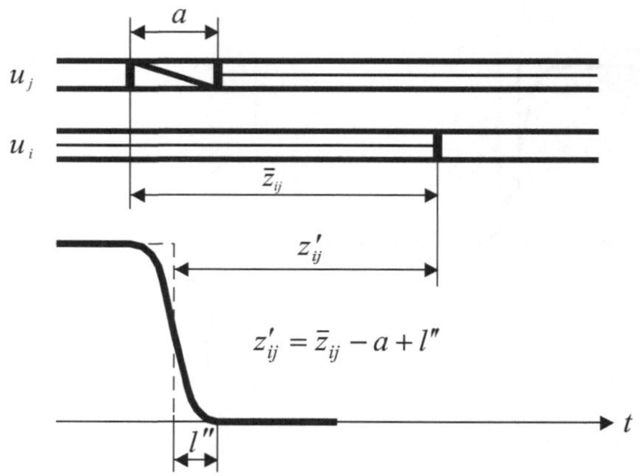

Figure AV.5

When intergreen time between two vehicle streams $\bar{z}_{ij} \le 0$, then the minimal effective intergreen time is calculated using the following expression (Fig. AV.6):

$$z'_{ij} = \bar{z}_{ij} + a - l. \tag{AV.11}$$

By substituting standard values for $a = 3\,\text{s}$ and $l' = l'' = 1\,\text{s}$, i.e., $l = 2\,\text{s}$, in expressions (AV.8) – (AV.11), the following expressions for z'_{ij} are obtained:

a) In the case of conflict between two vehicle traffic streams:

$$z'_{ij} = \bar{z}_{ij} - 1. \tag{AV.12}$$

b) In the case of conflict between pedestrians and vehicles (pedestrians lose the right-of-way):

$$z'_{ij} = \bar{z}_{ij} + 1. \tag{AV.13}$$

c) In the case of conflict between pedestrians and vehicles (pedestrians get the right-of-way):

$$z'_{ij} = \overline{z}_{ij} - 2 . \qquad \text{(AV.14)}$$

d) In the case of conflict between two vehicle traffic streams when $z'_{ij} \leq 0$:

$$z'_{ij} = \overline{z}_{ij} + 1 . \qquad \text{(AV.15)}$$

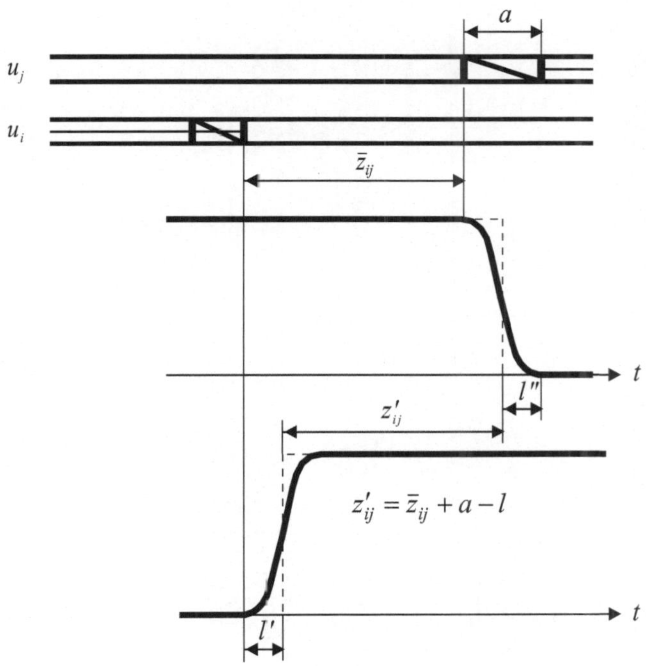

Figure AV.6

Appendix VI

DETERMINATION OF THE CONTROL VECTORS TRANSITION GRAPH

1. DETERMINATION OF THE SET OF FEASIBLE CONTROL VECTORS

The graph of control vectors transition, G_s, in the case relation R_s is symmetric, can be determined by finding all maximal cliques of compatibility graph G_g, and the control vectors transition graphs corresponding to these cliques. Graph G_s is determined as the union of these graphs. This procedure is very convenient because there exist many algorithms for finding cliques of a graph, so that the existing procedures can be easily implemented in this case.

In the case graph G_s is nonoriented, the set of feasible control vectors can be determined by finding the control vectors giving the right-of-way to a maximal number of signal groups. These vectors are used for generating all other feasible control vectors. The control vectors that give the right-of-way to the maximal number of signal groups correspond to the maximal (dominant) cliques of compatibility graph G_g.

The procedure for determination of all feasible control vectors comprises two steps:

- Extracting maximal cliques from graph G_g and obtaining corresponding control vectors,
- Generating all other feasible vectors using the vectors obtained in the previous step.

1.1. Determining maximal cliques of graph G_g and the set of maximal control vectors

A clique of a graph (Appendix I) is the set of nodes of any complete subgraph of the graph. If the clique is not contained in any other clique, it is called a *maximal* or *dominant* clique.

If the graph of signal groups compatibility,

$$G_g = (\mathfrak{D}_a, C_g) = (\mathfrak{D}_a, \Gamma_g), \tag{AVI.1}$$

is considered, where Γ_g is one-to-one mapping of set $\mathfrak{D}_a$ to its partitive set $\mathcal{P}(\mathfrak{D}_a)$, i.e.,

$$\Gamma_g : \mathfrak{D}_a \to \mathcal{P}(\mathfrak{D}_a), \tag{AVI.2}$$

then $\overline{\mathfrak{D}}_{am}^{\pi}$ is a maximal clique of the compatibility graph if

$$((\forall D_p \in \overline{\mathfrak{D}}_{am}^{\pi})(\forall D_q \in \overline{\mathfrak{D}}_{am}^{\pi}))\ D_q \in \Gamma_g D_p, \quad (p, q \in \mathscr{P}) \tag{AVI.3}$$

and if this clique is not contained (strictly) in any other clique.

A graph can have several maximal cliques. For solution of traffic problems it is necessary to find control vectors corresponding to all maximal cliques. Hence, it is necessary to find all maximal cliques, i.e., members of the set:

$$\overline{\mathfrak{D}}_{am} = \{\overline{\mathfrak{D}}_{am}^{1}, \overline{\mathfrak{D}}_{am}^{2}, \ldots, \overline{\mathfrak{D}}_{am}^{\pi}, \ldots, \overline{\mathfrak{D}}_{am}^{\Pi}\}, \tag{AVI.4}$$

where Π is the number of maximal cliques of graph G_g. The set of indices of maximal cliques in set $\overline{\mathfrak{D}}_{am}^{\pi}$, i.e., the index set is $\overline{\Pi} = \{1, 2, \ldots, \pi, \ldots, \Pi\}$.

The complete subgraph whose set of nodes is maximal clique $\overline{\mathfrak{D}}_{am}^{\pi}$ is denoted by $G_{gm}^{\pi} = (\overline{\mathfrak{D}}_{am}^{\pi}, \Gamma_g)$.

There are many algorithms for finding maximal cliques. These algorithms can be classified, mainly, in two groups. The first group consists of algorithms based on properties of Boolean algebra equations. One of these, Magu's algorithm, is described in a book by A. Caufmann (1975) [52]. Algorithms of the second group are based on "backtracking" procedures. Among these, the Bron and Kerbosh algorithm, developed in 1973 [16], is often used. Some other algorithms of this type are described in references [11], [12], and [63].

The problem of finding the set of maximal cliques was studied by Stoffers [77], who was the first to point out that control vectors giving the right-of-way to maximal numbers of compatible signal groups can be determined by extracting maximal cliques of the compatibility graph. His algorithm is of the "backtrack" type, and it is similar to Bron and Kerbosh algorithm.

Authors who studied the problem of determining optimal control vector sequences, after Stoffers, like Isabel Tully (1976) [85], paid great attention to finding cliques of compatibility graphs. However, the problem of finding cliques of compatibility graphs is not so difficult to deserve special algorithms developed just for this purpose. Compatibility graphs usually have less than 20 nodes, and thus they belong to simpler graphs regarding the problem of finding all maximal cliques. The existing algorithms are quite satisfactory, particularly Bron and Kerbosh algorithm, the FORTRAN IV version of which is given in a doctoral thesis of Isabel Tully (1976) [85].

J.D. Murchland (1979) developed the program for finding cliques, as a subroutine in the program for generating the sequence of control vectors [64].

Three intersections are presented in Figures AVI.1, AVI.2, and AVI.3, together with their graphs of signal groups compatibility, and all complete graphs with the set of nodes being maximal cliques of the compatibility graphs. Signal groups for intersections in Figures AVI.2 and AVI.3 contain a single traffic stream each.

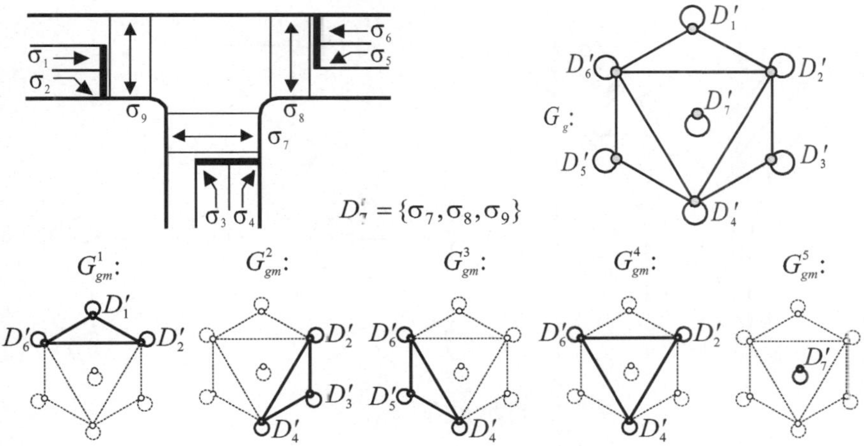

Figure AVI.1

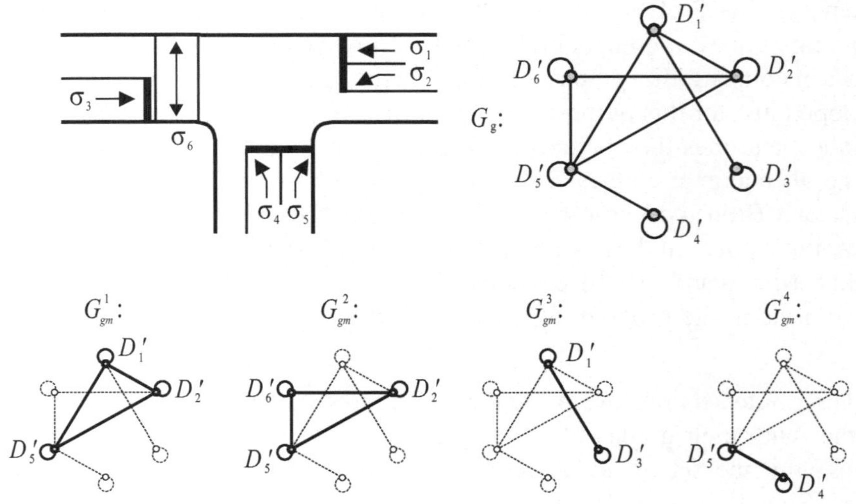

Figure AVI.2

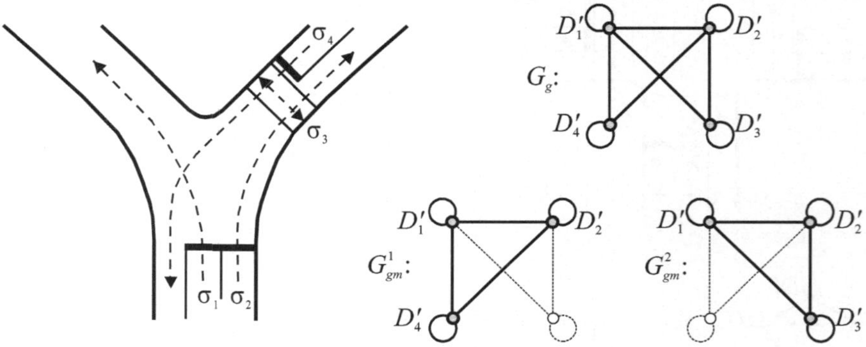

Figure AVI.3

As already mentioned, in order to determine all control vectors with maximal number of signal groups that can simultaneously have the right-of-way, first all maximal cliques have to be extracted from the compatibility graph.

Components of control vector $\mathbf{u}_a^\pi$, which gives the right-of-way to the maximal number of signal groups and corresponds to maximal clique $\overline{\mathfrak{D}}_{am}^{\pi}$, can be obtained as follows:

$$u_{ap}^{\pi} = \begin{cases} 1, \text{ if } D'_p \in \mathfrak{D}_{am}^{\pi} \\ 0, \text{ if } D'_p \notin \mathfrak{D}_{am}^{\pi} \end{cases}, \quad (\pi \in \overline{\Pi},\ p \in \mathscr{P}). \tag{AVI.5}$$

Such control vectors are called *maximal control vectors.*

The sets of signal groups that make maximal cliques for the example in Fig. AVI.2 are:

$$\overline{\mathfrak{D}}_{am}^{1} = \{D'_1, D'_2, D'_5\},$$

$$\overline{\mathfrak{D}}_{am}^{2} = \{D'_2, D'_5, D'_6\},$$

$$\overline{\mathfrak{D}}_{am}^{3} = \{D'_1, D'_3\},$$

$$\overline{\mathfrak{D}}_{am}^{4} = \{D'_4, D'_5\}.$$

Therefore, the maximal control vectors are:

$$\mathbf{u}_a^1 = (1,1,0,0,1,0)^T,\ \mathbf{u}_a^2 = (0,1,0,0,1,1)^T,$$

$$\mathbf{u}_a^3 = (1,0,1,0,0,0)^T,\ \mathbf{u}_a^4 = (0,0,0,1,1,0)^T.$$

1.2. Determining the set of all feasible control vectors

Each signal plan also includes control vectors giving the right-of-way to a number of signal groups, which is less than the maximal. Because of that, it is necessary to determine, also, other feasible control vectors (not only the maximal ones). These other feasible control vectors are obtained by starting with maximal vectors and reducing the number of signal groups having the right-of-way.

The set of all feasible vectors, $\mathbf{U}_f$, is obtained as a union of maximal control vectors and all subsets of control vectors obtained from maximal control vectors.

1.2.1. Finding all control vectors that correspond to a maximal control vector

Feasible control vectors are obtained from one maximal control vector as arrangements with repetitions of the number of elements having value 1 in maximal control vector $\mathbf{u}_a^\pi$ ($\pi \in \overline{\Pi}$) out of two elements [61].

The procedure of obtaining all feasible control vectors from control vector $\mathbf{u}_a^\pi$, which corresponds to maximal clique $\overline{\mathfrak{D}}_{am}^\pi$, i.e., function

$$P_{am} : \mathbf{U}_{am} \to \mathcal{P}(\mathbf{U}_f), \tag{AVI.6}$$

is described in the example below. $\mathcal{P}(\mathbf{U}_f)$ in expression (AVI.6) denotes a partitive set of the set of feasible control vectors.

Set $\mathbf{U}_{am}$ is the set of all maximal control vectors, i.e.,

$$\mathbf{U}_{am} = \{\mathbf{u}_a^1, \mathbf{u}_a^2, \ldots, \mathbf{u}_a^\pi, \ldots, \mathbf{u}_a^\Pi). \tag{AVI.7}$$

Each element of this set corresponds to a maximal clique.

Mapping P_{am} applied to $\mathbf{u}_a^1 = (1,1,0,0,1,0)^T$ gives:

$$\begin{aligned} P_{am}((1,1,0,0,1,0)^T) = \{&(1,1,0,0,1,0)^T, (1,1,0,0,0,0)^T, (1,0,0,0,1,0)^T, \\ &(0,1,0,0,1,0)^T, (1,0,0,0,0,0)^T, (0,1,0,0,0,0)^T, \\ &(0,0,0,0,1,0)^T, (0,0,0,0,0,0)^T\}. \end{aligned}$$

Subset $\mathbf{U}_{f\pi} \subset \mathbf{U}_f$ is obtained by applying P_{am} to $\mathbf{u}_a^\pi$, i.e.,

$$P_{am}(\mathbf{u}_a^\pi) = \mathbf{U}_{f\pi}. \tag{AVI.8}$$

1.2.2. The number of control vectors generated from one maximal control vector

The number of all feasible control vectors corresponding to one maximal clique, $\overline{\mathfrak{D}}_{am}^\pi$, i.e., to maximal control vector $\mathbf{u}_a^\pi$, is equal to the number of arrangements with repetitions of the number of elements in set $\overline{\mathfrak{D}}_{am}^\pi$ out of two elements. The number of all control vectors generated from maximal control vector $\mathbf{u}_a^\pi$ is, in fact, the cardinal number of set $P_{am}(\mathbf{u}_a^\pi)$. Since each component of $\mathbf{u}_a^\pi$ with value 1 can assume value 0, the number of feasible control vectors corresponding to $\mathbf{u}_a^\pi$ can be determined in the following way:

$$card\, P_{am}(\mathbf{u}_a^\pi) = \overline{V}_2^{h_\pi} = 2^{h_\pi}, \quad (\pi \in \overline{\Pi}), \tag{AVI.9}$$

where

$$h_{\pi} = card\,\overline{\mathfrak{D}}_{am}^{\pi} = \sum_{p=1}^{P} u_{ap}^{\pi}\,, \quad (\pi \in \overline{\Pi})\,. \tag{AVI.10}$$

For example, the number of control vectors corresponding to maximal cliques, for the intersection given in Fig. AVI.2, is:

$$\overline{V}_2^{h_1} = \overline{V}_2^{3} = card\,P_{am}(\mathbf{u}_a^1) = card\,P_{am}((1,1,0,0,1,0)^T) = 2^3 = 8\,,$$

$$\overline{V}_2^{h_2} = \overline{V}_2^{3} = card\,P_{am}(\mathbf{u}_a^2) = card\,P_{am}((0,1,0,0,1,1)^T) = 2^3 = 8\,,$$

$$\overline{V}_2^{h_3} = \overline{V}_2^{2} = card\,P_{am}(\mathbf{u}_a^3) = card\,P_{am}((1,0,1,0,0,0)^T) = 2^2 = 4\,,$$

$$\overline{V}_2^{h_4} = \overline{V}_2^{2} = card\,P_{am}(\mathbf{u}_a^4) = card\,P_{am}((0,0,0,1,1,0)^T) = 2^2 = 4\,.$$

1.2.3. Forming the set of all feasible control vectors

All feasible control vectors appear as elements of sets

$$P_{am}(\mathbf{u}_a^1), P_{am}(\mathbf{u}_a^2), \ldots, P_{am}(\mathbf{u}_a^{\pi}), \ldots, P_{am}(\mathbf{u}_a^{\Pi})\,.$$

Some control vectors, however, may belong to several sets. Therefore, the set of all feasible control vectors is determined as the union of sets $P_{am}(\mathbf{u}_a^1)$, $P_{am}(\mathbf{u}_a^2)$, … , $P_{am}(\mathbf{u}_a^{\pi})$, … , $P_{am}(\mathbf{u}_a^{\Pi})$, i.e.,

$$\mathbf{U}_f = \bigcup_{\pi \in \overline{\Pi}} P_{am}(\mathbf{u}_a^{\pi})\,. \tag{AVI.11}$$

For example, the set of all feasible control vectors, for the intersection presented in Fig. AVI.2, is defined by the following expression:

$$\begin{aligned}\mathbf{U}_f &= P_{am}(\mathbf{u}_a^1) \cup P_{am}(\mathbf{u}_a^2) \cup P_{am}(\mathbf{u}_a^3) \cup P_{am}(\mathbf{u}_a^4) \\ &= P_{am}((1,1,0,0,1,0)^T) \cup P_{am}((0,1,0,0,1,1)^T) \\ &\quad \cup P_{am}((1,0,1,0,0,0)^T) \cup P_{am}((0,0,0,1,1,0)^T)\,.\end{aligned}$$

where:

$$\begin{aligned}P_{am}(\mathbf{u}_a^1) = \{&(1,1,0,0,1,0)^T, (1,1,0,0,0,0)^T, (1,0,0,0,1,0)^T, \underline{(1,0,0,0,0,0)}^T, \\ &\underline{(0,1,0,0,1,0)}^T, \underline{(0,1,0,0,0,0)}^T, \underline{(0,0,0,0,1,0)}^T, \underline{(0,0,0,0,0,0)}^T\}\end{aligned}$$

$$\begin{aligned}P_{am}(\mathbf{u}_a^2) = \{&(0,1,0,0,1,1)^T, \underline{(0,1,0,0,1,0)}^T, (0,1,0,0,0,1)^T, \underline{(0,1,0,0,0,0)}^T, \\ &(0,0,0,0,1,1)^T, \underline{(0,0,0,0,1,0)}^T, (0,0,0,0,0,1)^T, \underline{(0,0,0,0,0,0)}^T\}\end{aligned}$$

$$P_{am}(\mathbf{u}_a^3) = \{(1,0,1,0,0,0)^T, \underline{(1,0,0,0,0,0)^T}, (0,0,1,0,0,0)^T, \underline{(0,0,0,0,0,0)^T}\}$$

$$P_{am}(\mathbf{u}_a^4) = \{(0,0,0,1,1,0)^T, (0,0,0,1,0,0)^T, \underline{(0,0,0,0,1,0)^T}, \underline{(0,0,0,0,0,0)^T}\}.$$

Therefore,

$$\begin{aligned}\mathbf{U}_f = \{&(1,1,0,0,1,0)^T, (1,1,0,0,0,0)^T, (1,0,0,0,1,0)^T, (1,0,0,0,0,0)^T,\\ &(0,1,0,0,1,0)^T, (0,1,0,0,0,0)^T, (0,0,0,0,1,0)^T, (1,0,1,0,0,0)^T,\\ &(0,0,1,0,0,0)^T, (0,1,0,0,1,1)^T, (0,1,0,0,0,1)^T, (0,0,0,0,1,1)^T,\\ &(0,0,0,0,0,1)^T, (0,0,0,1,0,0)^T, (0,0,0,1,1,0)^T, (0,0,0,0,0,0)^T\}.\end{aligned}$$

The underlined control vectors appear in several sets $P_{am}(\mathbf{u}_a^\pi)$, ($\pi \in \overline{\Pi}$).

1.2.4. The number of feasible control vectors

Since set $\mathbf{U}_f$ is the union of sets $P_{am}(\mathbf{u}_a^\pi)$, ($\pi \in \overline{\Pi}$), the number of feasible control vectors, i.e., the cardinal number of set $\mathbf{U}_f$ is determined using the inclusion–exclusion principle, according to the following expression [61]:

$$card\,\mathbf{U}_f = \sum_{\xi=1}^{\Pi} (-1)^{\xi-1} S_\xi[P_{am}(\mathbf{u}_a^1, \mathbf{u}_a^2, \ldots, \mathbf{u}_a^\Pi)], \tag{AVI.12}$$

where terms $S_\xi[P_{am}(\mathbf{u}_a^1, \mathbf{u}_a^2, \ldots, \mathbf{u}_a^\Pi)]$ represent the sum of cardinal numbers of all intersections ξ of different subsets that are elements of set $\{P_{am}(\mathbf{u}_a^1), P_{am}(\mathbf{u}_a^2), \ldots, P_{am}(\mathbf{u}_a^\Pi)\}$.

Thus, expressions for S_ξ, for different values of ξ, have the following form:

$$S_1[P_{am}(\mathbf{u}_a^1, \mathbf{u}_a^2, \ldots, \mathbf{u}_a^\Pi)] = \sum_{\pi=1}^{\Pi} card\, P_{am}(\mathbf{u}_a^\pi),$$

$$\begin{aligned}S_2[P_{am}(\mathbf{u}_a^1, \mathbf{u}_a^2, \ldots, \mathbf{u}_a^\Pi)] &= card\,(P_{am}(\mathbf{u}_a^1) \cap P_{am}(\mathbf{u}_a^2))\\ &+ card\,(P_{am}(\mathbf{u}_a^1) \cap P_{am}(\mathbf{u}_a^3))\\ &+ \ldots + card\,(P_{am}(\mathbf{u}_a^{\Pi-1}) \cap P_{am}(\mathbf{u}_a^\Pi),\end{aligned} \tag{AVI.13}$$

$$\ldots$$

$$S_\Pi[P_{am}(\mathbf{u}_a^1, \mathbf{u}_a^2, \ldots, \mathbf{u}_a^\Pi)] = card\left(\bigcap_{\pi=1}^{\Pi} P_{am}(\mathbf{u}_a^\pi)\right).$$

Cardinal numbers of sets $P_{am}(\mathbf{u}_a^{\pi})$ can be calculated using expression (AVI.9).

Cardinal numbers of intersections of sets, in expression (AVI.13), can be calculated as the number of arrangements with repetitions of h out of two elements. Symbol h represents the number of elements that are simultaneously equal to 1 in maximal control vectors corresponding to sets $P_{am}(\mathbf{u}_a^{s})$, $P_{am}(\mathbf{u}_a^{v})$,..., $P_{am}(\mathbf{u}_a^{w})$, which make the intersection whose cardinal number is to be found.

For example, when inspecting control vectors

$$\mathbf{u}_a^1 = (1,1,0,0,1,0)^T \text{, and}$$

$$\mathbf{u}_a^2 = (0,1,0,0,1,1)^T ,$$

it can be observed that $u_{a2}^1 = u_{a2}^2 = 1$ and $u_{a5}^1 = u_{a5}^2 = 1$, i.e., control vector $(0,1,0,0,1,0)^T$ belongs to sets $P_{am}(\mathbf{u}_a^1)$ and $P_{am}(\mathbf{u}_a^2)$.

Arrangements of 2 (2 components are simultaneously equal to 1) out of 2 elements (each component can take two values, either 1 or 0) correspond to control vectors:

$$(0,1,0,0,1,0)^T,\ (0,1,0,0,0,0)^T,\ (0,0,0,0,1,0)^T,\ (0,0,0,0,0,0)^T .$$

This means that both sets, $P_{am}(\mathbf{u}_a^1)$ and $P_{am}(\mathbf{u}_a^2)$, will contain these control vectors. Their number is $\overline{V}_2^2 = 2^2 = 4$.

The cardinal number of the intersection of sets $P_{am}(\mathbf{u}_a^{s})$, $P_{am}(\mathbf{u}_a^{v})$,..., $P_{am}(\mathbf{u}_a^{w})$ can be calculated by expression:

$$card\,(P_{am}(\mathbf{u}_a^{s}) \bigcap P_{am}(\mathbf{u}_a^{v}) \bigcap, \ldots, \bigcap P_{am}(\mathbf{u}_a^{w})) = \overline{V}_2^{h_{s,v,\ldots,w}} = 2^{h_{s,v,\ldots,w}} ,$$

where

$$h_{s,v,\ldots,w} = card\,(\overline{\mathfrak{D}}_{am}^{s} \bigcap \overline{\mathfrak{D}}_{am}^{v} \bigcap \ldots \bigcap \overline{\mathfrak{D}}_{am}^{w}) = \sum_{p=1}^{\mathrm{P}} u_{ap}^{s} u_{ap}^{v} \ldots u_{ap}^{w} . \quad \text{(AVI.14)}$$

The expressions for functions S_{ξ} can now be written as follows:

$$S_1[P_{am}(\mathbf{u}_a^1, \mathbf{u}_a^2, \ldots, \mathbf{u}_a^{\Pi})] = \sum_{\pi=1}^{\Pi} 2^{h_{\pi}} ,$$

$$S_2[P_{am}(\mathbf{u}_a^1, \mathbf{u}_a^2, \ldots, \mathbf{u}_a^{\Pi})] = 2^{h_{1,2}} + 2^{h_{1,3}} + \ldots + 2^{h_{\Pi-1,\Pi}} , \quad \text{(AVI.15)}$$

$$\ldots$$

$$S_{\Pi}[P_{am}(\mathbf{u}_a^1, \mathbf{u}_a^2, \ldots, \mathbf{u}_a^{\Pi})] = 2^{h_{1,2,\ldots,\Pi-1,\Pi}} .$$

The set of feasible control vectors, $\mathbf{U}_f$, is determined for the intersection presented in Fig. AVI.2. The maximal control vectors are:

$$\mathbf{u}_a^1 = (1,1,0,0,1,0)^T,\ \mathbf{u}_a^2 = (0,1,0,0,1,1)^T,$$
$$\mathbf{u}_a^3 = (1,0,1,0,0,0)^T,\ \mathbf{u}_a^4 = (0,0,0,1,1,0)^T.$$

First, exponents of base 2 have to be found:

$$h_1 = \sum_{p=1}^{6} u_{ap}^1 = 3,\ h_2 = \sum_{p=1}^{6} u_{ap}^2 = 3,\ h_3 = \sum_{p=1}^{6} u_{ap}^3 = 2,\ h_4 = \sum_{p=1}^{6} u_{ap}^4 = 2,$$

$$h_{1,2} = \sum_{p=1}^{6} u_{ap}^1 u_{ap}^2 = 2,\ h_{1,3} = \sum_{p=1}^{6} u_{ap}^1 u_{ap}^3 = 1,\ h_{1,4} = \sum_{p=1}^{6} u_{ap}^1 u_{ap}^4 = 1,$$

$$h_{2,3} = \sum_{p=1}^{6} u_{ap}^2 u_{ap}^3 = 0,\ h_{2,4} = \sum_{p=1}^{6} u_{ap}^2 u_{ap}^4 = 1,\ h_{3,4} = \sum_{p=1}^{6} u_{ap}^3 u_{ap}^4 = 0,$$

$$h_{1,2,3} = \sum_{p=1}^{6} u_{ap}^1 u_{ap}^2 u_{ap}^3 = 0,\ h_{1,2,4} = \sum_{p=1}^{6} u_{ap}^1 u_{ap}^2 u_{ap}^4 = 1,$$

$$h_{1,3,4} = \sum_{p=1}^{6} u_{ap}^1 u_{ap}^3 u_{ap}^4 = 0,\ h_{2,3,4} = \sum_{p=1}^{6} u_{ap}^2 u_{ap}^3 u_{ap}^4 = 0,$$

$$h_{1,2,3,4} = \sum_{p=1}^{6} u_{ap}^1 u_{ap}^2 u_{ap}^3 u_{ap}^4 = 0.$$

Functions S_ξ assume the following values:

$$S_1 = \sum_{\pi=1}^{4} 2^{h_\pi} = 2^3 + 2^3 + 2^2 + 2^2 = 24,$$

$$S_2 = 2^{h_{1,2}} + 2^{h_{1,3}} + 2^{h_{1,4}} + 2^{h_{2,3}} + 2^{h_{2,4}} + 2^{h_{3,4}} = 2^2 + 2^1 + 2^1 + 2^0 + 2^1 + 2^0 = 12,$$

$$S_3 = 2^{h_{1,2,3}} + 2^{h_{1,2,4}} + 2^{h_{1,3,4}} + 2^{h_{2,3,4}} = 2^0 + 2^1 + 2^0 + 2^0 = 5,$$

$$S_4 = 2^{h_{1,2,3,4}} = 2^0 = 1.$$

The application of formula AVI.11 gives:

$$card\,\mathbf{U}_f = \sum_{\xi=1}^{4} (-1)^{\xi-1} S_\xi = 24 - 12 + 5 - 1 = 16.$$

The relations in the set of signal groups and the obtained set of feasible control vectors, $\mathbf{U}_f$, enable formulation of mathematical definitions for relations and constraints that have to be satisfied by any feasible signal plan.

2. FORMING THE GRAPH OF CONTROL VECTORS TRANSITION, G_s

Sets of control vectors $\mathbf{U}_{f\pi}$ ($\pi \in \overline{\Pi}$) generated from control vectors that correspond to maximal cliques, i.e., from maximal control vectors, can be used to form the graph of control vectors transition, G_s.

As already mentioned, each maximal control $\mathbf{u}_a^\pi$ is mapped by function P_{am} to set $\mathbf{U}_{f\pi} \subset \mathbf{U}_f$, i.e.,

$$P_{am}(\mathbf{u}_a^\pi) = \mathbf{U}_{f\pi}.$$

Each control vector that is an element of set $\mathbf{U}_{f\pi}$ gives the right-of-way to a smaller number of signal groups than the maximal control vector $\mathbf{u}_a^\pi$. Each pair of control vectors that is an element of set $\mathbf{U}_{f\pi}$ belongs to relation R_s. Therefore, the graph of control vectors transition, having $\mathbf{U}_{f\pi}$ as the set of nodes,

$$G_{sm}^\pi = (\mathbf{U}_{f\pi}, \Gamma_{s\pi}), \quad (\pi \in \overline{\Pi}), \tag{AVI.16}$$

is a complete graph, with the property

$$\Gamma_{s\pi} : \mathbf{U}_{f\pi} \to \mathcal{P}(\mathbf{U}_{f\pi}), \quad (\pi \in \overline{\Pi}),$$

where

$$(\forall \mathbf{u}^l \in \mathbf{U}_{f\pi}) \wedge (\forall \mathbf{u}^n \in \mathbf{U}_{f\pi}) \Rightarrow \mathbf{u}^l \in \Gamma_{s\pi}\mathbf{u}^n.$$

The graph of control vectors transition can now be defined as:

$$\begin{aligned} G_s &= (\mathbf{U}_f, \Gamma_s) = (\mathbf{U}_f, R_s) \\ &= \bigcup_{\pi=1}^{\Pi} G_{sm}^\pi = \bigcup_{\pi=1}^{\Pi} (\mathbf{U}_{f\pi}, \Gamma_{s\pi}) = \bigcup_{\pi=1}^{\Pi} (\mathbf{U}_{f\pi}, R_{s\pi}) \end{aligned}, \tag{AVI.17}$$

where

$$\mathbf{U}_f = \bigcup_{\pi=1}^{\Pi} \mathbf{U}_{f\pi} \text{ and } R_s = \bigcup_{\pi=1}^{\Pi} R_{s\pi}.$$

The set of edges of graph G_s is R_s, and $R_{s\pi}$ is the set of edges of graph G_{sm}^{π}.

The compatibility graph subgraphs having maximal cliques as the set of nodes, and corresponding graphs G_{sm}^{π} are presented in Fig. AVI.4 for the intersection given in Fig. AVI.2. The union of graphs G_{sm}^{π}, ($\pi = 1,2,3,4$) gives the control vectors transition graph, G_s, shown in Fig. II.17.

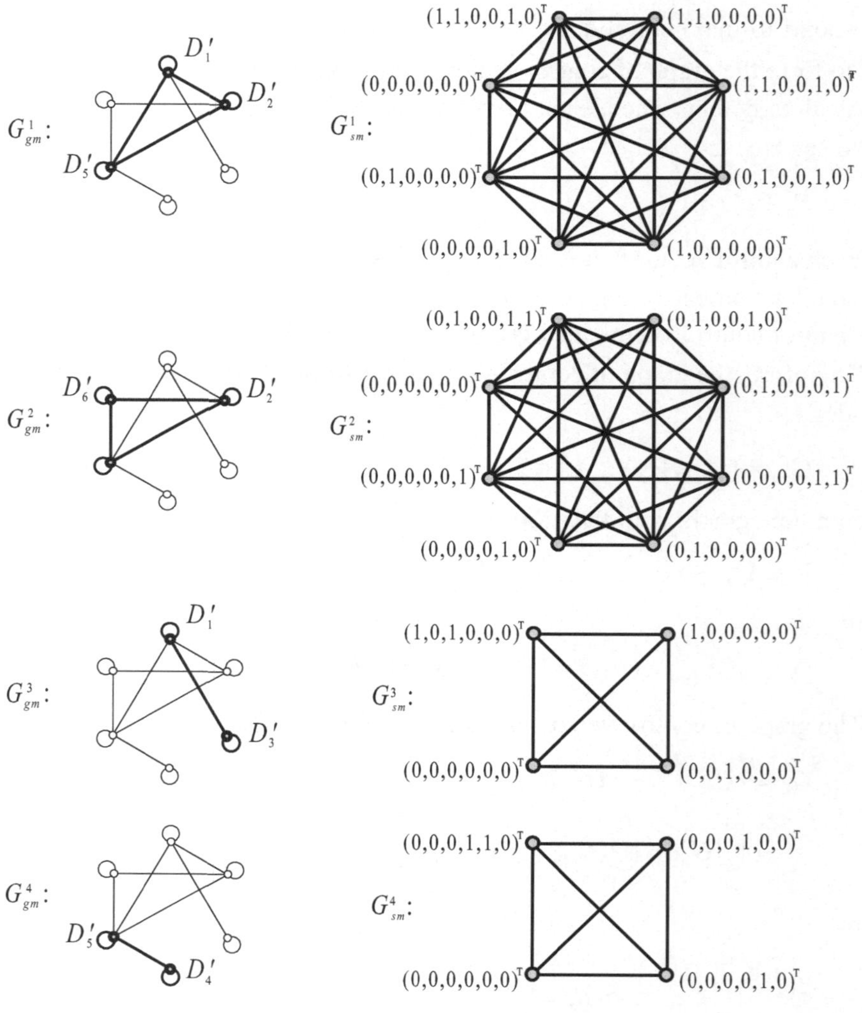

Figure AVI.4

Appendix VII

Description of STECSOT Program (STructurE and Cycle Split Optimization Technique)

1. THE STRUCTURE OF THE PROGRAM

STECSOT program consists of the main program and nine subprograms of SUBROUTINE type. The structure of the program is presented in Fig. AVII.1

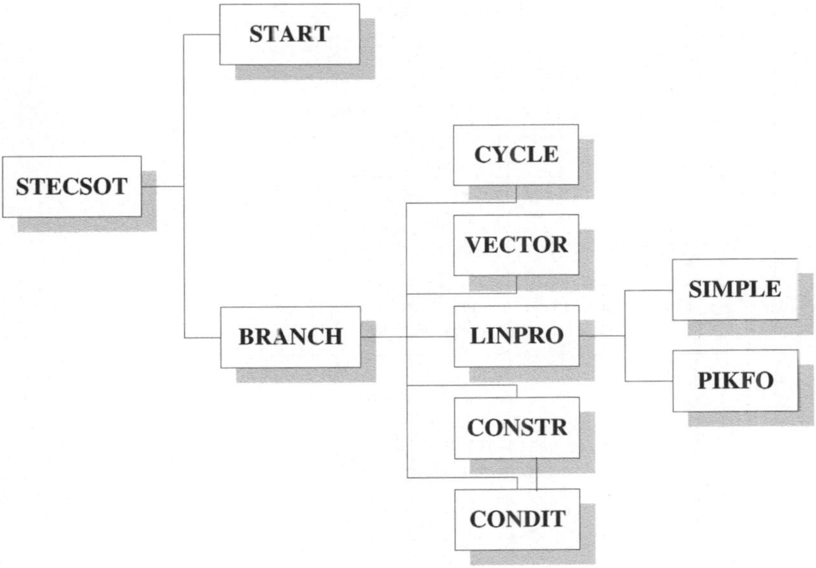

Figure AVII.1.

2. PROGRAM DESCRIPTION

2.1. Main Program

Input data are read at the beginning of the main program, and after that subroutines START and BRANCH are called. The flowchart of the main program is given in Fig. AVII.2.

Subroutine START arranges input data and prepares them for use in the BRANCH subroutine. Subroutine BRANCH realizes branch-and-bound procedure. The result of BRANCH subroutine is collection $\overline{\mathscr{U}}^*$, the elements of which are singleton subsets containing single elements, i.e., optimal signal plans. These results are returned to the main program in which results are printed.

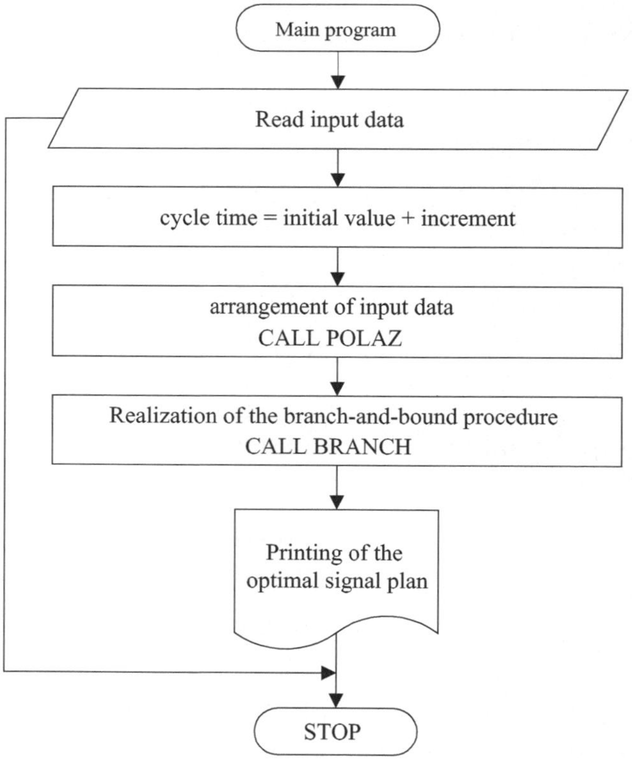

Figure AVII.2.

2.2. START Subroutine

START subroutine is called from the main program. This subroutine calculates flow volumes for each control vector, finds the control vector with maximal volume, calculates the initial bound, and determines coefficients of the criterion function. Calculated data are returned to the main program.

The flow chart of START subroutine is given in Fig. AVII.3.

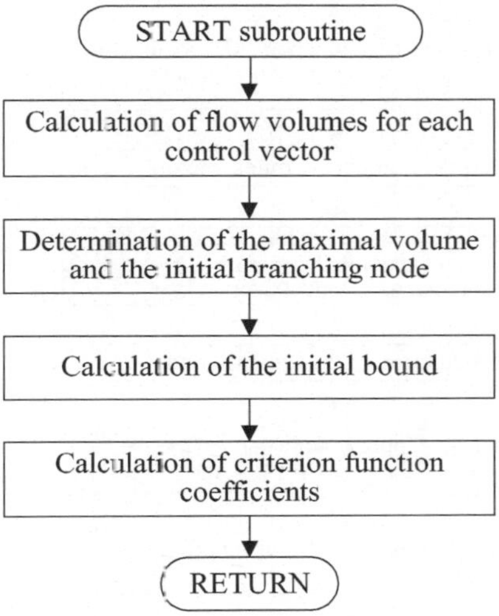

Figure AVII.3.

2.3. BRANCH Subroutine

BRANCH subroutine is called from the main program, and it realizes the branch-and-bound procedure by calling subprograms CYCLE, VECTOR, CONSTR, LINPRO, and CONDIT. The results of BRANCH subroutine, i.e., data about optimal signal plans, are returned to the main program.

The flow chart of BRANCH subroutine is given in Fig. AVII.4.

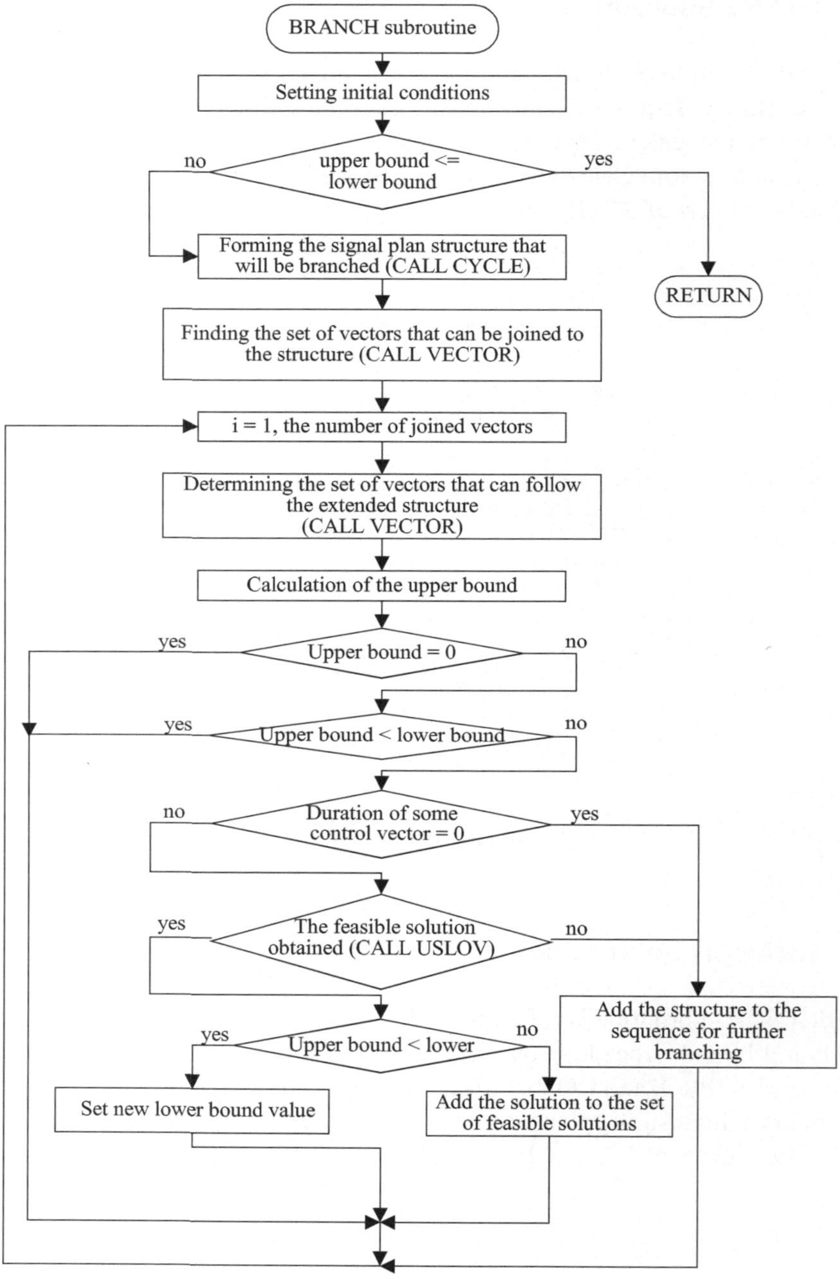

Figure AVII.4.

2.4. VECTOR Subroutine

VECTOR subroutine is called from BRANCH subroutine to determine elements of vector ISLED(I), which represents $\beta_1(\bar{u}^n)$. The elements of vector ISLED(I) can join the signal plan structure given by matrix MPOMOC(I,J) if the following constraint is satisfied:

$$\sum_{k=1}^{K}(u_p^k + u_p^{k(\text{mod}\,K)+1})(\text{mod}\,2) \le 2\,, \quad (p \in \mathscr{P} = \{1,2,\ldots,\text{NBRSG}\})\,,$$

where NBRSG is the number of signal groups. VECTOR subroutine provides, also, the information about the number of elements in this set, K100. Variables ISLED(I) and K100 are returned to the BRANCH subroutine.

The flowchart of VECTOR subroutine is given in Fig. AVII.5. Variable NBRUV contains the number of control vectors, and matrix MGRAF(I,J) represents the graph of control vectors transition.

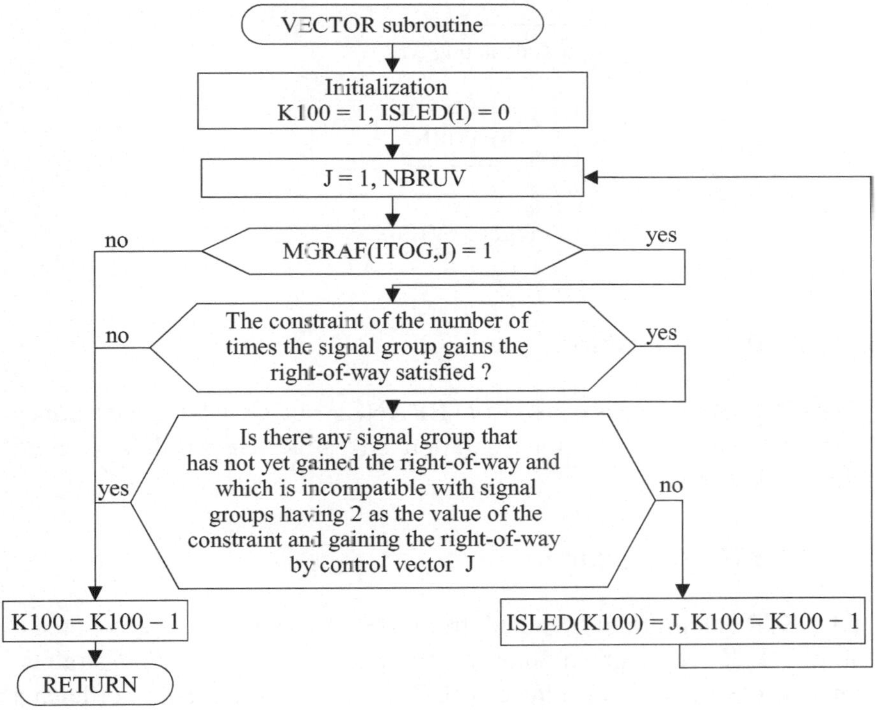

Figure AVII.5.

2.5. CYCLE Subroutine

CYCLE subroutine is called from BRANCH subroutine to form the structure, ISVI(I), of signal plan MPOMOC(I,J), which has to be branched. These variables, and the number of control vectors included in the sequence, KLL, are returned to BRANCH subroutine.

The flow chart of CYCLE subroutine is given in Fig. AVII.6.

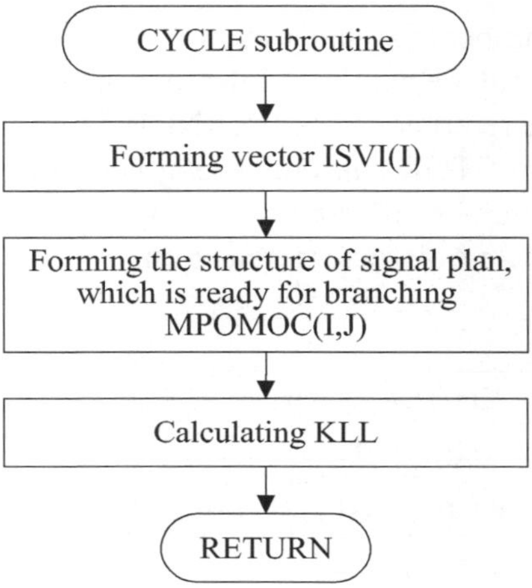

Figure AVII.6.

2.6. CONDIT Subroutine

CONDIT subroutine is called from BRANCH and CONSTR subroutines to determine whether the obtained solution is feasible. The result is returned to calling subprograms.

2.7. CONSTR Subroutine

CONSTR subroutine is called from BRANCH subroutine to create the minimal effective green time constraints and intergreen constraints. Redundant constraints are eliminated in CONSTR subroutine, and coefficients of criterion function are calculated, as well.

The flow chart of CONSTR subroutine is given in Fig. AVII.7.

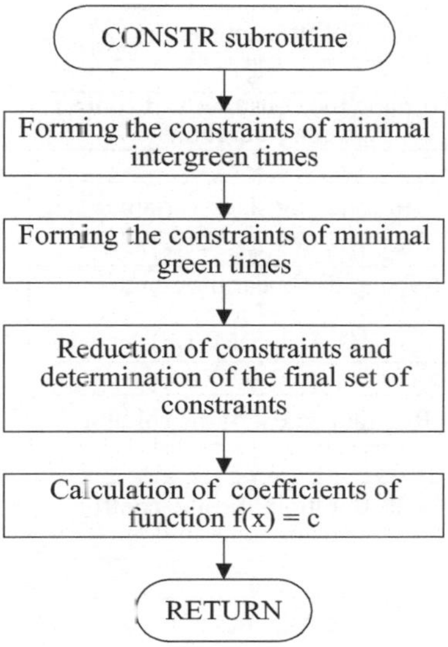

Figure AVII.7.

2.8. LINPRO Subroutine

LINPRO subroutine is called from BRANCH subroutine to solve linear programming problems. It introduces slack variables, forms the constraint that the sum of control vector durations has to be equal to the cycle time, and creates the model:

$$F(X) = C(X)$$

$$AX = B, \ X \geq 0,$$

and calls SIMPLE subroutine to solve the linear programming problem. The results obtained from SIMPLE subroutine are then rearranged, and the criterion value is calculated. The decision to stop further branching of some structure is also made by LINPRO when PICFO subroutine, called from LINPRO, established that duration of some control vector, determined as zero in LINPRO, will not be changed by introducing subsequent control vectors in the structure.

The flow chart of LINPRO subroutine is given in Fig. AVII.8.

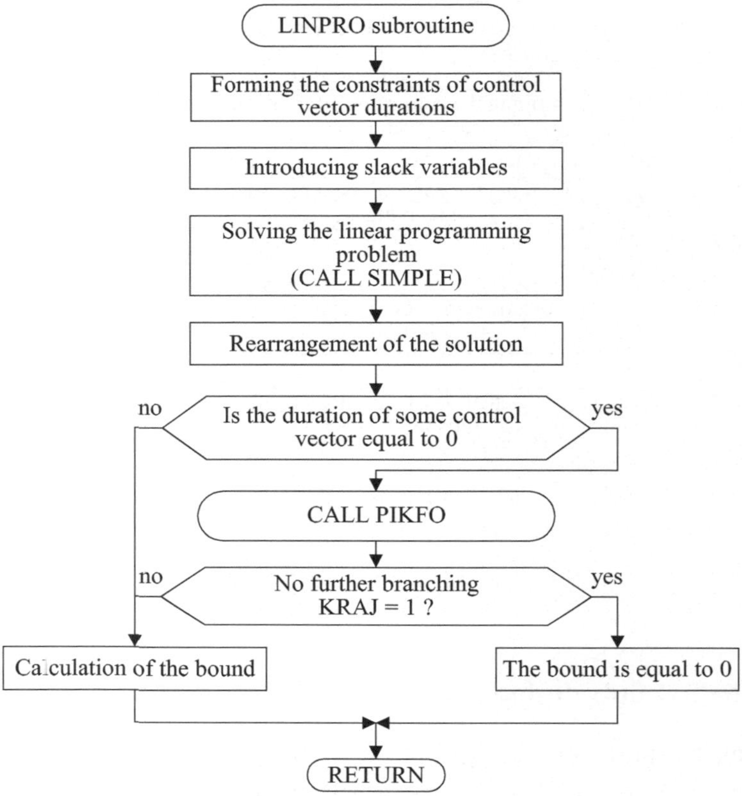

Figure AVII.8.

2.9. PIKFO Subroutine

PIKFO subroutine is called from LINPRO if the duration of some control vector in the structure created so far is equal to 0. PIKFO subroutine checks the values this vector could assume after further branching. If it remains 0, the indicator KRAJ assumes value 0, otherwise 1.

2.10. SIMPLE Subroutine

SIMPLE subroutine is a standard program for solving linear programming problems.

Appendix VIII

THE PROOF OF DELAY FUNCTION CONVEXITY

1. THE PROOF OF DELAY FUNCTION CONVEXITY IN CASE THE CYCLE TIME IS NOT GIVEN

A function is convex if and only if its Hesse matrix is positively semidefinite for any value of variables in the feasible domain [91].

A symmetric matrix is positive semidefinite if all of its principal minors are nonnegative. A minor is *principal* if its row indices and column indices are the same.

The function whose convexity has to be tested is the mathematical expectation of the total delay on an intersection, defined by the following expression (9.31):

$$J_c = 0.9 \sum_{p=1}^{P'} \sum_{e=1}^{e=E(p)} \left(a_{pe}^1 \left(\sum_{k=1}^{K} (1 - u_p^k) \tau^k \right)^2 + \frac{\bar{a}_{pe}^2 \left(\sum_{k=1}^{K} \tau^k \right)^3}{\left(\sum_{k=1}^{K} u_p^k \tau^k \right)^2 - \bar{a}_{pe}^3 \left(\sum_{k=1}^{K} \tau^k \right) \left(\sum_{k=1}^{K} u_p^k \tau^k \right)} \right).$$

Convex functions have the properties that their sum is a convex function, and a convex function multiplied by a constant also gives a convex function. Hence, function J_c is convex if all terms in expression (9.30) are convex. Each term consists of two terms, so that convexity of the whole expression can be established by investigating the convexity of each of them.

1.1. Investigation of convexity of the second term in delay function

If the second term in expression (9.31) is written as:

$$\delta'_{pe} = \frac{\bar{a}_{pe}^{2}\left(\sum_{k=1}^{K}\tau^{k}\right)^{3}}{\left(\sum_{k=1}^{K}u_{p}^{k}\tau^{k}\right)^{2} - \bar{a}_{pe}^{3}\left(\sum_{k=1}^{K}\tau^{k}\right)\left(\sum_{k=1}^{K}u_{p}^{k}\tau^{k}\right)}, \tag{AVIII.1}$$

and the following notations are introduced

$$\sum_{k=1}^{K}u_{p}^{k}\tau^{k} = g_{p},$$

$$\sum_{k=1}^{K}\tau^{k} = g_{p} + r_{p} = c,$$

then δ'_{pe} becomes:

$$\delta'_{pe} = \bar{a}_{pe}^{2}\frac{(g_{p}+r_{p})^{3}}{g_{p}^{2} - \bar{a}_{pe}^{3}(g_{p}+r_{p})g_{p}} = \bar{a}_{pe}^{2}\,\delta_{pe}. \tag{AVIII.2}$$

There are four possible positions of effective green and effective red time in the cycle. These cases are shown in Fig. 4.2. The convexity of δ'_{pe} will be investigated for all four cases. A general case will be considered, i.e., any signal group D'_{p} and traffic stream σ_{pe}, thus making indices p and $p(e)$ unnecessary. Therefore, it is necessary to investigate convexity of the expression

$$\delta = \frac{(g+r)^{3}}{g(g-\omega(g+r))} = \frac{(g+r)^{3}}{(1-\omega)g^{2} - \omega r g}, \tag{AVIII.3}$$

where

$$\sum_{k=1}^{K}\tau^{k} = g + r, \tag{AVIII.4}$$

$$\sum_{k=1}^{K}u_{p}^{k}\tau^{k} = g, \tag{AVIII.5}$$

$$\frac{q}{s} = \omega. \tag{AVIII.6}$$

The Hesse matrix, whose positive semidefiniteness is to be investigated, is defined as:

$$\nabla^2\delta = \begin{bmatrix} \frac{\partial^2\delta}{(\partial\tau^1)^2} & \frac{\partial^2\delta}{\partial\tau^1\,\partial\tau^2} & \cdots & \frac{\partial^2\delta}{\partial\tau^1\,\partial\tau^K} \\ \frac{\partial^2\delta}{\partial\tau^2\,\partial\tau^1} & \frac{\partial^2\delta}{(\partial\tau^2)^2} & \cdots & \frac{\partial^2\delta}{\partial\tau^2\,\partial\tau^K} \\ \vdots & \vdots & \cdots & \vdots \\ \frac{\partial^2\delta}{\partial\tau^K\,\partial\tau^1} & \frac{\partial^2\delta}{\partial\tau^K\,\partial\tau^2} & \cdots & \frac{\partial^2\delta}{(\partial\tau^K)^2} \end{bmatrix}. \tag{A VIII.7}$$

Function δ depends on g and r, and they depend on variables τ^k. It should be noted that g and r do not depend on same time variables. If r depends on some τ^k variables, then g does not depend on them, and vice versa.

Values of variables u_p^l and u_p^s (for any signal group D'_p) can be the same, i.e., $u_p^l = u_p^s = 1$ or $u_p^l = u_p^s = 0$, or they can be different, i.e., $u_p^l \neq u_p^s$. The set of indices of time intervals during which signal group D'_p is controlled by green signal indication is defined as:

$$\mathcal{K}' = \{k \mid u_p^k = 1,\ k \in \mathcal{K}\}. \tag{AVIII.8}$$

The set of indices of time intervals during which signal group D'_p is controlled by red signal indication is defined as:

$$\mathcal{K}'' = \{k \mid u_p^k = 0,\ k \in \mathcal{K}\}. \tag{AVIII.9}$$

The first derivatives of function δ are determined according to the following expressions:

a) $l \in \mathcal{K}'$

$$\frac{\partial\delta}{\partial\tau^l} = \frac{\partial\delta}{\partial g}\cdot\frac{\partial g}{\partial\tau^l} + \frac{\partial\delta}{\partial r}\cdot\frac{\partial r}{\partial\tau^l} = \frac{\partial\delta}{\partial g},$$

since $\frac{\partial g}{\partial\tau^l} = 1$, $\frac{\partial r}{\partial\tau^l} = 0$.

b) $s \in \mathcal{K}''$

$$\frac{\partial\delta}{\partial\tau^s} = \frac{\partial\delta}{\partial g}\cdot\frac{\partial g}{\partial\tau^s} + \frac{\partial\delta}{\partial r}\cdot\frac{\partial r}{\partial\tau^s} = \frac{\partial\delta}{\partial r},$$

since $\frac{\partial g}{\partial\tau^s} = 0$, $\frac{\partial r}{\partial\tau^s} = 1$.

The second derivatives of function δ are determined according to the following expressions:

a) $l \in \mathscr{K}'$, $s \in \mathscr{K}'$

a.1) $s = l$

$$\frac{\partial^2 \delta}{(\partial \tau^l)^2} = \frac{\partial}{\partial \tau^l}\left(\frac{\partial \delta}{\partial \tau^l}\right) = \frac{\partial}{\partial g}\left(\frac{\partial \delta}{\partial g}\right) \cdot \frac{\partial g}{\partial \tau^l} + \frac{\partial}{\partial r}\left(\frac{\partial \delta}{\partial g}\right) \cdot \frac{\partial r}{\partial \tau^l} = \frac{\partial^2 \delta}{\partial g^2}$$

$$= \frac{2r^2(g+r)}{((1-\omega)g^2 - \omega g r)^3}(3((1-\omega)g^2 - \omega g r) + (\omega(g+r))^2)$$

(AVIII.10)

a.2) $s \neq l$

$$\frac{\partial^2 \delta}{\partial \tau^l \partial \tau^s} = \frac{\partial}{\partial \tau^s}\left(\frac{\partial \delta}{\partial \tau^l}\right) = \frac{\partial}{\partial g}\left(\frac{\partial \delta}{\partial g}\right) \cdot \frac{\partial g}{\partial \tau^s} + \frac{\partial}{\partial r}\left(\frac{\partial \delta}{\partial g}\right) \cdot \frac{\partial r}{\partial \tau^s} = \frac{\partial^2 \delta}{\partial g^2}$$

The value of the second derivative, $\frac{\partial^2 \delta}{\partial \tau^l \partial \tau^s}$, in this case is the same as in a.1).

b) $l \in \mathscr{K}'$, $s \in \mathscr{K}''$

$$\frac{\partial^2 \delta}{\partial \tau^l \partial \tau^s} = \frac{\partial}{\partial \tau^s}\left(\frac{\partial \delta}{\partial \tau^l}\right) = \frac{\partial}{\partial \tau^s}\left(\frac{\partial \delta}{\partial g}\right) = \frac{\partial^2 \delta}{\partial g^2} \cdot \frac{\partial g}{\partial \tau^s} + \frac{\partial^2 \delta}{\partial g \partial r} \cdot \frac{\partial r}{\partial \tau^s} = \frac{\partial^2 \delta}{\partial g \partial r}$$

$$= \frac{-2gr(g+r)}{((1-\omega)g^2 - \omega g r)^3}(3((1-\omega)g^2 - \omega g r) + (\omega(g+r))^2)$$

(AVIII.11)

Since $\frac{\partial^2 \delta}{\partial \tau^l \partial \tau^s} = \frac{\partial^2 \delta}{\partial \tau^s \partial \tau^l}$, the value of this derivative is the same as obtained for $l \in \mathscr{K}''$ and $s \in \mathscr{K}'$.

c) $l \in \mathcal{K}''$, $s \in \mathcal{K}''$

c.1) $s = l$

$$\frac{\partial^2 \delta}{(\partial \tau^s)^2} = \frac{\partial}{\partial \tau^s}\left(\frac{\partial \delta}{\partial \tau^s}\right) = \frac{\partial}{\partial \tau^s}\left(\frac{\partial \delta}{\partial r}\right)$$

$$= \frac{\partial}{\partial g}\left(\frac{\partial \delta}{\partial r}\right) \cdot \frac{\partial g}{\partial \tau^s} + \frac{\partial}{\partial r}\left(\frac{\partial \delta}{\partial r}\right) \cdot \frac{\partial r}{\partial \tau^s} = \frac{\partial^2 \delta}{\partial r^2}$$

$$= \frac{2g^2(g+r)}{((1-\omega)g^2 - \omega g r)^3}(3((1-\omega)g^2 - \omega g r) + (\omega(g+r))^2) \qquad \text{(AVIII.12)}$$

c.2) $s \neq l$

$$\frac{\partial^2 \delta}{\partial \tau^s \partial \tau^l} = \frac{\partial}{\partial \tau^l}\left(\frac{\partial \delta}{\partial \tau^s}\right) = \frac{\partial}{\partial \tau^l}\left(\frac{\partial \delta}{\partial r}\right)$$

$$= \frac{\partial}{\partial g}\left(\frac{\partial \delta}{\partial r}\right) \cdot \frac{\partial g}{\partial \tau^l} + \frac{\partial}{\partial r}\left(\frac{\partial \delta}{\partial r}\right) \cdot \frac{\partial r}{\partial \tau^l} = \frac{\partial^2 \delta}{\partial r^2}$$

The value of the second derivative, $\frac{\partial^2 \delta}{\partial \tau^l \, \partial \tau^s}$, in this case is the same as in c.1).

From above expressions it is obvious that there exist only three different values of the second derivatives of function δ. These three values can be distributed in Hesse matrix in four different ways. Therefore, for all four cases the positive semidefiniteness of Hesse matrix has to be proven.

If the function

$$\varphi = \varphi(g,r) = \frac{2(g+r)(3(g^2 - \omega g(g+r)) + (\omega(g+r))^2)}{g^3(g - \omega(g+r))^3} \qquad \text{(AVIII.13)}$$

is introduced, then the expressions for second derivatives become:

a) $l \in \mathcal{K}'$, $s \in \mathcal{K}'$

$$\frac{\partial^2 \delta}{(\partial \tau^l)^2} = \frac{\partial^2 \delta}{\partial \tau^l \, \partial \tau^s} = r^2 \varphi \qquad \text{(AVIII.14)}$$

b) $l \in \mathcal{K}'$, $s \in \mathcal{K}''$

$$\frac{\partial^2 \delta}{\partial \tau^l \, \partial \tau^s} = \frac{\partial^2 \delta}{\partial \tau^s \, \partial \tau^l} = -g r \varphi \qquad \text{(AVIII.15)}$$

c) $l \in \mathscr{K}''$, $s \in \mathscr{K}''$

$$\frac{\partial^2 \delta}{(\partial \tau^s)^2} = \frac{\partial^2 \delta}{\partial \tau^s \, \partial \tau^l} = g^2 \varphi \tag{AVIII.16}$$

Function $\varphi(g,r)$ is positive because the expression

$$g - \omega(g+r) = \frac{gs - qc}{s}$$

is positive in the region where constraint is valid. This constraint is the flow balance constraint, and if satisfied, the intersection is not saturated.

For the first case of effective green and effective red position in the cycle shown in Fig. 4.2, Hesse matrix has the following form:

$$\nabla^2 \delta = \left[\begin{array}{cccc|cccc|cccc}
g^2\varphi & g^2\varphi & \dots & g^2\varphi & -gr\varphi & -gr\varphi & \dots & -gr\varphi & g^2\varphi & g^2\varphi & \dots & g^2\varphi \\
g^2\varphi & g^2\varphi & \dots & g^2\varphi & -gr\varphi & -gr\varphi & \dots & -gr\varphi & g^2\varphi & g^2\varphi & \dots & g^2\varphi \\
\vdots & \vdots & \vdots & \vdots & \vdots & \vdots & \vdots & \vdots & \vdots & \vdots & \vdots & \vdots \\
g^2\varphi & g^2\varphi & \dots & g^2\varphi & -gr\varphi & -gr\varphi & \dots & -gr\varphi & g^2\varphi & g^2\varphi & \dots & g^2\varphi \\
\hline
-gr\varphi & -gr\varphi & \dots & -gr\varphi & r^2\varphi & r^2\varphi & \dots & r^2\varphi & -gr\varphi & -gr\varphi & \dots & -gr\varphi \\
-gr\varphi & -gr\varphi & \dots & -gr\varphi & r^2\varphi & r^2\varphi & \dots & r^2\varphi & -gr\varphi & -gr\varphi & \dots & -gr\varphi \\
\vdots & \vdots & \vdots & \vdots & \vdots & \vdots & \vdots & \vdots & \vdots & \vdots & \vdots & \vdots \\
-gr\varphi & -gr\varphi & \dots & -gr\varphi & r^2\varphi & r^2\varphi & \dots & r^2\varphi & -gr\varphi & -gr\varphi & \dots & -gr\varphi \\
\hline
g^2\varphi & g^2\varphi & \dots & g^2\varphi & -gr\varphi & -gr\varphi & \dots & -gr\varphi & g^2\varphi & g^2\varphi & \dots & g^2\varphi \\
g^2\varphi & g^2\varphi & \dots & g^2\varphi & -gr\varphi & -gr\varphi & \dots & -gr\varphi & g^2\varphi & g^2\varphi & \dots & g^2\varphi \\
\vdots & \vdots & \vdots & \vdots & \vdots & \vdots & \vdots & \vdots & \vdots & \vdots & \vdots & \vdots \\
g^2\varphi & g^2\varphi & \dots & g^2\varphi & -gr\varphi & -gr\varphi & \dots & -gr\varphi & g^2\varphi & g^2\varphi & \dots & g^2\varphi
\end{array}\right] \tag{AVIII.17}$$

$$\nabla^2 \delta = gr\varphi \left[\begin{array}{cccc|cccc|cccc}
\alpha & \alpha & \dots & \alpha & -1 & -1 & \dots & -1 & \alpha & \alpha & \dots & \alpha \\
\alpha & \alpha & \dots & \alpha & -1 & -1 & \dots & -1 & \alpha & \alpha & \dots & \alpha \\
\vdots & \vdots & \vdots & \vdots & \vdots & \vdots & \vdots & \vdots & \vdots & \vdots & \vdots & \vdots \\
\alpha & \alpha & \dots & \alpha & -1 & -1 & \dots & -1 & \alpha & \alpha & \dots & \alpha \\
\hline
-1 & -1 & \dots & -1 & \frac{1}{\alpha} & \frac{1}{\alpha} & \dots & \frac{1}{\alpha} & -1 & -1 & \dots & -1 \\
-1 & -1 & \dots & -1 & \frac{1}{\alpha} & \frac{1}{\alpha} & \dots & \frac{1}{\alpha} & -1 & -1 & \dots & -1 \\
\vdots & \vdots & \vdots & \vdots & \vdots & \vdots & \vdots & \vdots & \vdots & \vdots & \vdots & \vdots \\
-1 & -1 & \dots & -1 & \frac{1}{\alpha} & \frac{1}{\alpha} & \dots & \frac{1}{\alpha} & -1 & -1 & \dots & -1 \\
\hline
\alpha & \alpha & \dots & \alpha & -1 & -1 & \dots & -1 & \alpha & \alpha & \dots & \alpha \\
\alpha & \alpha & \dots & \alpha & -1 & -1 & \dots & -1 & \alpha & \alpha & \dots & \alpha \\
\vdots & \vdots & \vdots & \vdots & \vdots & \vdots & \vdots & \vdots & \vdots & \vdots & \vdots & \vdots \\
\alpha & \alpha & \dots & \alpha & -1 & -1 & \dots & -1 & \alpha & \alpha & \dots & \alpha
\end{array}\right] = gr\varphi \begin{bmatrix} \alpha & -1 & \alpha \\ -1 & \frac{1}{\alpha} & -1 \\ \alpha & -1 & \alpha \end{bmatrix} \tag{AVIII.18}$$

where

$$\alpha = \frac{g}{r}.$$

Matrix $\nabla^2\delta$ is symmetric. This matrix is positive semidefinite if all of its principal minors are nonnegative [99]. Principal minors are determinants of submatrices with the same indices of rows and columns. Therefore, it is necessary to determine signs of all principal minors.

- Principal minors of first order – diagonal elements

  All diagonal elements have value

$$\alpha = \frac{g}{r} > 0 \text{ or } \frac{1}{\alpha} = \frac{r}{g} > 0$$

  and all are positive.

- Principal minors of second order

  Principal minors of second order, having different column and row indices, are:

$$M_1 = \begin{vmatrix} \alpha & -1 \\ -1 & \frac{1}{\alpha} \end{vmatrix} = 0,\ M_2 = \begin{vmatrix} \frac{1}{\alpha} & -1 \\ -1 & \alpha \end{vmatrix} = 0,$$

$$M_3 = \begin{vmatrix} -1 & \alpha \\ \frac{1}{\alpha} & -1 \end{vmatrix} = 0,\ M_4 = \begin{vmatrix} -1 & \frac{1}{\alpha} \\ \alpha & -1 \end{vmatrix} = 0.$$

  Their value is 0, as well as the value of other principal minors of second order having the same column and row indices.

- Principal minors of order higher than 2

  All principal minors of orders higher than 2 have at least two identical columns and rows, and because of that their value is 0.

For the other three cases of effective green and effective red position in the cycle it is easily established, in the same way as for the first case, that the corresponding Hesse matrix is positive semidefinite.

1.2. Investigation of convexity of the first term in delay function

If the first term in formula (9.31) is expressed as:

$$\beta'_{pe} = a^1_{pe}(\sum_{k=1}^{K}(1-u^k_p)\tau^k)^2, \tag{AVIII.19}$$

and the following notation is introduced

$$\sum_{k=1}^{K}(1-u^k_p)\tau^k = r^2_p,$$

then expression (AVIII.19) becomes:

$$\beta'_{pe} = a^1_{pe} r_p. \tag{AVIII.20}$$

According to the property of convex functions that the product of a convex function and a constant is also a convex function, it is enough to investigate convexity of function

$$\beta = r^2$$

to conclude whether function β'_{pe} is convex. The result of investigation holds for any $p \in \mathscr{P}'$, so that index p will be neglected in further text.

Since the convex function of a linear function is convex, and r is linear function of variables $\tau^1, \tau^2, \ldots, \tau^K$, it is necessary to investigate convexity of function $\beta = \beta(r)$. Hesse matrix in this case becomes:

$$\nabla^2\beta = \left[\frac{\partial^2\beta}{\partial r^2}\right] = [2].$$

It is obvious that this function is convex.

Since the sum of convex function is a convex function, and the product of the convex function and a constant is also a convex function, it can be concluded that criterion function J_c is a convex function.

2. THE PROOF OF DELAY FUNCTION CONVEXITY IN CASE THE CYCLE TIME IS GIVEN

The function whose convexity has to be tested (9.29) is the sum of terms (9.28):

$$\mathrm{M}v_p = 0.9 \sum_{e=1}^{e=\mathrm{E}(p)} \left(a_{pe}^1 \left(\sum_{k=1}^{\mathrm{K}} (1-u_p^k)\tau^k \right)^2 + \frac{a_{pe}^2}{\left(\sum_{k=1}^{\mathrm{K}} u_p^k \tau^k \right)^2 - a_{pe}^3 \left(\sum_{k=1}^{\mathrm{K}} u_p^k \tau^k \right)} \right).$$

It is enough to investigate convexity of only one term, i.e., of $\mathrm{M}v_p$.

Introducing symbols *r* and *g* for linear functions of τ^k, i.e.,

$$\sum_{k=1}^{\mathrm{K}} (1-u_p^k)\tau^k = r, \text{ and}$$

$$\sum_{k=1}^{\mathrm{K}} u_p^k \tau^k = g,$$

expression (9.28) becomes:

$$\mathrm{M}v_p = 0.9 \sum_{e=1}^{e=E(p)} \left(a_{pe}^1 r^2 + \frac{a_{pe}^2}{g^2 - a_{pe}^3 g} \right). \qquad \text{(AVIII.21)}$$

Since convex function of a linear function is also convex, then function $\mathrm{M}v_p$ is convex regarding variables $\tau^1, \tau^2, \ldots, \tau^{\mathrm{K}}$ if it is convex in regard to variables *g* and *r*.

Convexity of the first term in expression (AVIII.21) is proven in AVIII.1. The second term is convex if the following expression is convex:

$$\gamma = \frac{1}{g^2 - \varepsilon g}, \qquad \text{(AVIII.22)}$$

where $\varepsilon = a_{pe}^3$.

Since γ is a function of variable *g* only, Hesse matrix in this case becomes:

$$\nabla^2 \gamma = \left[\frac{\partial^2 \gamma}{\partial g^2} \right],$$

and it is necessary only to check $\frac{\partial^2 \gamma}{\partial g^2} \geq 0$ for feasible values of g.

The second derivative of function γ is given by expression:

$$\frac{\partial^2 \gamma}{\partial g^2} = \frac{2(3g(g-\varepsilon)+\varepsilon^2)}{(g(g-\varepsilon))^3}. \tag{ΠVIII.23}$$

This expression is positive if $g-\varepsilon>0$, i.e., if $g>\varepsilon$.

Since this expression is valid for any $p \in \mathscr{P}'$ and $e \in \mathscr{E}(p) = \{1,2,\ldots,e,\ldots,\mathrm{E}(p)\}$, i.e.,

$$g_p > \frac{c q_{pe}}{s_{pe}}, \quad (p \in \mathscr{P}', e \in \mathscr{E}(p)),$$

function γ will be convex if effective green times of all signal groups satisfy the condition that saturation degrees on all approaches to the intersection are less than 1.

It is shown that both terms of function $\mathrm{M}v_p$ are convex, which means that the function given by expression (9.29) is also convex.

REFERENCES

[1] Adams, W.F., Road traffic consideration as a random series, *J. Instn. Civ. Engrs.*, 4, 121, 1936.

[2] Akcelik, R., *Traffic Signals: Capacity and Timing Analysis*, Australian Road Research Board, Research Report ARR 123, 1981.

[3] Allsop, E.R., SIGSET—A computer program for calculating traffic signal settings, *Traffic Engineering and Control*, v.13, n.2, 58, 1971.

[4] Allsop, R.E., Optimisation of timings of traffic signals, in *Proc. AIRO Conf.*, Guida Editori, Napoli, 103, 1983.

[5] Allsop, R.E., Signal control at individual junctions: stage-based approach, in *Concise Encyclopaedia of Traffic & Transportation Systems* (Ed. Papageorgiou, M.), Pergamon Press, 478, 1991.

[6] Allsop, R.E., Treatment of opposed turning movements in traffic signal calculations, *Transportation Research*, 11 (6), 405, 1977.

[7] Allsop, R.E., Delay at a fixed traffic signal—I: theoretical analysis, *Transportation Science*, 6 (3), 260, 1972.

[8] Balas, E., A note on the branch-and-bound principle, *Opns. Res.*, 16, 442, 1968.

[9] Berge, C., *Théorie des graphes et ses application*, Dunod, Paris, 1958.

[10] Bielli, M., Ambrosino, G., Boero, M., *Artificial Intelligence Applications to Traffic Engineering*, VSP, Utrecht, The Nederland, 1994.

[11] Bierstone, E., Cliques and Generalized Cliques in a Finite Linear Graph, unpublished report, University of Toronto, 1967.

[12] Bonner, R.E., On some clustering techniques, *IBM J. Res. Develop.,* 8, 22, 1964.

[13] Браиловский, Н. О., Грановский Б.И., Управление движением транспортных средств, *Транспорт*, Москва, 1975.

[14] Branston, D., Some factors affecting the capacity of signalised intersections, *Traffic Engineering and Control*, 20, 390, 1979.

[15] Branston, D., Van Zuylen, H., The estimation of saturation flow - effective green time and passenger car equivalents at traffic signals by multiple linear regression, *Transp. Res.*, 12, 47, 1978.

[16] Bron, C., Kerbosch, J., Finding all cliques of an undirected graph, *Communications of the Association for Computing Machinery*, 16, (9), 575, September 1973.

[17] Burrow, I.J., OSCADY: A computer program to isolated traffic signal junctions, *TRRL Report*, 105, 1987.

[18] Camus, R., D'Amore, A., and Ukovich, W., Design of optimal fixed-time traffic controls for multi-junction nodes, in *Proc. of the II IAESTED Symposium on Modeling, Identification and Control*, March 2-5, 1982, Davos, Switzerland, Acta Press, 281.

[19] Cvetković, D., Simić, S., *Kombinatorika—Klasična i Moderna*, Naučna knjiga, Beograd, 1990.

[20] Cvetković, D., Milić, M., *Teorija grafova i njene primene*, Naučna knjiga, Beograd, 1977.

[21] Darroch, J.N., On the traffic light queue, *Ann. Math. Stst.*, 35, 38, 1964.

[22] Dekkers, A.J. et al., Berechnung der Signalfolge von Verkehrsampeln, *Philips' Technische Rundschau*, 12 (11), 1966.

[23] Desoer, C.A., *Notes for a Second Course on Linear Systems*, Van Nostrand Reinhold Company, New York, 1970.

[24] Evans, A.K. (editor), *Traffic Engineering Handbook*, Institute of Traffic Engineers, Connecticut, 2nd edition, 1950.

[25] Dougherty, E.R., Giardina, C.R., *Mathematical Methods for Artificial Intelligence and Autonomous Systems*, Prentice Hall, Englewood Cliffs, NJ, 1988.

[26] Forchhammer, N., Poulsen, L., Traffic Streams in Relation to Traffic Lights - the Operative Conflict Matrix, L.M. Ericsson technical note, unpublished, 1968.

[27] Gallivan, S., Heydecker, B.G., Optimising the Control Performance of Traffic Signals at a Single Junction, presented at 15th Annual Conference of the Universities Transport Study Group, Imperial College, London, 1983.

[28] Garfinkel, R.S., Nemhauser, G.L., The set-partitioning problem: set covering with equality constraints, *Operations Research*, 17, 848, 1969.

[29] Gartner, N.H., Demand-responsive traffic signals control research, *Transportation Res.*-A, 19A, 5/6, 369, 1985.

[30] Gazis, D.C., Potts, R.B., The oversaturated intersection, in *Proceedings of the 2nd International Symposium on the Theory of Road Traffic Flow*, London, 1963 (published by OECD, Paris, 1965).

[31] Gazis, D.C., *Traffic Science*, John Wiley & Sons, New York, 1970.

[32] Гаврилов, А.А., Моделирование дорожног движения, *Транспорт*, Москва, 1980.

[33] Guberinić, S., Mitrović Minić, S., Lazić, B., Signalna grupa: definicija, algoritmi, in *Proc. SYM-OP-IS '90*, 519, Kupari, 1990.

[34] Guberinić, S., Mitrović Minić, S., Signal group: definitions and algorithms, *Yugoslav Journal of Operations Research*, 3 (2), 219, 1993.

[35] Guberinić, S. et al., Signalni plan kojim se izjednačuju stepeni zasićenja vozačkih saobraćajnih struja, in *Proc. SYM-OP-IS 2000*, 467, 5–7 December, Beograd, 2000.

[36] Guberinić, S., Šenborn, G., Lazić, B., Određivanje skupa signalnih grupa za upravljanje saobraćajem na izolovanoj raskrsnici, *Saobraćaj*, 36, 459, 1989.

[37] Guberinić, S.Đ., Optimal control at a saturated intersection, in *Proceeding of JUREMA* 23 (5), Zagreb, 1978.

[38] Guberinić, S.Đ., Nova metoda optimizacije jedne klase višestepenih procesa i njena primena u automatskom upravljanju saobraćajnim procesom, Ph.D. thesis (University of Belgrade, Electrotechnical Faculty), 1990.

[39] Guberinić, S.Đ., Šenborn, G., Reljić, S., Maximization of traffic flow through the intersection by branch-and-bound bethod, in *Proceedings of the Conference on Control in Transportation Systems*, Baden-Baden, 153, 1983.

[40] Haight, F.A., *Mathematical Theories of Traffic Flow*, Academic Press, New York, 1963.

[41] Hewton, J.T., The metropolitan Toronto signal system, in *Proceeding of Symposium on Area Control of Road Traffic*, London, 1967 (Institute of Civil Engineers).

[42] Hope, L.M., Krystek, R., Some factors effecting left-turn capacity and delay at signalised intersections, in *Intersections without Traffic Signals*, editor Brilon W., Springer-Verlag, Berlin Heidelberg, 1988.

[43] Hunt, P.B. et al., SCOOT—a traffic-responsive method of coordinating traffic signals, *TRRL Report* LR 1014, Crowthorne, England, 1981.

[44] Hutchinson, T.P., Delay at a fixed time traffic signal -II: numerical comparison of some theoretical expressions, *Transpn Sci.*, 6, 286, 1972.

[45] Improta, G., Cantarella, G.E., Control system design for an individual signalised junction, *Transpn Res.*, 18B, 147, 1984.

[46] Inose, H., Amada, T., *Road Traffic Control*, Sangyo Tosho Ltd., 1972.

[47] Калужнин, А.А., *Введение в общую алгебру*, Наука, Москва, 1973.

[48] Kendall, G.D., Some problems in the theory of queues, *Journal of the Royal Statistical Society*, series B (methodological), XIII (2), 1951.

[49] Kimber, R.M., Hollis, E.M., Traffic queues and delays at road junctions, *TRRL Report*, 909, 1979.

[50] Kimber, R.M., Semmens, M.C., An experiment to investigate saturation flow at traffic signal junctions, *Traffic Engineering and Control*, 23, 110, 1982.

[51] Klir, G.J., Yuan, B., *Fuzzy Sets and Fuzzy Logic—Theory and Applications*, Prentice Hall, Inc., Upper Saddle River, NJ, 1995.

[52] Кофман, А., *Введение в прикладную комбинаторику*, Наука, Москва, 1975.

[53] Kronborg, P., Davidsson, F., Development and field trials of the new SOS algorithm for optimising signal control at isolated intersections,

in *Proceedings of the Eight International Conference on Road Traffic Monitoring and Control*, April 23–25, IEE, Savoy Place, London, UK, 1996.

[54] Land, A.H., Doig, A.G., An automatic method for solving discrete programming problems, *Econometrica*, 28, 497, 1960.

[55] Leonard, D.R., Tough, J.B., Baguley, P.C., CONTRAM: A traffic assignment model for predicting flows and queues during peak periods, Department of the Environment, *TRRL Report* LR 841, Crowthorne, England, 1978.

[56] Little, J.D.C. et al., An algorithm for the travelling salesman problem, *Opns. Res.*, 11, 972, 1963.

[57] McNeil, D.R., A solution to the fixed-cycle traffic light problem for compound Poisson arrivals, *J. Appl. Prob.*, 5, 624, 1968.

[58] Mesarović, M.D., Takahara, Y., *General Systems Theory: Mathematical Foundations*, Academic Press, 1975.

[59] Michalopoulos, P.G., Meyer, S.E., Settings for an isolated oversaturated intersection, *Traffic Engineering and Control*, 21 (12), 580, 1980.

[60] Miller, A.J., Settings for fixed-cycle traffic signals, *Opnal. Res. Quart.*, 14, 373, 1963.

[61] Mitrinović, D.S., Mihajlović, D., *Linearna algebra, analitička geometrija, polinomi*, IP "Građevinska knjiga," Beograd, 1962.

[62] Mitten, L., Branch-and-bound methods: general formulation and properties, *Opns. Res.*, 18, 24, 1987.

[63] Mulligan, G.D., Corneil, D.G., Corrections to Bierstone's algorithm for generating cliques, *J. Assoc. Comput. Mach.*, 19, 244, 1972.

[64] Murchland, J.D., Traffic Signal Sequence Generating Program SQGN: technical details, Transport Studies Group, University College, London, 1979.

[65] Murchland, J.D., Tully, I.M.Z., A Program for Generating Stage Sequences for a Traffic Signal Controller: SQGN, Traffic Studies Group, University College, London, 1976.

[66] Nakahara, T., Yumoto, N., Tanaka, A., Multi-criterion area traffic control system with feedback features, in *Proc. IFAC/IFIP 1st Int. Symposium on Traffic Control* (paper No. 3.9), June 1970.

[67] Newell, G.F., Statistical analysis of the flow of highway traffic through a signalised intersection, *Q. Appl. Math.*, 13, 353, 1956.

[68] Papageorgiou, M., *Application of Automatic Control Concepts to Traffic Flow Modeling and Control*, Springer-Verlag, Berlin Heidelberg, 1983.

[69] Pavel, G., *Planen von Signalanlagen für den Strassenverkehr*, Kirschbaum Verlag, Bonn-Bad Godesberg, Mai 1974.

[70] Печерский, М.П., Хорович, Б.J., Автоматизированые системы управления дорожным движением в гордах, *Транспорт*, Москва, 1979.

[71] Rinnooy, Kan, A.H.G., *On Mitten's Axioms for Branch-and-Bound*, Graduate School of Management, Delft, The Netherlands, 1975.

[72] Robertson, D.I., TRANSYT: A traffic network study tool, Ministry of Transport, *RRL Report* LR 253 (Road Research Laboratory), Crowthorne, England, 1969.

[73] Robertson, D.I., Vincent, R.A., Bus priority in a network of fixed time signals, Department of the Environment, *TRRL Report* 666 (Transport and Road Research Laboratory), Crowthorne, England, 1975.

[74] Salkin, H.M., *Integer Programming*, Addison Wesley, 1975.

[75] Serafini, P., Ukovich, W., A mathematical model for the fixed-time traffic control problem, *European Journal of Operational Research*, 42, pp. 152–165, 1989.

[76] Singh, M.G., Tamura, H., Modelling and hierarchical optimisation for oversaturated urban road traffic networks, *Int. J. Control*, 20 (6), 1974.

[77] Stoffers, E.K., Scheduling of traffic lights—A new approach, *Transportation Research*, 2, 199, 1968.

[78] Stojanović, S., Guberinić, S., Cena koštanja čekanja vozila u jedinici vremana, *Saobraćaj u gradovima*, 4, 9, 1977.

[79] Šegović, M. et al., Uticaj "filtriranja" jednih saobraćajnih struja kroz druge na kapacitet, in *Proc. SYM-OP-IS 2000*, 491, 5–7 december, Belgrade, 2000.

[80] Šenborn, G., Hijerarhijski struktuirani sistemi za automatsko upravljanje saobraćajem u gradu, in *Upravljanje sistemima*, Naučna knjiga, Beograd, 1986.

[81] Tarnoff, C.J., MacGowan, C.J., A report on urban traffic control research conducted by the Federal Highway Administration, *Control in Transportation Systems*, Proceedings of the IFAC/IFIP/IFORS, Third International Symposium, Columbus, OH, 1976.

[82] Teply, S. et al., *Canadian Capacity Guide for Signalised Intersections*, Second edition, Institute of Transportation Engineers, District 7—Canada: Institute of Transportation Engineers, Washington D.C., 1995.

[83] Trach, M., *Signalised Intersection Design*, Wydawmictwa Komunikacji i Lacznosci, Warsaw, 1988.

[84] Troutbeck, R.J., Current and future Australian practices for the design of unsignalised intersections, in *Intersections without Traffic Signals*, editor Brilon W., Springer-Verlag, Berlin Heidelberg, 1988.

[85] Tully, I.M.Z., Synthesis of Sequences for Traffic Signal Controllers Using Techniques of the Theory of Graphs, OUEL Report 1189/77, University of Oxford, 1977.

[86] Вентцель, Е.С., *Исследование операций*, Советское Радио, Москва, 1972.

[87] Vincent, R.A., Real-Time Traffic Control in Urban Areas - a Review, presented on International Summer School on Urban System, Dubrovnik, Yugoslavia, 1980.

[88] Webster, F.V., Cobbe, B.M., Traffic Signals, road research technical paper 56, HMSO, London, 1966.

[89] Webster, F.V., Traffic Signal Settings, road research technical paper 39, HMSO, London, 1958.

[90] Zadeh, L.A., The concept of state in system theory, in *Views of General Systems Theory*, edited by Mesarović M.D., John Wiley & Sons, 1964.

[91] Zlobec, S., Petrić J., *Nelinearno programiranje*, Naučna knjiga, Beograd, 1989.

[92] Zuylen, Henk van, Acyclic Traffic Controllers, traffic studies group, University College, London, 1976.

INDEX

N

O

P

Q

R

S

T

U

V